The Development of Behavior: Comparative and Evolutionary Aspects

Garland Series
In Ethology

Additional Books
in Ethology

The Development of Behavior: Comparative and Evolutionary Aspects

GORDON M. BURGHARDT

Department of Psychology
University of Tennessee
Knoxville, Tennessee

MARC BEKOFF

Department of Environmental,
  Population and Organismic
  Biology
University of Colorado
Boulder, Colorado

Garland STPM Press
New York & London

15 14 13 12 11 10 9 8 7 6 5 4 3 2 1

Library of Congress Cataloging in Publication Data

Main entry under title:

The Development of behavior.

   (Garland series in ethology)
   "Based on a symposium that took place June 6–7, 1977, at
the annual meeting of the Animal Behavior Society at
Pennsylvania State University, State College, Pa."
   Includes bibliographies and index.
   1. Animals, Habits and behavior of—Congresses. I.
Burghardt, Gordon M., 1941-II. Bekoff, Marc. III. Animal
Behavior Society. IV. Series.

QL750.D39   591.5   77-87206 ISBN 0-8240-7015-1

Printed in the United States of America

Table of Contents

Preface

This volume is based on a symposium that took place June 6–7, 1977, at the annual meeting of the Animal Behavior Society at Pennsylvania State University, State College, Pa. Our goal was to bring together a group of scientists actively studying the development of behavior from a variety of approaches in a diverse array of organisms. Each person invited was requested to emphasize comparative and evolutionary aspects of development. Although there were the usual difficulties in fitting together schedules, almost everyone invited agreed enthusiastically to participate. We aimed at a mixture of seasoned and relatively junior investigators.

Unlike most published symposia on behavioral development, this work deals with all classes of vertebrates as well as selected invertebrates (mollusks and insects). This choice is dictated because more than lip service must be paid to species other than mammals and birds if a broad understanding of development is ever to be attained. And the best people to present current and accurate comparative data are those working directly with these varied beasts. The second way in which the present volume differs from most previous ones is in the diverse theoretical and methodological approaches of its authors. We tried to avoid the often easy and facile rationalizations that have led to overly narrow volumes in the past. This is not to say we did not exercise judgment and selectivity, nor that we are unaware that all the authors are from North America (a necessity, due to financial constraints). While the volume is not meant to be a complete sourcebook, it is important to stress that the overlap between contributors is minimal and, with the extensive bibliographies, this volume should be useful as a text and reference.

We are grateful to Ed Price for his invaluable programming efforts. The authors and discussants were remarkably faithful. We thank Stuart Altmann, Colin Beer, and Eckhard Hess for their comments at the end of particular sessions. Donald Griffin, Ronald Oppenheim, and Leonard Rosenblum were also to comment in person but were unable to attend. The first two, however, submitted written comments that were considered by the authors concerned. This was likewise the case with all the discussants who attended. Eckhard Hess's comments are included herein after the chapter by H. Hoffman. We especially wish to thank the outside reviewers who graciously provided critical comments on several chapters. The Department of EPO Biology, University of Colorado, Boulder;

the Department of Psychology, University of Tennessee, Knoxville; and D. R. Griffin's laboratory at The Rockefeller University provided financial and secretarial help necessary for organizing and implementing the symposium and the published volume. The senior editor also wants to acknowledge the support of the John Simon Guggenheim Memorial Foundation during this period. Garland Publishing, Inc., particularly through the efforts of George Narita, greatly facilitated prompt publication.

G.M.B.

M.B.

Koblenz am Rhein
West Germany
August 29, 1977

Introduction

Ethology is concerned with four aspects of behavior which overlap extensively: phylogeny, ontogeny, causation, and function. In addition, one of the hallmarks of ethology is the comparative approach to the study of behavior, in which behavioral phenotypes are treated as structures and frequent attempts are made to reconstruct evolutionary pathways that may shed light on behavioral homologies (and analogies). Within the last fifteen years the variety of animals that have come under the careful scrutiny of both the aided and unaided eye has increased almost exponentially. And, not unexpectedly, such rapid growth has been accompanied by "growing pains." One of the most cumbersome offshoots of the heightened interest in previously ignored species is the emergence of all-encompassing theories and models based on often gratuitous assumptions about and simplifications of the actual behavior of the animals under study. Although new trends of thought in behavioral biology are firmly grounded in evolutionary theory, there is a tendency to play down the actual processes by which outcomes are attained; overt behavioral patterns and their variability are often ignored. Of course, theories and models *are* needed, but "grand systems" rarely survive the accumulation of broad-scale data. Currently, there is a need for more data and critical evaluations of existing theory. Only when solid comparative data are in will we be able to assess the cross-species applicability of general models and theories of behavioral evolution and ontogeny.

As suggested above, ontogeny is a very important area of study in ethology. It is also an area where a dichotomy, even an opposition, between evolution and individual experience has endured for many years. This has been most unfortunate. Despite the litany intoned by many prominent scientists, we do not find the nature-nurture (or instinct-learning, etc.) debate inherently useless or even dangerous. It must not be forgotten that the initial impact of classical ethology was based largely on its persuasive recitation of findings that seemed to favor "instinctive" rather than environmentally induced interpretations. But while it is universally agreed that all behavior, just as all structure, has a developmental (ontogenetic) history, so do developmental processes have an evolutionary history. Thus we felt the time propitious to bring together diverse material relating phylogeny and ontogeny in behavior.

Before each of the six sections into which the chapters are divided, we have commented briefly on each contribution and provided additional information as appropriate; elaborate comment is unnecessary. But we do wish to make several points concerning the present status of the field and our own personal reflections.

(1) Although some of the chapters (e.g., Altmann's) follow individual animals for considerable periods, longitudinal studies are still too rare. Many important questions will not be answered until such studies have been made.

(2) Variability within and among behavior patterns at different ages (both in different species and between individuals of the same species) must be carefully assessed with the new observational and analytical tools becoming available. One can speculate, for example, equally facilely on why neonates should be more *or* less variable than adults. Genetic and experiential factors should both be attended to, along with biochemical, maturational, and neurological aspects.

(3) Invertebrates need much more consideration by vertebrate-oriented theorists (cf. the present contribution by A. Bekoff). It is all too convenient to dismiss invertebrates as "irrelevant." The seminal role played and still being played by *Drosophila* in genetics and evolution (not to say behavior, as in Ringo's chapter) and the equally important role played in neurophysiology by mollusks, from the giant squid axon to the sea slug, strongly suggest that behavioral scientists should not arrogantly dismiss animals they casually squash at departmental picnics. On the one hand, invertebrates may be simpler and so facilitate elucidation of behavioral processes (and their neural bases) and development of methods as yet not available with vertebrates. On the other hand, processes once considered too advanced for invertebrates, such as apparent symbolic communication in bees, have been discovered; the implications of these are staggering. Workers assessing experiential factors in insect development will often find them important (e.g., see chapter by Topoff and Mirenda), whereas genetic factors will be established with irrefutable clarity (e.g., chapter by Hoy and Casaday).

(4) Direct studies of the evolution of behavior are notoriously rare, since behavior, it is often emphasized, does not fossilize. But of course it often does, or rather, the effects of behavior on substrates (tracks) or prey (bones) are evident. Imaginative use of fossils allows reasonable reconstructions not tied exclusively to the study of extant species. Paleoethology and ontogeny are not outlandish bedfellows (Berg's chapter).

(5) In choosing extant species to study, selections based on evolutionary and ecological criteria are often more useful than the normally encountered criteria of convenience and economy. Diversity in behavior is great in nature, and this fact should be reflected in our studies. The influence and acceptance (even the romance) of ethology lie largely in

reopening our eyes to the richness of life, so apparent to nineteenth-century naturalists yet lost with the rise of behaviorism in psychology and mechanism in biology.

(6) Reproductive fitness and other precise notions about how to measure adaptation and the process of natural selection are of great importance to ethology. Sociobiological thinking pervaded much of the 1977 XVth International Ethological Conference in Bielefeld, West Germany. As the same conference also stressed ontogeny, the necessity and timeliness of the deliberate convergence of ontogenetic and phylogenetic studies are apparent. Even in humans, the genetic-environmental question can be handled (e.g., see chapter by Plomin and Rowe).

(7) Theory, particularly testable models, are of critical heuristic importance. But theorists do need to be aware of actual facts and the degree of simplification of their assumptions, just as much as empirical workers need to be aware of current *useful* theory (see also Fagen's chapter).

All in all, we hope that the present volume advances the field of behavioral ontogeny by bringing together representative studies on diverse species, all of which, like ourselves, are faced with "problems" of development.

Gordon M. Burghardt
Marc Bekoff

Section One

TWO COMPLEMENTARY APPROACHES: FOSSILS AND PHYSIOLOGY

In this section two seemingly disparate themes are joined. Indeed they *are* disparate, but together they represent exciting directions that students of behavioral ontogeny should consider. What *can* fossils tell us about the development of behavior in extinct species? Carl Berg shows us how predatory behavior and its development in mollusks can be studied quantitatively by the analysis of shells. Anne Bekoff, on the other hand, approaches shells differently, going through them to observe and experiment with the developing chick embryo. Her studies give clear evidence of the existence of innate coordination patterns, although she is well aware of the complexities involved. Using chicken embryos, she has furthered the mounting evidence supporting "Lorenzian" notions about fixed action patterns, endogenous and spontaneous movements, maturation, and so forth.

1

Chapter One

Development and Evolution of Behavior in Mollusks, with Emphasis on Changes in Stereotypy

CARL J. BERG, JR.

Museum of Comparative Zoology
Harvard University
Cambridge, Massachusetts 02138

Recently, there has been a marked increase in the interest shown in mollusks as subjects for studies of the development and evolution of behavior. Most of the work has dealt with larval behavior and the inducement of metamorphosis (e.g., Hadfield, 1977; Harris, 1973, 1975; Kreigstein, Castellucci and Kandel, 1974; Strenth and Blankenship, 1977) or neurophysiological correlates of behavior (e.g., Kriegstein, 1977a, b; Willows and Dorsett, 1975). However, there are only a few papers describing postmetamorphic development of adult behavior patterns and the evolution of behavior. It is these that I plan to review, concentrating on rates of behavioral development and stereotypy of behavior patterns. In addition, I will present some of my own work on the development and evolution of predatory behavior in gastropods.

In general, behavior patterns concerned with locomotion and non-specialized feeding, such as grazing of algae, develop gradually after metamorphosis. For gastropods, metamorphosis usually consists of the change from a planktonic veliger which uses its ciliated velar lobes for both swimming and filter feeding, to a benthic crawling juvenile which uses its newly developed radula for rasping algae or animal tissues. Once the velar cilia are shed or resorbed, the juvenile can no longer swim; it must crawl, using its well-developed foot. Locomotor patterns develop slowly, with the animals first moving by ciliary action and later by waves of muscular contraction of the foot (Berg, 1972; Harris, 1973; Kriegstein,

Castellucci and Kandel, 1974). With strombid gastropods, a unique discontinuous leaping form of locomotion develops one week past metamorphosis, but even after six weeks there are still quantitative differences in locomotion between juveniles and adults (Berg, 1972).

Grazing by gastropods also develops gradually, with an appetitive orienting response developing first, followed by an incomplete biting and swallowing consummatory response. Only after metamorphosis, two days later, is the adult feeding pattern complete (Kriegstein, Castellucci, and Kandel, 1974). Few studies have been detailed enough to report the transitory phase and simply claim that the animals are fully capable of grazing immediately after metamorphosis.

Behavior patterns related to predation and escape from predation appear more abruptly, often with little or no apparent practice, soon after metamorphosis. The development of escape responses seems closely attuned to the size at which the animals are first preyed upon. Small strombid gastropods (*Strombus maculatus*) did not respond with the complete adult escape response to predaceous cone-snail stimulus until they were 2 mm in shell length and at least three weeks old (Berg, 1972). Once these criteria of maturation were met, the strombids performed the characteristic backward "flip" response to cones. Repeated presentations of the cones did not lower either the size or age criteria. Small cones did not prey upon *Strombus maculatus* less than 2 mm in length in the laboratory, but would feed upon larger ones. The development of the escape response appears related to the maturation of neural components (Berg, 1972). A similar case may exist with the nudibranch, *Tritonia diomedia*, which showed a marked escape-swimming response to being touched by the predatory starfish, *Pycnopodia helianthoides*. This response did not develop until the nudibranchs were about two months old (6–10 mm). Juveniles of the same age but which grew more slowly (only 3–6 mm) did not swim (G. Audesirk, personal communication). Unfortunately there are no data on the size of first predation by the starfish. The drop-off burial response of freshwater pulmonate snails to odors from crushed conspecifics is not present when the snails crawl from their egg cases (the free-swimming stage is absent in these snails) but develops later during the first few weeks after hatching (Snyder and Snyder, 1971). They found that the response to odor from predatory turtles is present in juvenile snails prior to hatching. However, these responses are maintained by the juvenile snails only until they reach the size at which they are no longer preyed upon. This size varies for the odors of different species of turtle and is correlated with the prey-catching ability of each of the different species. When the snails have reached a refuging size they no longer respond to that particular odor. Again, the escape response is closely attuned to ecological constraints of predation. There is no information on the mechanism by which these behavior patterns are turned on and off, but David Prior (personal communication), working with the surf clam, *Spisula solidissima*, describes an "ontogenetic switch" for the flip escape response that these clams show to predaceous starfish. When the clams are small (2–5 cm) they respond to a combination of chemical and tactile cues from the starfish by first withdrawing into the shell and then kicking at the substratum with their foot. Large clams (12–18 cm) respond by withdrawing, but only very occasionally kick. The change in responsiveness apparently is due to the decrease in the relative sizes of the areas of the siphons covered by starfish receptor cells. In large clams not enough cells are simultaneously stimulated by tactile cues to elicit the response. It is most likely that there is a gradual decrease in responsiveness as the animals get larger. Large clams probably live deep enough in the substratum to avoid starfish predation by simply withdrawing into their shell.

Cephalopods (i.e., cuttlefish, octopus and squid) perform their typical escape responses of swimming, inking, and changing colors immediately after hatching (Hanlon, 1977; Messenger, 1963; Wells, 1958, 1962; Wells and Wells, 1970).

Studies of the development of predatory behavior in mollusks are limited to the work with cephalopods (e.g., Hanlon, 1977; Joll, 1976; Messenger, 1963; Trantor and Augustine, 1973; Wells, 1958, 1962; Wells and Wells, 1970) and with those gastropods which bore through the shell of their prey (Berg, 1976; Carriker, 1957; Wood, 1968). Few of these papers have quantitative observations permitting anything but general comparisons of changes in variability in responses. The exception is Well's (1958, 1962) excellent work on the development of prey capture by the cuttlefish, *Sepia officinalis*. Juvenile cuttlefish appear to recognize their prey (live mysid shrimp) innately and attack it in an adult-like fashion. During the first week of posthatching development there is no change in the manner of attacking or aiming ability, but the time latency between first noticing the prey and its capture declines rapidly, irrespective of capture success, and is therefore probably not due to individual learning, but rather just the effect of repetition of the act of attacking. And most interestingly, the range of objects that are attacked widens. *Sepia* that never attacked food items before are more selective about what they attack than are the experienced animals, i.e., there is an increase in the variability of prey items with respect to experience.

The earlier work of Wells (1958, 1962) with cuttlefish is supported by considerable evidence: the observations of Bradley (1974), Hanlon (1977), Itami *et al.* (1963), Joll (1976), Thomas and Opresko (1973), Trantor and Augustine (1973), and Wells and Wells (1970) on six species of octopus; Choe (1966), Choe and Ohshima (1963), LaRoe (1971), and Ohshima and

Choe (1961) on two species of squid; and Choe (1966), Choe and Ohshima (1963), and Ohshima and Choe (1961) on three species of cuttlefish. In all cases the authors report that after a nonfeeding period varying in length for each species, newly hatched juveniles attack prey in an adult manner and show definite selection of live moving prey, primarily crustaceans, which make up the normal diets of adults. It is unfortunate that there have been no further quantitative studies with the cephalopods on the development of behavior.

The work on predatory gastropods is very similar to that on the cephalopods in being simply descriptions of the behavior of juveniles and their prey preferences. Carriker (1957) concentrates on predator orientation reactions in the American oyster drill *Urosalpinx cinerea* and notes that newly hatched drills show a definite positive chemotactic response to external metabolites from prey clams. Wood (1968) discusses prey preference in detail. He believes that the hatchlings undergo "ingestive conditioning," which determines later prey selection. He claims this process is similar to classical imprinting described by Lorenz and the "food imprinting" described by Burghardt and Hess (1966) for snapping turtles. Although quantitative data were not presented, Wood (1968) goes on to suggest that juvenile oyster drills change and improve their techniques for boring into each of their prey species. This has also been suggested by Fischer-Piette (1935) for European oyster drills.

I decided to test this hypothesis by a series of rearing studies using another group of marine snails which also bore into their prey. A distinct advantage of working with boring gastropods is that the borehole provides an easily quantifiable and permanent record of behavior patterns involved in prey orientation and boring. I worked with the Naticidae, the common moon snails on mud flats (Berg, 1975, 1976). These snails prey upon bivalves and gastropods by grabbing the prey in their large foot, manipulating and orienting it into the preferred position, and then drilling a hole through the prey's hard calcareous shell. This hole probably is made using a combination of chemical and mechanical means (Berg, 1976) in a manner similar to that of the oyster drills mentioned above (Carriker and Van Zandt, 1972). Each species of predator bores into each of its prey species in a distinct area (Ansell, 1960; Berg, 1975; Berg and Porter, 1974) usually near the center of the shell. I developed methods of measuring the position of the borehole with respect to the major anterior-posterior and dorsal-ventral axes of the bivalve shells (Fig. 1) and with respect to both the anterior-posterior axis and the axis of coiling for gastropod shells (Fig. 2) (Berg and Nishenko, 1975; Berg and Porter, 1974). By measuring the position of the borehole with respect to the total length of the axis and presenting it as a proportionality, shells of different

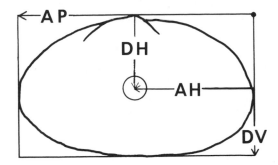

FIG. 1. *Methods for measuring borehole position in a bivalve shell. AP = anterior edge to posterior edge distance; AH = anterior edge to center of borehole distance; DV = dorsal edge to ventral edge distance; DH = dorsal edge to center of borehole distance. Ratios: AH/AP and DH/DV.*

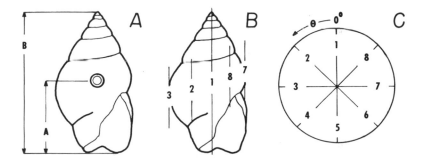

FIG. 2. *Methods for measuring borehole position in a gastropod shell. A. Linear measurements: A = distance to borehole; B = total length; hole ratio = A/B. B. Circular distribution measurements: numbers indicate sector designations. C. Anterior view of sector designations, increasing in a counterclockwise direction from 1 = 0° in the plane of the columella.*

sizes and shapes can be compared. What is important to this discussion is not the position of the hole *per se,* but rather changes in the sterotypy of the position which I measured during postmetamorphic development of these snails.

Stereotypy is defined as the fidelity with which the positioning of the borehole is repeated by members of the population of snails being studied. Stereotypy will be measured as the inverse of variability $(S = 1/V)$ and vice versa $(V = 1/S)$(Schleidt, 1974). I will be using two measures of variability: (1) The coefficient of variation, defined as the ratio between the standard deviation and the mean ($V = s/\bar{x}$) and expressed as variability in percent of the mean ($V \times 100$). This measure will be used for linear measurements; and (2) the value r will be used for circular distributions where $r = \sqrt{(\cos \alpha)^2 + (\sin \alpha)^2}$ is a measure of the concentration of values about the mean direction (Batschlet, 1965). In this case variability (V_r) is equal to the inverse of $r(V_r = 1/r)$. The magnitude of the values of V and V_r cannot be compared. Probabilities of no significant differences between coefficients of variation (V) were tested, using the formula

$$ c = \frac{V_1 - V_2}{S_{V_1} + S_{V_2}} $$

where

$$ S_{V_1} = \frac{V_1}{\sqrt{2n}} \sqrt{1 + 2(\frac{V_1}{100})^2} $$

and c approximates a t distribution for $n_1 + n_2 - 2$ d.f. (Sokal and Rohlf, 1973, p. 137).

Because I am discussing variability of locations on a surface area, I will use the product of the variabilities from each axis as my relative measure of total variability of borehole position depending on whether I am discussing bivalve or gastropod prey, respectively: $V_t = V_a \times V_d$ or $V_t = V_a \times V_r$.

In the first experiment (Berg, 1976) I reared specimens of *Natica gualtieriana* by collecting them in the plankton as free-swimming veligers, placing them in small individual rearing chambers, and presenting each with a single prey snail. Before metamorphosis, these snails float in the water column and have neither a radula nor a proboscis; therefore it is very unlikely that they have any premetamorphic experience at prey capture and manipulation. Upon first presentation the predators captured the prey, manipulated it, and bored through the shell in a fully adult-like and proficient manner. The first borehole was well formed (Fig. 3) and there was no evidence of incomplete boreholes or random rasping. The general quality of the borehole and the proficiency of boring seemed to decrease during later postmetamorphic development. A comparison of the variability in the positions of the first boreholes with those of the 20th boreholes shows (Table 1) an increase in the variability ($1/r$) of position around the shell along the coiling axis but a decrease ($p < .05$) in the

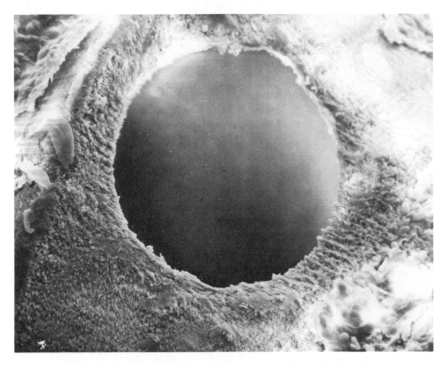

FIG. 3. *Scanning electron photomicrograph of the first bore-hole made by a newly metamorphosed* Natica gualtieriana. *Borehole diameter is approximately 125 μm.*

variability with respect to the anterior-posterior axis of the shell (V_a). The longitudinal axis of the gastropod prey shell is probably the major cue in prey orientation. The prey shell is "balanced," perpendicular to the anterior-posterior axis of the predator's foot, and held there as the prey shell is rotated around its own axis. Looking at total variability, V_t, it again appears that as the animals mature and have more experience at prey orientation and boring, the less proficient and more variable those behaviors become.

My second experiment was designed to partition out the relative effects of maturation and experience on the development of predatory behavior in another naticid species, *Lunatia triseriata* (Berg, 1975). This species undergoes all of its larval development within the egg capsule. The young emerge from the capsule as fully formed juvenile snails; there is no planktonic stage. Upon hatching, each snail was placed in its own petri dish, containing seawater, sand, and a small clam. It was difficult to collect clams small enough to feed the snails. One group was fed whole-

TABLE 1. Changes in the variability of borehole position for *Natica gualtieriana*.

COEFFICIENT	FIRST PREY	TWENTIETH PREY	Δ VARIABILITY
r	(7) .847	(8) .306	↑
V_a	(13) 8.17	(10) 4.48	↓
$V_t = \dfrac{1}{r} \times V_a$	9.65	14.64	↑

Sample size is given in parentheses. r is a measure of the concentration of values about the mean direction for circular distributions. The higher the value of r, i.e. the closer r approaches unity, the less variability in the sample. V_a is a measure of the coefficient of variation ($V = \dfrac{s}{\bar{x}} \times 100$) and the lower the value, the less variability there is in the sample.

intact clams; the other was offered only the soft clam meat. At 32 days after hatching each of the control-group snails was given an intact clam, their first hard object to manipulate. There was no obvious difference between the two groups, but my final sample sizes were small ($n = 7, = 2$) because I discovered that these animals are voracious but obligate carnivores. They did not feed upon the clam meat that I presented. With both groups, the first prey encountered was captured, manipulated, and bored in a fully competent and proficient manner. Variability in borehole placement in the first prey bored by the experimental group (those given prey immediately upon hatching) can be compared with variability shown by adults. It should be pointed out, however, that because the prey given to the juveniles is too small for adults to feed upon, cross-species comparison must be made (Table 2). There is no reason to expect that the amount of variability in borehole placement should be solely a function of species bored rather than of ontogenetic changes within the predators themselves. There is no significant difference in variability of borehole placement in the two species preyed upon by adults (Table 2). As with *N. gualtieriana*, variability decreases ($p < .001$) between juveniles and adults for measurements along the anterior-posterior axis (V_a) but increases ($p < .001$) for the other axis (V_d). And again, overall variability (V_t) increases markedly from the juveniles' first boreholes to those of adults.

Upon metamorphosis, naticid gastropods are competent predators, boring into their prey in a species-typical and -stereotyped fashion. The position of the borehole is a function of behavioral acts of prey orientation, some of which are more stereotyped than others, perhaps reflecting differences in the selective advantages they confer. With development (maturation and experience combined) overall orientation becomes more

TABLE 2. Changes in the coefficient of variation ($V = \frac{s}{\bar{x}} \times 100$) for *Lunatia triseriata* boring into bivalve prey.

COEFFICIENTS	FIRST PREY	ADULT PREY		Δ VARIABILITY
	Gemma gemma	*Mercenaria mercenaria*	*Mya arenaria*	
$V_a = \dfrac{AH}{AP}$	(7) 24.7	(10) 12.9	(12) 13.3	↓
$V_d = \dfrac{DH}{DV}$	13.1	46.4	56.3	↑
$V_t = V_a \times V_d$	323.6	598.6	748.8	↑

Sample size is given in parentheses ( ). Ratios are the distance from anterior edge of the shell to the center of the borehole (*AH*) divided by the total anterior-posterior distance (*AP*) and for the distance from the dorsal edge to the center of the borehole (*DH*) divided by the dorsal-ventral distance (*DV*).

variable and the general quality of the borehole declines. These changes are probably related most directly to the increase in the predators' size. With a larger and stronger foot, the animals have a greater ability to feed upon prey with a wider range of shapes and sizes. Variability in orientation would allow for the handling of these different types of prey. In addition, as the predator gets larger, the efficiency with which he bores is probably less energetically important. Thus, an increase in variability of response could be conceived of as being adaptive. This is in contrast to nut opening by squirrels (Eibl-Eibesfeldt, 1951), which show progressive efficiency in opening nuts, and to vertebrate predators, which generally become less variable in orientation of the killing bite (Mueller, 1974).

Comparison with other data on changes in variability in behavior during ontogeny emphasizes the frailty of the common notion that behavior patterns always become less variable during ontogeny due to the effects of maturation and experience (Bekoff, Chapter 9, and 1977a; Schleidt and Shalter, 1973; Wiley, 1973). It should also be noted that the variability shown above for orientation of predatory behavior is greater than that reported for some behavior patterns serving for communication (see reviews by Barlow, 1977; Bekoff, 1977a, b; Schleidt, 1974) but much less than that for behavior patterns related to grooming or care of the body surface (Bekoff, Chapter 9). Without knowing the form of the distribution of these different characteristics, we must be extremely cautious in evaluating differences (Schleidt, 1974). More rigorous quantitative comparisons of ontogenetic data are definitely called for.

Because mollusks have hard calcareous shells which fossilize readily, we have the unique capability of using borehole position as a measure of predatory behavior to compare changes in the variability of behavior patterns through their evolution. Probably the most common assumption would be that behavior patterns would become less variable, especially where they assume communication functions. This is part of the general process of ritualization. For predatory behavior, we would expect that predator and prey behavior should evolve together, perhaps causing major changes in behavior through time. To test this hypothesis we measured the distribution of boreholes attributable to naticids in fossil gastropod shell assemblages (Berg and Nishenko, 1975, submitted for publication) using the same techniques developed for ontogenic studies. Shells from collections in the U.S. National Museum and the American Museum of Natural History were measured and only samples of uniform and well-documented geological age were used. We then compared borehole positions in gastropod shells over the approximate 100-million-year span of naticid boring (Sohl, 1969). The distribution pattern of boreholes around the shell was initially random ($r = .17$; $p > .05$) with a slight tendency to be centered over the aboral surface (Fig. 4). By the Paleocene, boring became restricted to the adoral surface in a nonrandom fashion

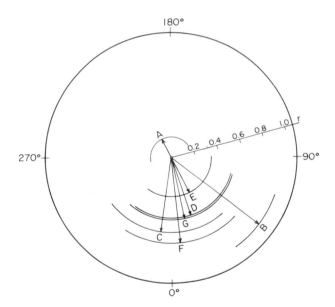

FIG. 4. *Grouped mean circular distributions of naticid boreholes in prey snail shells. Position of vector indicates mean angle; vector length indicates* r *value for strength of distribution with* r = 1.0, *the strongest nonrandom distribution. Arc segments indicate angular deviations of mean angles: A = Cretaceous; B = Paleocene; C = Eocene; D = Miocene; E = Pliocene; F = Pleistocene; G = Recent.*

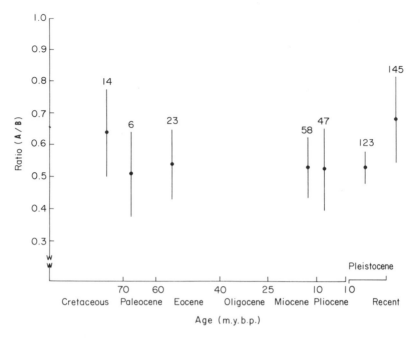

FIG. 5. *Grouped mean hole ratios (A/B) of naticid boreholes in prey snail shells. Age in millions of years before present. Numbers indicate sample size; dots = mean values; lines − ± standard deviation of the mean.*

($r = .93$; $p < .01$). This orientation has remained to the present. Also, distribution on the longitudinal shell axis (Fig. 5) was variable ($V = 21.3$) in the Cretaceous but became more stereotyped ($p < .001$) by Recent times ($V = 9.1$ and 7.9 [$p > .05$] for two Recent subpopulations.) Because of differences among predator species in borehole positioning in each prey species, we looked most closely at three sets of data in which the boreholes can be attributed to the same species of predator, *Lunatia heros,* and all are in nassariid prey of similar shape. These samples span approximately 20 million years, but the Pleistocene sample was taken within 60 km of the Recent sample. These data (Table 3) indicate no apparent change in the actual position of the borehole nor in the variability of the angular deviation. There does appear to be a decrease ($p < .001$) in variability along the longitudinal shell axis (V_a) between the Miocene and Recent times. Through evolutionary time the behavior is becoming more stereotyped, suggesting some selection pressure. I know of no other studies of the evolution of stereotypy of behavior patterns. Comparative studies of modern animals cannot give this type of data.

TABLE 3. Change in the variability of borehole position for *Lunatia heros* boring into nassariid prey.

COEFFICIENTS	MIOCENE	PLEISTOCENE	RECENT	Δ VARIABILITY
n	20	10	64	—
Mean angle	355°	346.6°	337.6°	—
r	.87	.89	.83	≈
Mean ratio: $\dfrac{AH}{AP}$	.49	.50	.54	—
V_a	15.7	12.3	9.1	↓
$V_t = (\dfrac{1}{r} \times V_a)$	18.05	13.82	10.96	↓

r is a measure of the concentration of angular values about the mean direction. The closer r comes to 1, the less variable is the sample. V_a is a measure of the coefficient of variation. The lower the value, the less variable is the sample. n indicates sample size.

There are some studies on the evolution of behavioral patterns in mollusks, and they should be mentioned here. All are comparative studies dealing mainly with the evolution of gastropod escape responses. Snyder (1967) describes the alarm reaction of freshwater snails (pulmonate gastropods) to substances released when conspecifics are crushed. Because of the widespread distribution of alarm reactivity in the freshwater snails, he suggests that the response may be a "primitive" behavior pattern. I have found the unique behavior patterns involved in escape responses of marine snails (prosobranch gastropods) to be shared by all members of a large taxonomic group, the superfamily Strombacea (Berg, 1974). There I suggested, on the basis of the sharing the distinctive behavior patterns involved in locomotion, shell righting, and escape, and on the basis of a common descent from an apporhaid-like ancestor, that the behavior of the strombids was more conservative than their widely varying shell morphology. Combined with the naticid work described above, I believe that gastropod behavior patterns related to predator-prey interactions are conservative, i.e., have changed comparatively little through evolution. Heavy selection pressure, by a predator for example, should cause behavior patterns to quickly become stereotyped and remain so—unless, of course, the selection pressure again changes rapidly.

Willows and Dorsett (1975), working on the neurological basis of escape responses in two species of marine nudibranchs (opisthobranch gastropods), describe what they believe to be a transitional state in the evolution of escape responses. Here, two closely related species, *Tritonia diomedia* and *T. hombergi,* show distinct differences between central neuronal arrangements, thresholds of stimuli needed to elicit stereotyped swimming escape response, and elements of response duration over repeated trials. Whether the behavior patterns are being lost or gained by *T. hombergi* is not known. It would be interesting to do developmental studies of the behavior and neurophysiology of both species to see how the escape response develops. Indeed, similar neuroethological studies with prosobranch gastropods whose behavior is well documented (e.g., strombids and naticids) would be of considerable comparative interest. However, prosobranchs have less convenient nervous systems for this kind of study. At the moment nudibranchs are ideally suited for studies of the neurophysiological basis for the evolution and development of behavior.

References

Ansell, A. D. 1960. Observations on predation of *Venus striatula* (Da Costa) by *Natica alderi* (Forbes). *Proc. Malacol. Soc. London* 34:157–164. Plate 9.

Barlow, G. W. 1977. Modal action patterns. *In How animals communicate,* ed., T. A. Sebeok, pp. 94–125. Bloomington: University of Indiana Press.

Batschelet, E. 1965. *Statistical methods for the analysis of problems in animal orientation and certain biological rhythms.* Washington: American Institute of Biological Sciences. pp. 1–57.

Bekoff, M. 1977a. Social communication in canids: evidence for the evolution of a stereotyped mammalian display. *Science* 197:1097–1099.

Bekoff, M. 1977b. Quantitative studies of three areas of classical ethology: social dominance, behavioral taxonomy, and behavioral variability. pp. 1–46, *In Quantitative methods in the study of animal behavior.* ed., B. A. Hazlett, New York: Academic Press.

Berg, C. J., Jr. 1972. Ontogeny of the behavior of *Strombus maculatus* (Gastropoda: Strombidae). *Am. Zool.* 12:427–443.

Berg, C. J., Jr. 1974. A comparative ethological study of strombid gastropods. *Behaviour* 51:274–322.

Berg, C. J., Jr. 1975. A comparison of adaptive strategies of predation among naticid gastropods. *Biol. Bull.* 149:420–421.

Berg, C. J., Jr. 1976. Ontogeny of predatory behavior in marine snails (Prosobranchia: Naticidae). *Nautilus* 90:1–4.

Berg, C. J., Jr., and S. Nishenko. 1975. Stereotypy of predatory boring behavior of Pleistocene naticid gastropods. *Paleobiology* 1:258–260.

Berg, C. J., Jr., and S. Nishenko. 1975. Paleoethology and the evolution of predatory behavior in naticid gastropods. (Submitted for publication.)

Berg, C. J., Jr., and M. E. Porter. 1974. A comparison of predatory behavior among the naticid gastropods *Lunatia heros, Lunatia triseriata* and *Polinices duplicatus. Biol. Bull.* 147:469–470.

Bradley, E. A. 1974. Some observations of *Octopus joubini* reared in an inland aquarium. *J. Zool.* 173:355–368.

Burghardt, G. M., and E. H. Hess. 1966. Food imprinting in the snapping turtle, *Chelydra serpentina*. *Science* 151:108–109.

Carriker, M. R. 1957. Preliminary study of behavior of newly hatched oyster drills, *Urosalpinx cinerea* (Say). *J. E. Mitchel Soc.* 73:328–351.

Carriker, M. R., and D. van Zandt. 1972. Predatory behavior of a shell-boring muricid gastropod. *In Behavior of marine animals,* eds., H. E. Winn and B. L. Olla, pp. 157–244. New York: Plenum.

Choe, S. 1966. On the eggs, rearing, habits of the fry, and growth of some cephalopoda. *Bull. Mar. Sci.* 16:330–348.

Choe, S., and Y. Ohshima. 1963. Rearing of cuttlefishes and squids. *Nature* 197:307.

Eibl-Eibesfeldt, I. 1951. Beobachtungen zur Fortpflanzungsbiologie und Jungendentwicklung des Eichhörnchens (*Sciurus vulgaris* L.). *Z. Tierpsychol.* 8: 370–400.

Fischer-Piette, E. 1935. Historie d'une moulière. Observations sur une phase de desequilibre faunique. *Bull. Biol. France Belgique* 69:153–177.

Hadfield, M. G. 1977. Chemical interactions in larval settling of a marine gastropod. *In NATO conference on marine natural products,* eds., D. J. Faulkner and W. H. Fenical, pp. 403–413. New York: Plenum.

Hanlon, R. T. 1977. Laboratory rearing of the Atlantic reef octopus, *Octopus briareus* Robson, and its potential for mariculture. *Proc. Eighth Ann. Meet. World Mariculture Soc., Jan. 1977.* (In press).

Harris, L. G. 1973. Nudibranch associations. *Curr. Top. Comp. Pathobio.* 2:213–315.

Harris, L. G. 1975. Studies on the life history of two coral-eating nudibranchs of the genus *Phestilla. Biol. Bull.* 149:539–550.

Itami, K., Y. Izawa, S. Maeda, and K. Nakai. 1963. Notes on the laboratory culture of the octopus larvae. *Bull. Jap. Soc. Sci. Fish.* 29:514–520.

Joll, L. M. 1976. Mating, egg-laying and hatching of *Octopus tetricus* (Mollusca: Cephalopoda) in the laboratory. *Mar. Biol.* 36:327–333.

Kriegstein, A. R. 1977a. Development of the nervous system of *Aplysia californica. Proc. Natl. Acad. Sci. U.S.A.* 74:375–378.

Kriegstein, A. R. 1977b. Stages in the post-hatching development of *Aplysia californica. J. Exp. Zool.* 199:275–288.

Kriegstein, A. R., V. Castellucci, and E. R. Kandel. 1974. Metamorphosis of *Aplysia californica* in laboratory culture. *Proc. Nat. Acad. Sci. U.S.A.* 71:3654–3658.

LaRoe, E. T. 1971. The culture and maintenance of the loliginid squids *Sepioteuthis sepioidea* and *Doryteuthis plei. Mar. Biol.* 9:9–25.

Messenger, J. B. 1963. Behaviour of young *Octopus briareus* Robson. *Nature* 197:1186–1187.

Mueller, H. C. 1974. The development of prey recognition and predatory behaviour in the American Kestrel *Falco sparverius. Behaviour* 49:313–324.

Ohshima, Y., and S. Choe. 1961. On the rearing of young cuttlefish and squid. *Bull. Jap. Soc. Sci. Fish.* 27:979–986.

Schleidt, W. M. 1974. How "fixed" is the fixed action pattern? *Z. Tierpsychol.* 36:184–211.

Schleidt, W. M., and M. D. Shalter. 1973. Stereotypy of a fixed action pattern during ontogeny in *Coturnix coturnix coturnix. Z. Tierpsychol.* 33:35–37.

Snyder, N. F. R. 1967. An alarm reaction of aquatic gastropods to intraspecific extract. *Cornell Univ. Agric. Exp. Stat., Memoir* 403:1–122.

Snyder, N. F. R., and H. A. Snyder. 1971. Defenses of the Florida apple snail *Pomacea paludosa. Behaviour* 40:175–215.

Sohl, N. F. 1969. The fossil record of shell boring by snails. *Am. Zool.* 9:725–734.

Sokal, R. R., and F. J. Rohlf. 1973. *Biometry.* San Francisco: Freeman. 776 pp.

Strenth, N. F., and J. E. Blankenship. 1977. Laboratory culture and metamorphosis of larval *Aplysia brasiliana* Rang (Gastropoda, Opisthobranchia). *Bull. Am. Malacol. Union.* 1976:48.

Thomas, R. F., and L. Opresko. 1973. Observations on *Octopus joubini:* four laboratory reared generations. *Nautilus* 87:61–65.

Tranter, D. J., and O. Augustine. 1973. Observations on the life history of the blue-ringed octopus *Hapalochlaena maculosa. Mar. Biol.* 18:115–128.

Wells, M. J. 1958. Factors affecting reactions to *Mysis* by newly hatched *Sepia. Behaviour* 13:96–111.

Wells, M. J. 1962. Early learning in *Sepia. Zool. Soc. Lond. Symp.* 8:149–169.

Wells, M. J., and J. Wells. 1970. Observations on the feeding, growth rate and habits of newly settled *Octopus cyanea. J. Zool.* 161:65–74.

Willows, A. O. D., and D. A. Dorsett. 1975. Evolution of swimming behavior in *Tritonia* and its neurophysiological correlates. *J. Comp. Physiol.* 100:117–133.

Wiley, R. H. 1973. The strut display of male sage grouse: A "fixed" action pattern. *Behaviour* 47:129–152.

Wood, L. 1968. Physiological and ecological aspects of prey selection by the marine gastropod *Urosalpinx cinerea* (Prosobranchia: Muricidae). *Malacologia* 6:267–320.

Chapter Two

A Neuroethological Approach to the Study of the Ontogeny of Coordinated Behavior

ANNE BEKOFF

*Department of Environmental, Population
and Organismic Biology
University of Colorado
Boulder, Colorado 80309*

Introduction

Our understanding of the form, and in some cases the neural bases, of complex coordinated behaviors such as courtship displays, aggressive displays, and vocalizations has greatly increased in recent years. Reviews of studies of such behaviors in adults are provided by Barlow (1977), Bekoff (1977a), Schleidt (1974), and Hoyle (1970, 1975). Most studies have dealt with the relatively discrete and relatively sterotyped motor patterns which seem to make up the natural "units" of behavior. Often called "fixed," or "modal" (see Barlow, 1977) action patterns, these co-ordinated behaviors were first recognized by Whitman (1898), Heinroth (1910), and Lorenz (1950) as playing a central role in behavior.

Many analyses of coordinated behaviors in adult vertebrates have tended to focus on the stereotypy or variability of these motor patterns (e.g., Stamps and Barlow, 1973; Wiley, 1973, 1975; Schleidt, 1974; Crews, 1975). A different approach focusing on analyses of the neural bases for complex coordinated behaviors was first developed by von Holst (1937; 1939). Although Weiss (1941) argued that neurobiologists ought to give more attention to the problem of how the central nervous system pro-duces coordinated behavior, and Tinbergen, in his classic book *The Study of Instinct* (1951) emphasized the importance of neurobiological analyses of complex coordinated behaviors to the understanding of classical ethological issues, little further progress was made until recently.

Beginning in the early 1960s, the "neuroethological" approach was very successfully employed by invertebrate neurobiologists. One of the most significant results of these studies was the first proof that a behavior could be produced by a central pattern generator. Wilson (1961) showed that the pattern of wing muscle contractions seen during locust flight is not dependent on sensory feedback. This is not to say that sensory input is not used during normal flight produced by the central pattern generator. Recently, elegant experiments by Wendler (1974) and Burrows (1975) have indicated that input from locust wing sensory receptors plays a role in stabilizing the amplitude of wing movements and in reducing variability in period length. In several other studies it has been shown that in addition to raising the general excitability level, sensory input may modulate the output of the central pattern generator so as to compensate for intrinsic asymmetries in output, for bodily damage, and for unpredictable environmental variables (Wilson, 1968; Pearson, 1972; Grillner, 1975; Pearson and Duysens, 1976). On the other hand, it should also be noted that central pattern generators can modulate sensory input (e.g., Russell, 1971; Kennedy, Calabrese, and Wine, 1974; Murphey and Palka, 1974; Forssberg, Grillner, and Rossignol, 1975; Grillner, Rossignol, and Wallén, 1976).

Evidence for central pattern generators, and some information on the nature of these neural circuits, has subsequently been found for numerous other behaviors in a variety of invertebrates (e.g., the swimming-escape response in the marine mollusk, *Tritonia,* Dorsett, Willows, and Hoyle, 1973, and Willows, Dorsett, and Hoyle, 1973; courtship in the grasshopper, *Gomphocerippus rufus,* Elsner, 1973; feeding in the snail, *Helisoma trivolvis,* Kater, 1974; swimming in the leech, *Hirudo medicinalis,* Kristan and Calabrese, 1976; Friesen, Poon, and Stent, 1976). For reviews see Hoyle (1975), Eibl-Eibesfeldt (1975), Brown (1975), and Fentress (1976). The success of this approach in dealing with issues of interest to ethologists is due to its emphasis on studying the function of the nervous system during the performance of selected, adaptive behaviors which are part of the animal's normal behavioral repertoire.

The very real possibility of recording intracellularly from the central neurons involved in a particular behavior has made invertebrates particularly attractive to neuroethologists (Hoyle, 1970, 1975). However, while this possibility is somewhat more remote in vertebrate studies, there is still a great deal to be learned about the neural bases of coordinated behaviors without having to rely on intracellular recordings from identified neurons. Thus, following the lead of invertebrate neurobiologists, studies of the neural bases for coordinated behaviors in unanesthetized, behaving vertebrates have begun (for reviews, see DeLong, 1971; Konishi, 1971; Grillner, 1976; Herman *et al.,* 1976; Fentress, 1976). As in studies of invertebrates, one of the first important results of these studies was the demonstration of central pattern generating networks in vertebrates (e.g., cat walking, Engberg and Lundberg, 1969; newt walking, Székeley, Czéh, and Vörös, 1969; mouse grooming, Fentress, 1973; frog calling, Schmidt, 1974; fish swimming, Grillner, Perret, and Zangger, 1975).

While we are still far from completely understanding the neural bases for complex coordinated behavior even in adults, it is useful at this point to consider what can be learned from developmental studies. As Tinbergen (1951) pointed out, the study of development cannot be separated from the causal analysis of adult behavior. Adult behavior is, after all, the result of a developmental process. Moreover, complex behaviors (or the neural circuits underlying them) may actually be composed of simpler units which are more readily recognized or studied in immature animals. Thus, in addition to providing information about developmental mechanisms, ontogenetic studies may make significant contributions to our understanding of the neural bases of adult behaviors.

This chapter will deal with one of the major questions in the field of developmental neuroethology: How are the neural circuits underlying complex coordinated behaviors such as courtship displays, aggressive displays, and the like assembled? In this context, knowing *when* a coordinated behavior first begins to develop is critical because it is only by analyzing the whole sequence of ontogeny from beginning to end that we can begin to understand the rules governing the development of coordinated behavior.

Since, at present, we actually know very little about the neural mechanisms involved in the development of complex coordinated behaviors, I will discuss some relevant studies of somewhat simpler, cyclically repetitive behaviors, such as hatching in chick embryos and locomotion in both vertebrates and invertebrates. In doing so, I will emphasize the techniques and results which I feel will be most important in guiding future research aimed at elucidating the mechanisms underlying the development of more complex coordinated behaviors.

Levels of Coordination

Coordinated behavior results from a group of muscles working together to produce a particular movement or sequence of movements. Thus, the basic element of coordination is a pattern of muscle contractions. This pattern is, of course, the result of selective activation of the motor neuron pools innervating particular muscles which causes each muscle to contract in a specific spatiotemporal relationship to other muscles (Figure 1). This usually, but not always (see p. 26), results in a movement or series of movements that we can recognize as being "coordinated." When movements rather than the underlying muscle activity are analyzed, the characteristics usually associated with coordination are smoothness and a recognizable pattern, or sequence, of movements.

In order to deal with the issues of when and how coordinated behaviors are assembled during ontogeny, it is useful to recognize that coordination can occur at several levels of organization (Weiss, 1941; Tinbergen, 1951). For example, we can speak of:

1. Intramuscular coordination—the motor neurons innervating a single muscle fire together as a unit to cause a smooth contraction of that muscle. This level of coordination will not be considered further in this chapter. For further discussion of intramuscular coordination, see Basmajian (1967).

2. Intrajoint coordination—the various muscles acting to move a single joint contract in a coordinated spatiotemporal pattern. For example, during hatching in the chick embryo, ankle antagonists show a pattern of alternation (Fig. 1).

3. Interjoint coordination—muscles acting at different joints within the same limb contract in a defined spatiotemporal pattern so that the joint movements are coordinated with one another and the limb as a whole can be said to be performing coordinated movement. For example, in the hatching chick embryo knee and ankle extensors co-contract. Knee and ankle flexors co-contract and extensors alternate with flexors (Fig. 1).

4. Interlimb coordination—muscles of different limbs contract in a defined pattern so that movements of one limb bear a recognizable relationship to the movements of another. A simple example would be the alternation of right and left hindlimbs of a cat during walking. A more complex example is seen in the coordination of leg and wing movements during courtship in a grasshopper (Elsner, 1973). This level can be extended to include coordination of other body parts as well, as in the coordination of head and leg movements during walking in the chick (Bangert, 1960).

It is important to note that the coordinated pattern of muscle contraction cannot be defined without reference to a particular behavior, since patterns of muscle contraction may change depending on the behavior being performed. In the extreme case, muscles which act as synergists during one behavior may act as antagonists during another (Wilson,

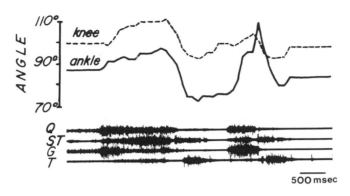

hatching: 20-day embryo

FIG. 1. *A composite figure showing a typical example of the behavioral and electrophysiological data which were collected simultaneously from an embryo during an episode of hatching behavior.* Top: *Instantaneous joint angle of knee and ankle plotted against time. Extension is defined as an increase in angle, flexion as a decrease.* Bottom: *The EMG record which was recorded simultaneously with the videotape record from which the joint angle data above were measured. The two records were correlated by means of a time calibration pulse which appeared synchronously on both records at 2.5 Hz (e.g., see* T *channel of EMG record). Alternation of the antagonists (*Q *and second burst in* ST; G *and* T*) and coactivation of knee and ankle extensors (*Q *and* G, *respectively) and of knee and ankle flexors (second burst in* ST *and* T, *respectively) are clearly evident.* Q = *quadriceps femoris;* ST = *semitendinosus;* G = *gastrocnemius lateralis;* T = *tibialis anterior. (From Bekoff, 1976.)*

1962; Kammer, 1968, 1970; Elsner, 1974; Ayers and Davis, 1977; Sherman, Novotny, and Camhi, 1977).

Methods for Studying the Development of Coordinated Behaviors

In order to study the development of a coordinated behavior, it is necessary to have a standard with which the behavior of the developing animal can be compared. Therefore, the first step of a developmental study is usually to characterize the adult (or fully developed) behavior. This has been done successfully for a variety of vertebrate and invertebrate behaviors. A few examples include: cat walking (Engberg and Lundberg, 1969); fish swimming (von Holst, 1937; Grillner and Kashin, 1976); locust flight (Wilson and Weis-Fogh, 1962); grasshopper courtship (Elsner, 1973); and cricket ecdysis (Carlson and Bentley, 1977). The next step is then to use the same technique to characterize the behavior of younger animals.

At this point I would like to emphasize the necessity for studying the development of movements performed during normal behavior, rather than isolated reflexes. Prior to the elegant work of Hamburger and his associates (for reviews, see Hamburger, 1963; 1968; 1973; Oppenheim, 1974), studies of the embryonic and early postnatal ontogeny of behavior often concentrated exclusively on reflex development (for review of these studies, see Carmichael, 1970). While such studies are valuable in plotting the course of development of functional reflex arcs and the time at which various sensory inputs become functional in producing motor responses, they cannot substitute for study of the development of the normal behavioral repertoire. This is due to the fact that many coordinated behaviors depend, at least in part, on central pattern generators rather than on peripheral (or reflex) pattern generation (see p. 20).

At present there are two basic approaches to characterizing movements for the purpose of studying the development of coordinated behaviors (other than vocalizations—for reviews of the extensive literature on development of vocalizations and the methods used to study them, see Nottebohm, 1975; Marler, 1976). The two approaches are (1) some form of behavioral observation, and (2) electromyogram (EMG) recordings of the activity of identified muscles. Each has advantages and disadvantages that determine which is the method of choice in any given instance. In ideal cases they may be combined (e.g., Engberg and Lundberg, 1969; Bekoff, 1976; Grillner and Kashin, 1976; Ayers and Davis, 1977; Carlson and Bentley, 1977).

For behavioral observations to be of use in a study of the ontogeny of a coordinated behavior, they must be made in a precise, quantitative manner. Therefore it is usually necessary to make a videotape or film record that can be analyzed on a frame-by-frame basis (e.g., Stamps and Barlow, 1973; Wiley, 1973; Bekoff, 1976; M. Bekoff, 1977b; Barlow, 1977). In favorable cases, movement of each joint of each limb (or body part) involved in a particular behavior can be quantified. For example, changes in joint angles can be measured (Goslow, Reinking, and Stuart, 1973; Bekoff, 1976) or additional information can be obtained using a method such as Eshkol-Wachmann movement notation (Golani, 1976).

Behavioral observations have the advantage that they can often proceed with minimal disturbance to the animal and can be used even in difficult field situations. Filmed records can be analyzed most successfully when the behavior of interest involves large amplitude movements. However, problems often arise when developing animals are the subject of study. For example, resolution may be a problem when filming embryos still within the egg or uterus. Moreover, in addition to the fact that embryos themselves are small, embryonic movements may also be of small amplitude, compounding the problem.

Another difficulty encountered in ontogenetic studies is that developing animals may undergo striking morphological changes during the study. For example, limb proportions (Fig. 2) or body size may change drastically and these changes may make comparisons of movements performed at different stages difficult (but see M. Bekoff, 1977b). Postural changes may also occur during development, particularly at birth or hatching, which may alter the appearance of a movement without changing the underlying coordinated sequence of muscle contractions (see p. 32). Moreover, weak muscles in developing animals may also prevent expression of coordinated movements when resistance is encountered, even though the muscles are activated in the appropriate pattern. The important point then, is that, particularly in developing animals, there might not be a 1-to-1 relationship between observed movement and underlying muscular contractions, or that even if there is a relationship, it might not be recognizable even after careful analysis of filmed or videotaped records.

An alternative method for characterizing coordinated behaviors during ontogeny, which avoids some of the problems just mentioned, involves recording electromyograms (EMGs) from individual, identified muscles.

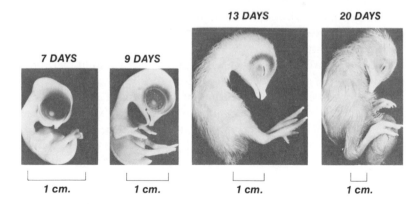

13 DAYS　　**20 DAYS**

7 DAYS　　**9 DAYS**

1 cm.　　1 cm.　　1 cm.　　1 cm.

FIG. 2. *Photographs of chick embryos at four different stages of development. Note the striking differences in hindlimb morphology.*

This technique provides a reliable record of the activity of individual muscles in that bursts of action potentials are recorded during each muscle contraction. No action potentials are recorded when the muscle is not contracting (Fig. 1; see Basmajian [1967] for further discussion of electromyography). By simultaneously recording from two or more muscles which are involved in producing the behavior of interest, a record of the sequence, or pattern of muscle activity is obtained (e.g., Wilson, 1961; Székely, Czéh, and Vörös, 1969; Osse, 1969; Martin and Gans, 1972; Elsner, 1973; Seiler, 1973; Tuttle and Basmajian, 1974; Bekoff, 1976). EMG records can be filmed from an oscilloscope, stored on tape, or recorded with a chart recorder and then later subjected to precise, quantitative analyses (Wyman, 1965; Gerstein and Perkel, 1972; Perkel et al., 1975; Bekoff, 1976).

A major advantage of the EMG recording technique is seen in cases in which behavioral observations might be deceptive, such as during small amplitude movements or weak muscle contractions, when observed movement may not correspond to the output of the neural pattern generating circuits. For example, behavioral observations of 7-day chick embryos did not detect coordinated ankle joint movements, while EMG recordings showed clearly that these movements were the result of coordinated activation of extensor and flexor muscles (Bekoff, Stein, and Hamburger, 1975; Bekoff, 1976). In another case, leg movements of newts, which on the basis of behavioral observations appeared coordinated (although no attempt was made to quantify individual joint angle changes) were shown to be the result of uncoordinated muscle contractions when EMG recordings were made (Czéh and Székely, 1970). Thus EMG recordings provide a clear and unambiguous record of the pattern of muscle contraction, and, therefore, of the motor neuron output, so that the developmental status of the neural circuitry underlying the production of a particular behavior can be assessed in many cases where behavioral observations are not adequate for this purpose.

In cases in which the movements are of large amplitude and are clearly stereotyped, when there is no reason to suspect that movement is not faithfully representing the output of the neural pattern generating circuit, EMG records may not be necessary, although they may be useful to confirm the behavioral observations. In addition, the on-off nature of EMG bursts may make multiple-channel EMG records easier to analyze than changes in joint angles or other behavioral measures when dealing with complex behaviors.

One disadvantage of EMG records is that, although within a single EMG channel there is a good correlation between the firing rate of action potentials and the strength of contraction of the muscle being recorded from (Basmajian, 1967; Osse, 1969), there is no parameter of the EMG record which can be used to make direct comparisons of the relative strength of muscle contractions between muscles or between preparations (Martin and Gans, 1972). This is due to the fact that parameters which might be used to judge the magnitude of the EMG, such as the height of the spikes or their firing rate, are strongly affected by electrode properties such as size and impedance, and by electrode placement. In addition, the effective movement produced by activation of one muscle depends strongly on such factors as response properties of the muscle itself (Partridge, 1966; Wilson and Larimer, 1968), initial length of the muscle (Bergel et al., 1972; Grillner, 1972; Goslow, Reinking, and Stuart, 1973) which is dependent on joint angle, and on other mechanical considerations depending on joint angles (Sherrington, 1910; Grillner, 1972). Concurrent activity in other muscles also influences the actual movement produced. For example, co-contraction of antagonists may be used to brake strong movements (Wilson and Weis-Fogh, 1962; Basmajian, 1967; Engberg and Lundberg, 1969) or to load the system so as to produce a springlike action (Hoyle, 1955; Brown, 1967; Godden, 1970). Thus it is not possible to reconstruct the movements produced by the muscle contractions accurately using EMG records alone. For this reason, combined EMG and film or videotape records provide more information than will either technique alone (e.g., Engberg and Lundberg, 1969; Osse, 1969; Bekoff, 1976).

In order to make EMG recordings, it is of course necessary to implant electrodes. If this were to interfere with, or alter, normal behavior then this technique would be of little value. However, fine wire electrodes (50 μm or less in diameter) and long, flexible leads can be used to allow relatively unrestricted movement of an animal. For example, Elsner (1973) has been able to implant more than 30 EMG electrodes in thoracic muscles of a single grasshopper without interfering with complex courtship behaviors. For dealing with embryos, suction electrodes pulled from Tygon tubing and suspended from polyethylene tubing have been found to be flexible enough to follow movements without damaging delicate embryonic chick muscles (Bekoff, 1976). EMG recordings have also been made from mammalian fetuses (Lewis and Basmajian, 1959; Ranney and Basmajian, 1960; Ånggärd, Bergström, and Bernhard, 1961; Bergström, Hellström, and Stenberg, 1962; Boëthius, 1967) and from insects during postembryonic development (Bentley and Hoy, 1970; Kutsch, 1971; Altman, 1975; Kammer and Rheuben, 1976).

One caution, which applies to both behavioral observations and EMG recordings when they are used to analyze the development of the neural circuitry underlying coordinated behaviors, is that if the output of the neural pattern generating circuit is inhibited, neither of these techniques will detect it. This is important to note because there are several cases known in which specific behavior patterns are inhibited during development (Bentley and Hoy, 1970; Camhi, 1974; deGroat *et al.,* 1975; Truman, 1976). Therefore, it is important to recognize that absence of a particular behavior from the repertoire of an organism at a particular stage of development does not necessarily mean absence of the neural pattern-generating circuit which produces that behavior.

Development of Intrajoint and Interjoint Coordination

Using these rather simple and straightforward techniques first to characterize precisely a well-developed coordinated behavior, and then to trace the ontogeny of that behavior has provided some important information on when various elements of coordinated behaviors first appear.

One of the most important results to emerge is that, at least in some cases, the elements of coordinated behavior patterns develop much earlier than behavioral observations would suggest. For example, in the chick embryo, the first movements which appear recognizably coordinated to the behavioral observer occur during tucking, a prehatching behavior which first appears at about 16–17 days of incubation (Hamburger and Oppenheim, 1967; Oppenheim, 1973). These movements are smooth and appear to involve coordination of wing, trunk, and head movements. Tucking movements and other coordinated prehatching movements, as well as hatching behavior itself, have all been classified as Type III movements (Hamburger and Oppenheim, 1967). Behaviorally they are quite distinct in appearance from the other classes of embryonic movements, Types I and II, which are the only types observed prior to the onset of tucking and which continue to appear in between pre-hatching behavior episodes up until hatching. Type I and Type II movements appear jerky and uncoordinated, in striking contrast to the Type III behaviors (Hamburger and Balaban, 1963; Hamburger *et al.,* 1965; Hamburger and Oppenheim, 1967). Specifically, coordination between body parts was not detected, although the possibility that intrajoint or interjoint coordination might be present was not excluded (Hamburger, 1968).

In order to trace the development of hatching behavior to its origins and to examine the possibility that coordinated movements might develop prior to 17 days, studies of intrajoint and interjoint coordination in the leg of the chick embryo were carried out (Bekoff, Stein, and Hamburger, 1975; Bekoff, 1976). A schematic diagram of the experimental setup is shown in Figure 3. Videotape records were made at all stages of

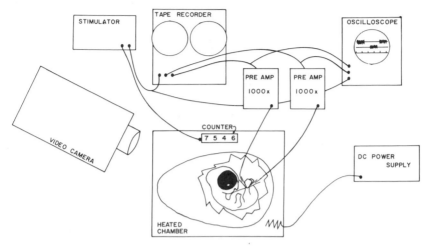

FIG. 3. *Schematic diagram of the experimental setup for simultaneously recording behavioral and EMG data from a spontaneously behaving chick embryo. A hole is made in the shell over the embryo and the egg is placed in a heated, humidified chamber. Extracellular suction electrodes are placed on individual, identified muscles and the recorded EMGs are amplified × 1000, viewed on the oscilloscope screen and also stored on magnetic tape for later filming and analysis. A stimulator is used to produce a time mark on one channel of the EMG record at a frequency of 2.5 Hz. This same signal drives a counter which is in the field of view of the videotape camera. The videotape records of behavior can then be synchronized on a frame-by-frame basis with the simultaneously recorded EMG records.*

development, from near the time at which the first leg movements could be detected (6–6½ days of incubation) through hatching. However, due to the small size of the early embryos and the small amplitude of many embryonic movements, frame-by-frame analysis of videotape records was not a satisfactory method for resolving changes in leg joint angles at very early stages. Thus, while qualitative observations of the embryonic movements could be made from the videotapes, in order to determine whether or not coordinated movement developed prior to the time at which behavioral observations could easily detect patterned movement, EMG recordings were made from identified knee and ankle muscles (Fig. 4). The pattern of activity of these muscles was determined during hatching and found to be relatively stereotyped (Fig. 1). Knee and ankle extensors were active synchronously, as were knee and ankle flexors. Flexors and extensors alternated. In addition, one bi-functional muscle,

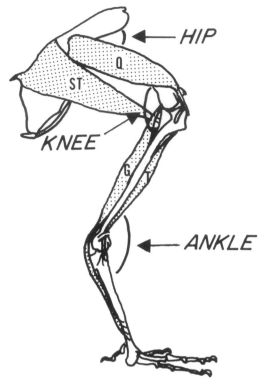

FIG. 4. *A diagram of the chick leg indicating the anatomical relationships of four of the leg muscles from which EMG recordings have been made. (From Bekoff, 1976.)*

semitendinosus, which is both a hip extensor and a knee flexor (Chamberlain, 1943; Bekoff, 1976) showed two bursts of activity, one during extension of the leg and the other during flexion. Thus a consistent spatiotemporal pattern of muscle activity could be identified. This was used to define intrajoint (ankle) and interjoint (knee-ankle) coordination.

Analyses of EMG records made during Type I behavior at various earlier stages of development showed that interjoint coordination was present at least as early as nine days of incubation and intrajoint coordination at least as early as seven days (Fig. 5). Simultaneous recordings have not yet been made from knee and ankle muscles in 7-day embryos so it is not known whether interjoint coordination is established that early.

These results establish that both intrajoint and interjoint coordination develop entremely early during the embryonic life of the chick.

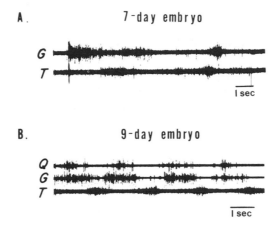

FIG. 5. *EMG records showing several cycles of activity during typical sequences of activation in* (A) *a seven-day embryo and* (B) *a nine-day embryo.*

Intrajoint coordination appears to develop even before reflex arcs are functional (Bekoff, Stein, and Hamburger, 1975) and therefore the underlying neural circuitry does not appear to require sensory feedback for its assembly. Intrajoint coordination of the wing movements involved in flight and stridulation in crickets (Bentley and Hoy, 1970), flight in locusts (Kutsch, 1971, 1974; Weber, 1972; Altman, 1975), and flight in moths (Kammer and Rhuben, 1976) also develop prior to the time at which the behavior is first seen and its development does not seem to be dependent on sensory input. The neural pattern generator underlying coordinated swimmeret movement in lobsters has also been shown to develop after experimental removal of sensory feedback (Davis, 1973).

The only EMG study of the development of intrajoint coordination which has been carried out on a mammal is a study of ankle antagonists in postnatal kittens (Scheibel and Scheibel, 1970). The earliest evidence of intrajoint coordination of the ankle muscles was seen at 12–14 days after birth. This is interesting in light of behavioral observations of kitten fetuses which suggested that interlimb coordination develops prenatally (Brown, 1915; Windle and Griffin, 1931). These results may suggest that intrajoint and interlimb coordination of hip muscles develop long before intrajoint (and interlimb) coordination of ankle muscles. In any case, results of quantitative EMG analyses of fetal limb coordination (intrajoint, interjoint, and interlimb) would undoubtedly be interesting.

What Next? Intrajoint and Interjoint Coordination

There is still much to be learned about the ontogeny of hatching behavior in the chick embryo. One unsolved question involves the time of development of interjoint coordination. If intrajoint coordination of ankle and knee develop separately, and interjoint coordination develops later, it would suggest that the pattern generator for one leg is composed of a set of circuits, one for each joint, which are tightly coupled during such behaviors as hatching.

Another question involves the relationship of hatching behavior of the embryo to posthatching behavior of the chick. A neural circuit capable of programming extension at all three leg joints in alternation with flexion of all three is used during hatching (Bekoff, 1976). This pattern is simple, but in fact is quite similar to the pattern seen in the cat hindlimb during walking (Engberg and Lundberg, 1969). Certainly the behavior of hatching, taken as a whole, differs markedly from walking. Head, trunk, and wing movements observed during hatching seem unsuited for walking. The posture of the embryo and the position of the legs during hatching differ drastically from the posthatched, walking chick (Fig. 6), and yet it may be that the same neural pattern-generating circuit for leg movements is used. It would be economical for the chick to use the

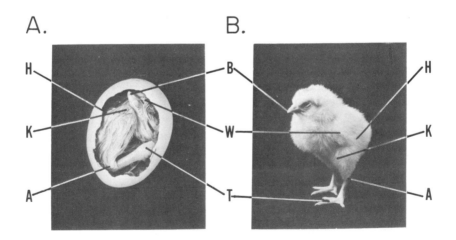

FIG. 6. *A. Photograph of a 20-day old embryo in the hatching position. B. Photograph of a one-day old hatched chick. Note the very different postures and leg positions.* B = *beak;* W = *wing;* H = *hip;* K = *knee;* A = *ankle;* T = *toes.*

circuit developed embryonically for hatching in post-hatching walking. In this context, recent studies in invertebrates are particularly interesting since it has been found that the same pattern generating circuits may be used for more than one behavior (Elsner, 1974; Camhi, 1977).

Some of my own unpublished EMG results indicate that intrajoint coordination at the ankle during walking in unrestrained week-old chicks is quite similar to that seen during hatching. Current work (in collaboration with M. Nusbaum) is designed to compare interjoint and interlimb coordination during walking and hatching. If these two behaviors turn out to share the same, or elements of the same, pattern generator this result will have important implications for our understanding of the nature of embryonic antecedents of postembryonic behavior since it will indicate that such antecedents need have little resemblence to the postnatal behavior. It will also suggest the power of the EMG recording technique in detecting the underlying similarities.

Development of Interlimb Coordination

A third unresolved question deals with the development of interlimb coordination. In the chick embryo, behavioral observations indicate that right and left legs alternate during hatching (Hamburger and Oppenheim, 1967; Oppenheim, 1973). Thus interlimb coordination develops embryonically. Foelix and Oppenheim (1973) mention that brief episodes of alternating flexions and extensions of the two legs are occasionally seen in 9-day embryos. However, Provine (1977) reports that consistent patterns of interlimb coordination are present only during the last week of incubation as was suggested previously by Hamburger and co-workers (Hamburger and Oppenheim, 1967; Hamburger, 1968; Hamburger, 1973). Between 15 days and hatching (21 days) an increase in synchronous wing movements and alternating leg movements was observed (Provine, 1977). These observations suggest that, in the chick, interlimb coordination develops after intrajoint and interjoint coordination. Based on behavioral observations Hughes and Prestige (1967) also suggested that interlimb coordination developed later than intrajoint and interjoint coordination in *Xenopus* tadpoles. Given the jerkiness and small amplitude of many early embryonic movements, unambiguously tracing interlimb coordination back to its earliest appearance in chick and amphibian embryos may require analyses of EMG records.

Behavioral observations of kitten fetuses have also suggested prenatal development of interlimb coordination (Brown, 1915; Windle and Griffin, 1931). A recent study of interlimb coordination in newborn rats during swimming using frame-by-frame analysis of videotapes has shown that phase coupling of the limbs is well-developed even as early as the first day after birth (Bekoff and Trainor, in preparation). Fentress (in Chapter 16) has reported a similar result for newborn mice. Thus, despite the fact that nonquantitative behavioral observations of rat fetuses did not reveal interlimb coordination (Narayanan, Fox, and Hamburger, 1971; Bekoff, unpublished observations), it seems likely that more precise behavioral observations or EMG recordings will reveal that interlimb coordination develops prenatally in rats. This is interesting since, under normal circumstances (i.e., if not placed in a swimming pool), coordination of all four limbs is not seen in rats until over a week after birth (Stelzner, 1971; Altman and Sudershan, 1975).

The only studies which have used EMG recordings to trace the development of interlimb coordination to its beginnings have been done on hindwing-forewing coordination in Orthopteran insects (Bently and Hoy, 1970; Kutsch, 1971; Altman, 1975). These studies found that interlimb coordination developed after intrajoint coordination.

Conclusions and Some Implications For Future Studies

Results in chicks and rats suggest that the neural circuits responsible for producing coordinated behavior may develop long before the coordinated behavior is readily recognizable to the casual observer. The most important message provided by these results is that studies whose aim is to trace a coordinated behavior back to its origins cannot fail to consider the possible relevance of antecedent behaviors which do not bear an obvious resemblance to the particular behavior of interest. That is, behaviors may appear to develop *de novo* when, in fact, the underlying neural circuitry had developed gradually. Using the proper techniques it should often be possible to follow this gradual development.

These conclusions undoubtedly apply to the development of more complex coordinated behaviors such as courtship displays, aggressive displays, etc., as well as to relatively simpler behaviors such as hatching and walking. In fact, Barlow (1977) points out that many displays have been derived from locomotory behavior and some displays are little more than ritualized locomotion. The relatively simpler behaviors, hatching and locomotion, have been the primary subject of study in the analysis of the neural bases for coordinated behavior to date because they are more amenable to analysis. They are relatively stereotyped, cyclically repetitive, and reasonably easy to elicit in the laboratory, even in semirestrained animals. While we do not yet have a complete understanding of how even these relatively simple behaviors are assembled during development, there are reassuring indications that analysis of such simple units is relevant to our ultimate goal of understanding more complex behaviors. Most significant is the recent finding that complex sequences of coordinated behavior such as cricket ecdysis result from the incorporation of many simpler units of behavior through the use of coordinating and controlling elements (Carlson and Bentley, 1977). Future studies will determine whether this is true also for vertebrates, but there seems little reason to doubt it *a priori.*

The importance of this result is that, if we must think of a complex coordinated behavior such as the greylag goose triumph ceremony (Fischer, 1965) as a single, unitary behavior, produced by a single, complex pattern generator, then the possibility of investigating that behavior and its neural basis seems quite remote at present. If, on the other hand, the display is found to consist of many separable units, then the possibility of analyzing each unit and then attempting to obtain information about the coordination and control of these units seems more reasonable. From a developmental point of view, the question of how the coordinated behavior comprising each unit is assembled can be studied first, and then the mechanisms by which the units become linked into adaptively significant sequences can be analyzed.

In summary, the EMG recording technique, especially when combined with quantitative behavioral observations, provides a powerful tool for studying the development of coordinated behavior. As emphasized above, the power of the EMG recording techniques lies in its ability to monitor the output of the central nervous system during ongoing, normal behaviors. In developmental studies, it allows the recognition of patterned motor output when behavioral observations are not adequate for this purpose. Its use thus represents an important step toward a neuroethological analysis of the ontogeny of coordinated behaviors. For example, the use of the EMG recording technique has shown that the neural circuitry underlying specific coordinated behaviors is, at least in some cases, assembled long before the behavior is expressed in a recognizable form (Bentley and Hoy, 1970; Bekoff, Stein, and Hamburger, 1975; Bekoff, 1976). In addition, it has provided evidence that at least some elements of these neural pattern generators normally develop in the absence of patterned sensory input. In the future, the use of this neuroethological approach has strong potential for providing a better understanding of (1) the mechanisms by which coordinated behaviors are assembled and (2) the relationship of early behaviors to the complex coordinated behaviors of adults.

References

Altman, J., and K. Sudarshan. 1975. Postnatal development of locomotion in the laboratory rat. *Anim. Behav.* 23:896–920.

Altman, J. S. 1975. Changes in the flight motor pattern during the development of the Australian plague locust, *Chortoicetes terminifera. J. Comp. Physiol.* 97:127–142.

Ayers, J. L., and W. J. Davis. 1977. Neuronal control of locomotion in the lobster, *Homarus americanus.* I. Motor programs for forward and backward walking. *J. Comp. Physiol.* 115:1–27.

Bangert, H. 1960. Untersuchungen zur Koordination der Kopf-und Beinbewegungen beim Haushuhn. *Z. Tierpsychol.* 17:143–164.

Barlow, G. W. 1977. Modal action patterns. In *How animals communicate,* ed., T. A. Sebeok, pp. 94–125. Bloomington: University of Indiana Press.

Basmajian, J. V. 1967. *Muscles alive: their functions revealed by electromyography.* Baltimore: Williams and Wilkins. 421 pp.

Bekoff, A. 1976. Ontogeny of leg motor output in the chick embryo: a neural analysis. *Brain Res.* 106:271–291.

Bekoff, A., P. S. G. Stein, and V. Hamburger. 1975. Coordinated motor output in the hindlimb of the 7-day chick embryo. *Proc. Natl. Acad. Sci. U.S.A.* 72:1245–1248.

Bekoff, M. 1977a. Quantitative studies of three areas of classical ethology: social dominance, behavioral taxonomy, and behavioral variability. In *Quantitative methods in the study of animal behavior,* ed., B. A. Hazlett, pp. 1–46. New York: Academic Press.

Bekoff, M. 1977b. Social communication in canids: evidence for the evolution of a stereotyped mammalian display. *Science* 197:1225–1227.

Bentley, D., and R. R. Hoy. 1970. Postembryonic development of adult motor patterns in crickets: a neural analysis. *Science* 170:1409–1411.

Bergel, D. H., M. C. Brown, R. G. Butler, and R. M. Zacks. 1972. The effect of stretching a contracting muscle on its subsequent performance during shortening. *J. Physiol.* 225:21P–22P.

Brown, J. L. 1975. *The evolution of behavior.* New York: W. W. Norton. 761 pp.

Brown, R. H. J. 1967. Mechanism of locust jumping. *Nature* 214:939.

Brown, T. G. 1915. On the activities of the central nervous system of the unborn foetus of the cat; with a discussion of the question whether progression (walking, etc.) is a "learnt" complex. *J. Physiol.* 49:208–215.

Burrows, M. 1975. Monosynaptic connexions between wing stretch receptors and flight motoneurons of the locust. *J. Exp. Biol.* 62:189–219.

Camhi, J. M. 1974. Neural mechanisms of response modification in insects. *In Experimental analysis of insect behavior,* ed., L. Barton Browne, pp. 60–86. New York: Springer Verlag.

Camhi, J. M. 1977. Behavioral switching in cockroaches: transformations of tactile reflexes during righting behavior. *J. Comp. Physiol.* 113:283–301.

Carlson, J. R., and D. Bentley. 1977. Ecdysis: neural orchestration of a complex behavioral performance. *Science* 195:1006–1008.

Carmichael, L. 1970. The onset and early development of behavior. In *Carmichael's manual of child psychology*. Vol. 1, pp. 447–563. ed., P. Mussen, New York: Wiley.

Chamberlain, F. W. 1943. Atlas of avian anatomy. *Mich. Agric. Exp. Stat. Memoir Bull.* No. 5.

Crews, D. 1975. Inter- and intraindividual variation in display patterns in the lizard, *Anolis carolinensis. Herpetologica* 31:37–47.

Czéh, G., and G. Székely. 1971. Muscle activities recorded simultaneously from normal and supernumerary forelimbs in *Ambystoma. Acta Physiol. Acad. Sci. Hung.* 40:287–301.

Davis, W. J. 1973. Development of locomotor patterns in the absence of peripheral sense organs and muscles. *Proc. Nat. Acad. Sci. U.S.A.* 70:954–958.

de Groat, W. C., J. W. Douglas, J. Glass, W. Simonds, B. Weimer, and P. Werner. 1975. Changes in somato-vesical reflexes during postnatal development in the kitten. *Brain Res.* 94:150–154.

Delong, M. 1971. Central patterning of movement. *Neurosci. Res. Program Bull.* 9:10–30.

Dorsett, D. A., A. O. D. Willows, and G. Hoyle. 1973. The neuronal basis of behavior in *Tritonia.* IV. The central origin of a fixed action pattern demonstrated in the isolated brain. *J. Neurobiol.* 4:287–300.

Eibl-Eibesfeldt, I. 1975. *Ethology, the biology of behavior.* New York: Holt, Rinehart and Winston. 625 pp.

Elsner, N. 1973. The central nervous control of courtship behavior in the grasshopper *Gomphocerippus rufus* L. (Orthoptera: acrididae). In *Neurobiology of invertebrates.* ed., J. Salanki, pp. 261–287, Budapest: Akad. Kiado.

Elsner, N. 1974. Neural economy: bifunctional muscles and common central pattern elements in leg and wing stridulation of the grasshopper *Stenobothrus rubicundus* Germ. (Orthoptera: Acrididae). *J. Comp. Physiol.* 89:227–236.

Engberg, I., and A. Lundberg. 1969. An electromyographic analysis of muscular activity in the hindlimb of the cat during unrestrained locomotion. *Acta Physiol. Scand.* 75:614–630.

Fentress, J. C. 1972. Development and patterning of movement sequences in inbred mice. In *The biology of behavior,* ed., J. A. Kiger, pp. 83–132. Oregon: Oregon State Univ. Press.

Fentress, J. C. 1973. Development of grooming in mice with amputated forelimbs. *Science* 179:704–705.

Fentress, J. C. 1976. *Simpler networks and behavior.* Sunderland, Mass.: Sinauer Associates. 403 pp.

Fischer, H. 1965. Das Triumphgeschrei der Graugans *(Anser anser). Z. Tierpsychol.* 22:247–304.

Foelix, R. F., and R. W. Oppenheim. 1973. Synaptogenesis in the avian embryo: ultrastructure and possible behavioral correlates. In *Studies on the development of behavior and the nervous system: behavioral embryology,* Vol. 1, ed., G. Gottlieb, pp. 103–139. New York: Academic Press.

Forssberg, H., S. Grillner, and S. Rossignol. 1975. Phase dependent reflex reversal during walking in chronic spinal cats. *Brain Res.* 85:103–107.

Friesen, W. O., M. Poon, and G. S. Stent. 1976. An oscillatory neuronal circuit generating a locomotory rhythm. *Proc. Natl. Acad. Sci. U.S.A.* 73:3734–3738.

Gerstein, G. L., and D. H. Perkel. 1972. Mutual temporal relationships among neuronal spike trains. Statistical techniques for display and analysis. *Biophys. J.* 12:453–473.

Godden, D. H. 1970. The neural basis for locust jumping. *Am. Zoologist* 9:1139–1140.

Golani, I. 1976. Homeostatic motor processes in mammalian interactions: a choreography of display. In *Perspectives in ethology,* Vol. 2. eds., P. P. G. Bateson and P. Klopfer, pp. 69–134. New York: Plenum Press.

Goslow, G. E., R. M. Reinking, and D. C. Stuart. 1973. The cat step cycle: hind limb joint angles and muscle lengths during unrestrained locomotion. *J. Morph.* 141:1–42.

Grillner, S. 1972. The role of muscle stiffness in meeting the changing postural and locomotor requirements for force development by the ankle extensors. *Acta Physiol. Scand.* 86:92–108.

Grillner, S. 1975. Locomotion in vertebrates: central mechanisms and reflex interaction. *Physiol. Rev.* 55:247–304.

Grillner, S., and S. Kashin. 1976. On the generation and performance of swimming in fish. In *Neural control of locomotion,* eds., R. M. Herman, S. Grillner, P. S. G. Stein and D. G. Stuart, pp. 181–201. New York: Plenum Press.

Grillner, S., and P. Zangger. 1975. How detailed is the central pattern generator for locomotion? *Brain Res.* 88:367–371.

Grillner, S., C. Perret, and P. Zangger. 1976. Central generation of locomotion in the spinal dogfish. *Brain Res.* 109:255–269.

Grillner, S., S. Rossignol, and P. Wallén. 1976. Phase dependent reflex reversal during swimming in the spinal dogfish. *Acta Physiol. Scand.* (In press).

Hamburger, V. 1963. Some aspects of the embryology of behavior. *Q. Rev. Biol.* 38:342–365.

Hamburger, V. 1968. IV. Emergence of nervous coordination. Origins of integrated behavior. *Dev. Biol. Suppl.* 2:251–271.

Hamburger, V. 1973. Anatomical and physiological basis of embryonic motility in birds and mammals. In *Studies on the development of behavior and the nervous system: behavioral embryology.* Vol. 1, ed., G. Gottlieb, pp. 52–76. New York: Academic Press.

Hamburger, V., and M. Balaban. 1963. Observations and experiments on spontaneous rhythmical behavior in the chick embryo. *Develop. Biol.* 7:533–545.

Hamburger, V., and R. Oppenheim. 1967. Prehatching motility and hatching behavior in the chick. *J. Exp. Zool.* 166:171–204.

Hamburger, V., E. Wenger, and R. Oppenheim. 1966. Motility in the chick embryo in the absence of sensory input. *J. Exp. Zool.* 162:133–160.

Hamburger, V., M. Balaban, R. Oppenheim, and E. Wenger. 1965. Periodic motility of normal and spinal chick embryos between 8 and 17 days of incubation. *J. Exp. Zool.* 159:1–14.

Heinroth, O. 1910. Beiträge zur Biologie, insbesondere Psychologie und Ethologie der Anatidien. *Verh. 5th Internatl. Ornithol. Kongr.* pp. 589–702.

Herman, R. M., S. Grillner, P. S. G. Stein, and D. G. Stuart, eds. 1976. *Neural control of locomotion.* New York: Plenum Press. 822 pp.

Hoyle, G. 1955. Neuromuscular mechanisms of a locust skeletal muscle. *Proc. Roy. Soc. Lond. B.* 143:343–367.

Hoyle, G. 1970. Cellular mechanisms underlying behavior-neuroethology. *Adv. Insect Physiol.* 7:349–444.

Hoyle, G. 1975. Identified neurons and the future of neuroethology. *J. Exp. Zool.* 194:51–74.

Hughes, A., and M. C. Prestige. 1967. Development of behaviour in the hindlimb of *Xenopus laevis. J. Zool.* 152:347–359.

Kammer, A. E. 1968. Motor patterns during flight and warm-up in *Lepidoptera. J. Exp. Biol.* 48:89–109.

Kammer, A. E. 1970. A comparative study of motor patterns during pre-flight warm-up in hawkmoths. *Z. Vergl. Physiol.* 70:45–56.

Kammer, A. E., and M. B. Rheuben. 1976. Adult motor patterns produced by moth pupae during development. *J. Exp. Biol.* 65:65–84.

Kater, S. B. 1974. Feeding in *Helisoma trivolvis:* The morphological and physiological bases of a fixed action pattern. *Am. Zoologist.* 14:1017–1036.

Kennedy, D., R. L. Calabrese, and J. J. Wine. 1974. Presynaptic inhibition: primary afferent depolarization in crayfish neurons. *Science* 186:451–454.

Konishi, M. 1971. Ethology and neurobiology. *Am. Scientist* 59:56–63.

Kristan, W. B. and R. L. Calabrese. 1976. Rhythmic swimming activity in neurons of the isolated nerve cord of the leech. *J. Exp. Biol.* 65:643–668.

Kutsch, W. 1971. The development of the flight pattern in the desert locust, *Schistocerca gregaria. Z. Vergl. Physiol.* 74:156–168.

Kutsch, W. 1974. The influence of the wing sense organs on the flight motor pattern in maturing adult locusts. *J. Comp. Physiol.* 88:413–424.

Lorenz, K. Z. 1950. The comparative method in studying innate behavior patterns. *Symp. Soc. Exp. Biol.* 4:221–268.

Marler, P. 1976. Sensory templates in species-specific behavior. In *Simpler networks and behavior,* ed., J. C. Fentress, pp. 314–329. Sunderland: Sinauer Associates.

Martin, W. F., and C. Gans. 1972. Muscular control of the vocal tract during release signalling in the toad *Bufo valliceps. J. Morph.* 137:1–27.

Murphey, R. K., and J. Palka. 1974. Efferent control of cricket giant fibers. *Nature* 248:249–251.

Narayanan, C. H., M. W. Fox, and V. Hamburger. 1971. Prenatal development of spontaneous and evoked activity in the rat *(Rattus norvegicus albinus). Behaviour* 40:100–134.

Nottebohm, F. 1975. Vocal behavior in birds. In *Avian biology.* Vol. 5, eds., D. Farner, J. R. King, and K. C. Parkes, pp. 289–332. New York: Academic Press.

Oppenheim, R. W. 1972. An experimental investigation of the possible role of tactile and proprioceptive stimulation in certain aspects of embryonic behavior in the chick. *Dev. Psychobiol.* 5:51–91.

Oppenheim, R. W. 1973. Prehatching and hatching behavior: a comparative and physiological consideration. In *Studies on the development of behavior and the nervous system: behavioral embryology.* Vol. 1, ed., G. Gottlieb, pp. 163–244. New York: Academic Press.

Oppenheim, R. W. 1974. The ontogeny of behavior in the chick embryo. In *Advances in the study of behavior.* Vol. 5, eds., D. S. Lehrman, R. A. Hinde, and E. Shaw, pp. 133–172. New York: Academic Press.

Osse, J. W. M. 1969. Functional morphology of the head of the perch (*Perca fluviatilis L.*): an electromyographic study. *Neth. J. Zool.* 19:289–392.

Partridge, L. D. 1966. Signal-handling characteristics of load moving skeletal muscle. *Am. J. Physiol.* 210:1178–1191.

Pearson, K. G. 1972. Central programming and reflex control of walking in the cockroach. *J. Exp. Biol.* 56:173–193.

Pearson, K. G., and J. Duysens. 1976. Function of segmental reflexes in the control of stepping in cockroaches and cats. In *Neural control of locomotion,* eds., R. M. Herman, S. Grillner, P. S. G. Stein, and D. G. Stuart, pp. 519–537. New York: Plenum Press.

Perkel, D. H., G. L. Gerstein, M. S. Smith, and W. G. Tatton. 1975. Nerve-impulse patterns: a quantitative display technique for three neurons. *Brain Res.* 100:271–296.

Provine, R. R. 1977. Emergence of between-limb synchronization in chick embryos. *Neurosci. Abstr.* 3:116.

Russell, I. J. 1971. The role of the lateral line efferent system in *Xenopus laevis. J. Exp. Biol.* 54:621–641.

Scheibel, M. E., and A. B. Scheibel. 1970. Developmental relationship between spinal motoneuron dendrite bundles and patterned activity in the hind limb of cats. *Exp. Neurol.* 29:328–335.

Schleidt, W. M. 1974. How "fixed" is the fixed action pattern? *Z. Tierpsychol.* 36:184–211.

Schmidt, R. S. 1974. Neural correlates of frog calling. Independence from peripheral feedback. *J. Comp. Physiol.* 88:321–333.

Seiler, R. 1973. On the function of facial muscles in different behavioral situations. A study based on muscle morphology and electromyography. *Am. J. Phys. Anthropol.* 38:567–572.

Sherman, E., M. Novotny, and J. M. Camhi. 1977. A modified rhythm employed during righting behavior in the cockroach *Gromphadorhina portentosa. J. Comp. Physiol.* 113:303–316.

Sherrington, C. S. 1910. Flexion-reflex of the limb, crossed extension reflex, and reflex stepping and standing. *J. Physiol.* 40:28–121.

Stamps, J. A., and G. W. Barlow. 1973. Variation and stereotypy in the displays of *Anolis aneus* (Sauria: Iguanide). *Behaviour* 47:67–94.

Stelzner, D. J. 1971. The normal postnatal development of synaptic endfeet in the lumbosacral spinal cord and of responses in the hind limbs of the albino rat. *Exp. Neurol.* 31:337–357.

Székely, G., G. Czéh, and G. Vörös. 1969. The activity pattern of limb muscles in freely moving normal and deafferented newts. *Exp. Brain Res.* 9:53–62.

Tinbergen, N. 1961. The study of instinct. Oxford: Clarendon Press, 228 pp.

Truman, J. W. 1976. Development and hormonal release of adult behavior patterns in silkmoths. *J. Comp. Physiol.* 107:39–48.

Tuttle, R., and J. V. Basmajian. 1974. Electromyography of *Pan gorilla:* an experimental approach to the problem of hominization. *Symp. 5th Cong. Intl. Primat. Soc.* 303–314.

von Holst, E. 1937. On the nature of order in the central nervous system. In *The behavioural physiology of animals and man.* Vol. 1. Translated from German by R. Martin. pp. 3–32. Coral Gables: University of Miami Press.

von Holst, E. 1939. Relative coordination as a phenomenon and as a method of analysis of central nervous functions. In *The behavioral physiology of animals and man.* Vol. 1. Translated from German by R. Martin. pp. 33–135. Coral Gables: University of Miami Press.

Weber, T. 1972. Stabilisierung des Flugrhythmus durch 'Erfahrung' bei der Feldgrille. *Naturwissenschaften* 59:366.

Weiss, P. 1941. Self-differentiation of the basic patterns of coordination. *Comp. Psychol. Monogr.* 17:1–96.

Wendler, G. 1974. The influence of proprioceptive feedback on locust flight coordination. *J. Comp. Physiol.* 88:173–200.

Whitman, C. O. 1898. Animal behaviour. Biol. Lect. Marine Biol. Lab., Woods Hole. 285–338.

Wiley, R. H. 1973. The strut display of the male sage grouse: a "fixed" action pattern. *Behaviour* 47:129–152.

Wiley, R. H. 1975. Multidimensional variation in an avian display: implications for social communication. *Science* 190:482–483.

Willows, A. O. D., D. A. Dorsett, and G. Hoyle. 1973. The neuronal basis of behavior in *Tritonia.* III. Neuronal mechanism of a fixed action pattern. *J. Neurobiol.* 4:255–285.

Wilson, D. M. 1961. The central nervous control of flight in a locust. *J. Exp. Biol.* 38:471–490.

Wilson, D. M. 1962. Bifunctional muscles in the thorax of grasshoppers. *J. Exp. Biol.* 39:669–677.

Wilson, D. M. 1968. Inherent asymmetry and reflex modulation of the locust flight motor pattern. *J. Exp. Biol.* 48:631–641.

Wilson, D. M., and J. L. Larimer. 1968. The catch property of ordinary muscle. *Proc. Natl. Acad. Sci. U.S.A.* 61:909–916.

Wilson, D. M., and T. Weis-Fogh. 1962. Patterned activity of coordinated motor units, studied in flying locusts. *J. Exp. Biol.* 39:643–667.

Windle, W. F., and A. M. Griffin. 1931. Observations on embryonic and fetal movements of the cat. *J. Comp. Neurol.* 52:149–188.

Wyman, R. 1965. Probabilistic characterization of simultaneous nerve impulse sequences controlling dipteran flight. *Biophys. J.* 5:447–471.

Section Two

AN INVERTEBRATE SAMPLER

Insects are the most successful animals in the world in terms of number of species and individuals. While insects have long interested students of natural history, epidemiology, and economics (e.g. crop damage), only a few species have been used in seminal behavioral research that has affected work on vertebrates. This is becoming less true as the advantages of working with these organisms become apparent. Here we present representative studies involving three of the most behaviorally influential groups: crickets, fruit flies, and ants.

Ronald Hoy and George Casaday continue the theme begun by Anne Bekoff in tracing, neurologically, the development and genetics of stridulation in crickets, expanding on already classic studies by David Bentley and Hoy. John Ringo details the development of behavior in *Drosophila*, until recently thought incapable of even the most elementary forms of learning. The original work of Ringo and his co-workers on the lek behavior of these flies shows the intimate connections between social behavior, ecological niche, life span, and development. Howard Topoff and John Miranda demonstrate the importance of social stimulation in the remarkably precocial behavior of slavemaking ants. Their elegant and clear-cut experimental studies are examples of the power of classical ethological methods of assessing stimulus control. Again, the relations between ecology, life style, and development are stressed.

Chapter Three

Acoustic Communication in Crickets: Physiological Analysis of Auditory Pathways

RONALD R. HOY and GEORGE B. CASADAY

Cornell University
Section of Neurobiology and Behavior
Langmuir Laboratory
Ithaca, New York 14853

Introduction

The study of species-specific acoustic communication is of interest to any observer of animal behavior, whether motivated by professional interests or personal curiosity. Besides vertebrates, crickets and their orthopteran relatives are among the most conspicuous users of sounds for communication. Well before there was a systematic science of animal behavior, cricket song was the subject of human attention. However, only in the past twenty-five years has it become evident that the detailed analysis of cricket songs would be informative for areas of biological science other than entomology. For students of evolution, ecology, ethology, and neurobiology, cricket acoustic behavior has been a rich source of information (Alexander, 1962; Huber, 1975); the fact that it serves the animal as a communication system for mate attraction makes this behavior especially attractive for attempts to integrate evolution of communication with behavioral and neurobiological mechanisms. In this chapter we will analyze some aspects of acoustic communication behavior in crickets, especially those which pertain to a general problem facing all animals, vertebrate and invertebrate, that employ species-specific signals for mate attraction: how do the sender of the signal and the conspecific receiver come to be "coupled" such that the message in the signal can be translated into effective action? From an evolutionary perspective this question concerns the coevolution of the signal generation and signal recognition mechanisms. From a neurobiological point of view it concerns the physiological mechanisms for generation and recognition of the signals. From the development perspective it concerns the assembly and "tuning" of the two systems. A truly satisfying analysis must include information from several disciplines: evolutionary theory, genetics, neurobiology, and development. This should come as no surprise to the serious student of animal behavior; the past decade has seen the emergence of fields which bear the neologistic titles *neuroethology* and *neurogenetics.* The problem of coupling sender and receiver is hardly solved, but we will present data that might convince the reader that a solution is possible.

In most cricket species only the male sings, and the songs are concerned with reproductive behavior, courtship of females, and aggression toward rival males. The male of many species defends a territory using a rivalry or aggressive song. From his territory he sings a long-range or calling song. A female can hear the calling song from at least several meters away, and, if she is ready to mate, she seeks out the male. The male then shifts to a courtship song, and the pair complete a mating ritual which can also involve tactile and chemical cues. In some species the male sings yet another song which keeps the female nearby after copulation. This all seems to be an effective species-isolating mechanism, judging from the rarity of hybrids in nature (Hill *et al.* 1972).

Much of the work we will discuss concerns the calling song of the Australian field crickets *Teleogryllus oceanicus* and *T. commodus* (Fig. 1). The top line of the figure (*A*) diagrams the song of *T. oceanicus*. It is made up of short pulses of an almost pure tone with a carrier frequency or pitch of about 4.5 kHz. Each pulse represents a single closing stroke of the wings. Four or five pulses make up the chirp section, which is followed by a series of doublet pulses called the trill section. The bottom line (*C*) is the song of the sympatric species *T. commodus.* Its chirp section is similar to that of the song of *oceanicus,* but the trills are groups of about 9 pulses rather than doublets. Also, the carrier frequency is 3.7 kHz rather than 4.5 kHz.

As long ago as 1913 Regen found that in *Gryllus campestris,* the calling song alone, without visual, chemical, or tactile cues, can attract females. He showed this by letting males sing to conspecific females over a telephone. The same is true for *oceanicus* and *commodus.* Each will preferentially approach a source of its conspecific song. We now know that the male and female are matched with respect to both characteristics of the calling song, the carrier frequency and the temporal pattern. The rest of this chapter will concern the mechanisms of this matching and the development of these mechanisms.

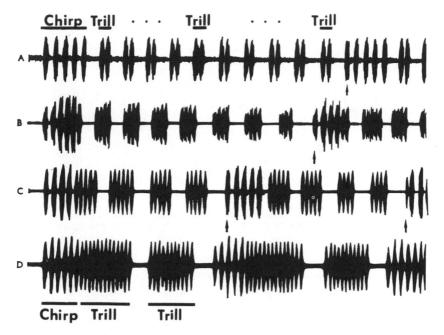

FIG. 1. *Temporal pattern of sound pulses in the calling songs of* Teleogryllus *and interspecific hybrids. These oscillograms show that the calling song is composed of two sections, a chirp followed by a series of trills that comprise a song phrase, the fundamental repeating unit of the call. The arrow indicates the end of one phrase and the beginning of the next one.* A: T. oceanicus. B: *hybrid T-1* (T. ocean.♀× T. comm.♂). C: *hybrid T-2* (T. comm.♀× T. ocean.♂). D: T. commodus. *The duration of the* T. oceanicus *phrase is about 2 s. From Bentley and Hoy, 1972.*

Carrier Frequency

The pulses of the calling song are made up of almost pure tones. If the female is to hear the calling song, her auditory system must be sensitive at that frequency. This is indeed the case, as shown in Figure 2. The stippled area shows the frequency distribution of energy in the calling song of *T. commodus* (Hoy, unpublished). The maximum for this species is at 3.7 kHz. The broken line is the threshold curve for a typical auditory interneuron (Loftus-Hills *et al.* 1971). So, the auditory system is sensitive at the pitch of the calling song. It is also possible to get a behavioral measure of sensitivity at various carrier frequencies. Hill (1974) artificially gener-

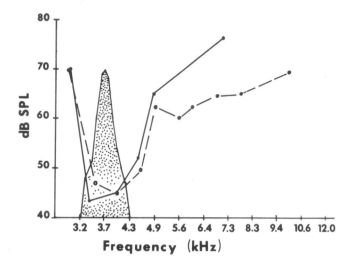

FIG. 2. *Sound energy in the song and auditory sensitivity of* Telegryllus commodus: stippled area, *sound energy distribution;* solid line, *threshold for an auditory interneuron (Loftus-Hills et al., 1971);* dashed line, *behavioral threshold (Hill, 1974).*

ated songs with the *commodus* temporal pattern, but with a range of carrier frequencies. He then tested females to determine the intensity necessary to cause a criterion proportion of the animals to approach a loudspeaker. The results indicated by the solid line show that females are most sensitive near the natural frequency.

This frequency match can be explained by rather simple physical properties of the sound-producing and sound-receiving organs. When the male strokes its wings together, a scraper on the edge of one wing strikes the teeth of the file on the other wing. The tooth strike rate depends on the speed at which the cricket slides the wings past each other. The tooth-strike rate determines the fundamental frequency of the song. Also, the wing has a specialized structure, the harp, which functions as a sharply tuned resonator, tuned at the carrier frequency. The resonator emphasizes the carrier frequency and filters out other frequency components. (For a more detailed review of these mechanisms see Michelsen and Nocke, 1974.)

Turning to the receiver: the ear or tympanum is located on the tibia of the foreleg. It is a thin membrane which allows sound to reach the auditory neurons inside the leg. Behind the tympanum is an air-filled cavity, the leg trachea. It has recently been shown (Paton, 1975) that this air-filled space is a resonator tuned to the species carrier frequency. Hill

(1977) has furthermore shown that only the part of the trachea in the leg segment bearing the tympanum is necessary for tuning.

So we can conclude that the sender and receiver are matched with respect to a carrier frequency mainly by the mechanical properties of both systems. This assures that the female can hear the male. However, we should emphasize that we know of no compelling evidence that females actually use differences in carrier frequency to discriminate between conspecific and heterospecific songs. This point remains an open question which deserves further work.

Temporal Pattern

Temporal patterns of cricket songs are sufficiently characteristic to be used for taxonomic purposes (Alexander, 1962). This suggests that they are exploited by the crickets themselves for species identification. Several investigators have studied this problem (Shuvalov and Popov, 1973; Walker, 1957; Zaretsky, 1972), but we will discuss only a recent result of Gerald Pollack (unpublished) in this laboratory. Virgin *T. oceanicus* females were given a choice between two loudspeakers. Both played songs at 5 kHz but with different temporal patterns. One was the *T. oceanicus* pattern and the other the *T. commodus* pattern. Of 37 animals tested, 24 chose the conspecific pattern and 13 chose the heterospecific. The difference is significant at the 0.025 level and shows that crickets can discriminate between calling songs on the basis of temporal pattern alone. The discrimination is far from perfect, but remember that the calling song is only the first step in a complex mate-selection process.

We have no proven explanation for this ability to discriminate. It is unlikely that the match in temporal pattern between sender and receiver is the result of simple mechanical properties as is the case for carrier frequency. Therefore we will present a hypothetical model which could explain the match in terms of mechanisms in the central nervous system (Fig. 3). The model assumes that both male and female have identical or homologous timing elements. Similar models have been proposed for communication in *Drosophila* (Ewing, 1969), birds (Marler, 1973), and even human speech (Liberman *et al.,* 1967). In this hypothetical scheme, timing elements in the male are excited by command fibers recently shown to cause singing (Bentley, 1977). These elements then generate the song pattern. In the female, the timing elements act as a filter or resonator for song recognition. The timing elements are made up of identical neurons in both male and female, and their development is directed by the same genes. Therefore, under all conditions, the sender and receiver will

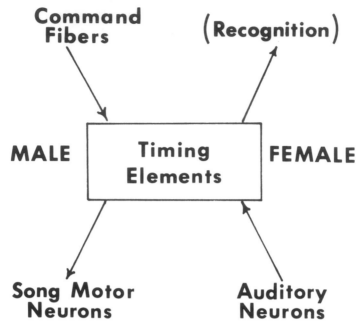

FIG. 3. *A model of the role of the nervous system in processing temporal information in acoustic communication in crickets. In the male, command fibers arising from the brain are seen to activate thoracic central pattern generators, called "timing elements" in this scheme. These central pattern generators (Bentley, 1969) presumably set the temporal pattern in the song rhythm by activating the song motor neurons. In the female, song pattern detection might be achieved by a scheme in which the auditory neurons of the ear send a precise "copy" of the temporal pattern of the acoustic stimulus to the thoracic centers. There, acoustic interneurons can either "filter" the input or relay it onto the brain. Either at the level of thoracic centers or the brain, the temporal pattern of the input can either be filtered according to the closeness of "fit" with timing properties in these neuronal filters, or the input can be matched against some sort of template which encodes the species temporal pattern; the closeness of fit in the matching again presumably results in "recognition." Presumably, communication coupling could be achieved between male and female if there are elements common to both in the nervous system, particularly among the "timing elements." It is further suggested that the temporal parameters of these timing elements are under genetic control by a common set of genes. We emphasize that this is a highly speculative model; little neurophysiological information exists to illuminate its major features, particularly the critical timing elements.*

be matched. A different interpretation may be found in the work of the von Helversens (1975a, b).

The model explains two major experimental results. The first of these is the behavior of hybrids between *oceanicus* and *commodus*. (We summarize results that have been reported at length elsewhere [Bentley and Hoy, 1972; Hoy and Paul, 1973; Hoy *et al.*, *1977*].) Hybrid males resulting from cross between *T. oceanicus* and *T. commodus* sing calling songs that are distinctly different from either parental type and are characteristic of the hybrid, although there is more variability among hybrid males' calls than among those of either parental type. Examples of these differences are illustrated in curves *B* and *C* of Figure 1. Statistical analysis of the temporal pattern of parental and hybrid calls reveals that the calls of both hybrid types (T-1 and T-2) are intermediate in pattern between those of the parents; this is interpreted (and supported by analysis of back-cross hybrids in Bentley, 1971) as evidence of polygenic control of song pattern. An additional finding is that one parameter of rhythm, the intertrill interval (the interval between doublet groups in *oceanicus* and between the longer trills in *commodus*) is sex-linked. That is, the call of each hybrid type tends to resemble that of the maternal parent more than that of the paternal parent. Mechanisms to account for this phenomenon are purely speculative at present and range from cytoplasmic factors in the egg to expression of genes on the maternal X chromosome. (Males are XO sex-determined and get their lone X chromosome from their mother.) When female hybrids were tested for either their song preference or their ability to discriminate among song types by measuring their phonotaxic preference, evidence for matching between the male sender and the female receiver again became evident. Hybrid females preferred the calls of their hybrid brothers to those of either parental species (Table 1). In fact, hybrid specificity occurs even between hybrid cross types themselves: females of cross type T-1 preferred the calls of T-1 males over those of T-2 hybrid males, and conversely, T-2 females preferred T-2 calls over T-1 calls (Table 2). This suggests a maternal sex linkage in song responsiveness of the female hybrids. Since females have two X chromosomes, the data are consistent with an explanation that maternal sex linkage is due to factors that control song responsiveness that reside in the egg cytoplasm or with genes that reside on the maternal X chromosome and are selectively expressed; at present the mechanism is a matter of pure speculation. Whatever the mechanism, these results suggest that the observed species-specific matching of call types that underlies call specificity in nature has a strong genetic basis. The preference of the various hybrid females for the corresponding hybrid males is consistent with homologous central nervous system mechanisms mediating song production in the male and song recognition in the female.

TABLE 1. Performance of female crickets walking on a Y-maze while being presented a calling song from a loudspeaker placed to the left or right of the animal.

CALLING SONG PLAYED FROM SPEAKER	NUMBER OF FEMALES MEETING 75% PERFORMANCE CRITERION	FEMALES MEETING CRITERION (%)
T. oceanicus on maze		
T. oceanicus	14/22	63.6
T. commodus	4/22	18
T. commodus on maze		
T. oceanicus	3/15	20
T. commodus	21/28	75
Hybrid (T−1) on maze		
T. oceanicus	3/11	27.3
T. commodus	8/21	38
Hybrid (T−1)	21/28	75

Performance was measured for 20 choices with sound coming from the right and 20 from the left; data selected from females that responded to sound coming from both right and left (after Hoy and Paul, 1973). Response criteria based on females responding to each speaker for at least 15 of 20 turns toward the active speaker.

TABLE 2. Phonotaxic performance of free-walking hybrid females when given a choice between reciprocal hybrid calling songs playing from two, simultaneously playing loudspeakers.

HYBRID TYPE PERFORMING DISCRIMINATION	RESPONSE TO T-1 CALLING SONG	RESPONSE TO T-2 CALLING SONG	RESPONSE TO SIBLING HYBRID TYPE
T-1	97	50	66%
T-2	40	125	76%

Columns record the total number of females of each hybrid type that responded to a particular hybrid song; females were not retested so that the data are based on many different females. Hybrid T-1 females were generated from the mating of *T. ocean.* females × *T. comm.* males; T-2 from *T. comm.* female × *T. comm.* male. Data after Hoy, Hahn, and Paul, 1977.

It is important to note that these genetic results apply to *crickets:* strikingly different results have been reported from similar analyses in acridid grasshoppers, where hybrid males do not necessarily sing hybrid songs but may sing songs that resemble one or the other parental type, and where hybrid females may show a preference for parental calls over hybrid ones (von Helverson and von Helverson, 1975a, b). It is of interest to note, however, that there is one feature of song inheritance that seems to be shared between crickets and grasshoppers: in both instances there is evidence for sex linkage of factors that affect species-specific communication; song behavior of hybrids tends to resemble that of their mother's species more than that of their father's species. The meaning of this is not clear at present but it is not likely to be without significance when considering the evolution of communication systems and their genetic bases.

A second observation which is easily explained by the model concerns the variation in pulse rate in the songs of some tree crickets as temperature varies (Walker, 1957). As the temperature of a male increases, the pulse rate of its song increases. Correspondingly, as the temperature of a female increases, the pulse rate which she prefers increases. For example, at 70° F females prefer a pulse rate near 55 pulses per second and males sing at about 55 pulses per second. There is a similar match at other temperatures. If identical systems mediate both song production and song recognition, this result would be expected.

The model we have described is attractive because of its simplicity and great explanatory power. It is made more believable by the fact that females do, although rarely, move their wings in a stridulatory manner. However, it remains a speculation, and we believe that it may never be proved or disproved on the basis of behavioral experiments alone. Therefore we have chosen to study the central nervous system to try to understand the mechanism of pattern recognition in neuronal terms.

Neurophysiological Analysis

We believe a neuron level analysis of this system may be possible because the cricket's nervous system is relatively small and simple. We know that we can identify some neurons as individuals in animal after animal and we are now trying to map all the neurons—large and small—of the auditory pathway.

In *T. oceanicus,* about 70 primary auditory neurons begin at the tympanal organ or ear in the tibia of the prothoracic legs (Fig. 4). The axons travel in the leg nerve to the prothoracic ganglion (T1), where all the fibers terminate in a discrete region, the acoustic neuropil (Fig. 5). The insect neuropil is highly structured. Its synaptic areas and tracts are quite analogous with the nuclei and tracts of the vertebrate central nervous system. This order is important for two reasons. First, it allows stereotactic placement of electrodes in specific regions, for example, the acoustic neuropil or even a particular part of it. Second, it lets us describe a

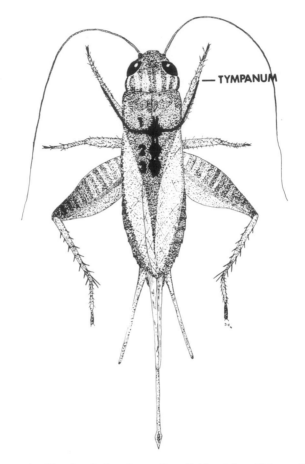

FIG. 4. *Sketch of the Australian field cricket,* Teleogryllus oceanicus, *showing the location of relevant parts of the nervous system of importance to the physiological studies described in the text. The tympanum is the site of "ear" of the cricket and is located on the animal's tibia. Nerve fibers from the ear terminate in the prothoracic ganglion (ganglion 1). Acoustic information is carried from ganglion 1 by acoustic interneurons, in an ascending direction to the brain, or descending, perhaps to the mesothoracic ganglion (2) or the metathoracic ganglion (3).*

neuron, not only in terms of general morphology, but also in terms of the location of its cell body, the tracts in which its fibers run, and the specific areas it interconnects.

To do the thorough mapping we are attempting requires a technique which reveals the physiology and morphology of both large and small

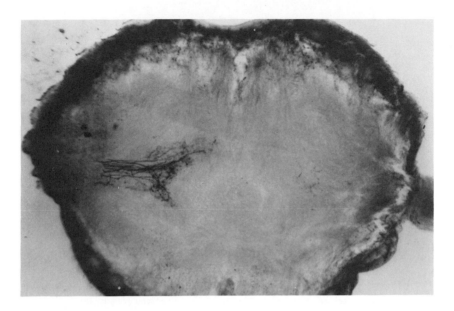

FIG. 5. *Histological section made longitudinally through the plane of the acoustic neuropile. Numerous acoustic receptor axons are seen to pass from the ear (tympanal organ) and terminate in the acoustic neuropile in the prothoracic ganglion. It is within this neuropile that microelectrodes are inserted in order to monitor and stain acoustic interneurons. The receptor axons are stained with cobalt sulfide; the preparation was enhanced by Timm's procedure, modified by Tyrer and Bell (1974).*

neurons. Thus we have chosen a rather new extracellular cobalt staining technique (Rehbein *et al.,* 1974). We probe the neuropil with a cobalt chloride–filled micropipette electrode. When we isolate a unit, we record from it and eject cobalt into the extracellular space. The cobalt enters the cell, spreads through it, and after histological treatment makes the cell visible. We will not attempt to summarize our data to date; rather, we will show a few examples to illustrate our approach to the problem.

Figure 6 is a photograph of a whole ganglion T1 with a stained interneuron. It has a large dendrite in the acoustic neuropil, an anterior axon which goes to the brain, and a contralateral soma which is out of the plane of focus. This cell is unique in each half of ganglion, and has a puzzling physiology (Fig. 7). It is spontaneously active and is inhibited rather than excited by sound at 5 kHz, near the calling song frequency. Furthermore, it does not code song. At this point we can only hope that

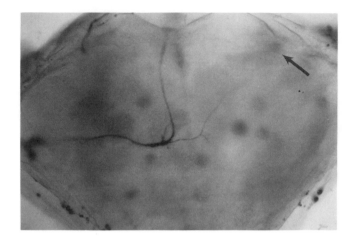

FIG. 6. *An interneuron whose activity appears to be suppressed during stimulation by 5 kHz tones. It is shown as it appears in a whole mount preparation after cobalt sulfide staining, fixation, and clearing. The neuron sends major projections in the area of the acoustic neuropile, to the left. It sends its axon anterior to the brain, toward the top of the page, and it is connected to its contralaterally located cell body* (arrow) *by a thin neurite, seen going off and out of focus to the right. (From Cassaday and Hoy, 1977.)*

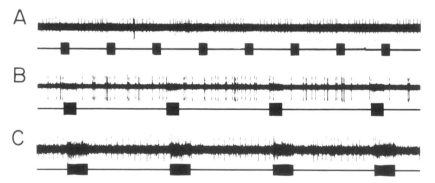

FIG. 7. *Activity pattern of the interneuron shown in Figure 6. Three different neurons (A, B, and C) of this type are shown here. In each case the upper record is an oscillogram of the neuron's discharge; the lower trace represents a stimulus tone of 5 kHz. Note that the apparently spontaneous activity is suppressed by acoustic stimulation. The time scale is indicated by the black bar at the bottom of the figure and stands for 500 msec. (From Cassaday and Hoy, 1977.)*

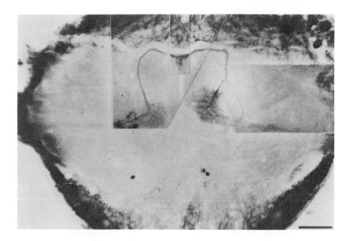

FIG. 8. *A photomontage illustrating the neuroanatomy of an interneuron which faithfully codes the temporal pattern of calling song. This is a cobalt-Timm's enhanced stain; sections were taken through the plane of the acoustic neuropile. Note that the neuron sends sizable projections to both bilateral acoustic neuropiles. Of particular interest is that this neuron does not have either an ascending or descending axon. (From Casaday and Hoy, 1977.)*

further work on this cell will give us some clue about its function.

Figure 8 is a composite photograph from horizontal sections of the ganglion. This large neuron connects the two sides of the ganglion and has no axon leaving T1. Of course, a cell connecting the two sides of the animal suggests a function in direction finding, although such a function has not yet been demonstrated. Physiologically, the unit codes the song pattern almost perfectly (Fig. 9). It produces a group of spikes for each pulse in the *oceanicus* song. Also, note that the cell maintains its firing rate throughout the segment of song. It can thus code the intensity of long-lasting stimuli, an important property if the cell is in fact involved in direction finding.

The last cell we want to discuss (Fig. 10) is quite important for two reasons. First, it is a very small cell, which we have been able to unambiguously stain twice. Its soma is only between 5 and 10 μm in diameter and its processes are only about 1 μm in diameter. So, we are encouraged to believe that we really can study very small cells in detail. Second, its axon projects, not to the brain, but posterior toward the second thoracic ganglion. This is where at least some components of the song pattern generator for singing are located. So this cell is potentially a link between

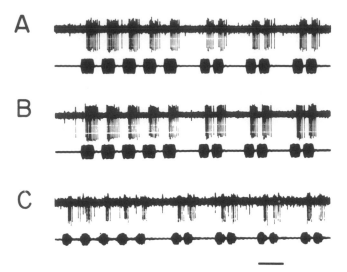

FIG. 9. *Physiological activity of the interneuron shown in Figure 10. Three different neurons (A, B, and C) are shown in their response to calling song. Top trace is the physiological response, bottom trace is a portion of calling song that is played to the preparation during the recording session. Note the precision of the coding of the temporal pattern of the song by the pattern of the neuron's discharge. (From Casaday and Hoy, 1977.)*

the sensory and motor systems. This link is exactly the sort which must exist if our hypothetical model for song recognition is correct.

We are speculating at this point, but we are encouraged by our results so far. We have definitely or tentatively identified about a dozen cells, and we hope that we are not too far from completing a map of ganglion T1. It really seems possible that with more work we might be able to find an explanation, in neuronal terms, for the matching of sender and receiver in this communication system.

Ontogeny

The indirect message of this review is that acoustic behavior in crickets is a model system for the total biological analysis of species-specific communication, ranging from global concerns such as taxonomy, evolution, and ecology to more proximate concerns such as physiological and anatomical analyses of the neural pathways involved in acoustic behavior; behavior genetic analysis can be seen to link the "ultimate" aspects with

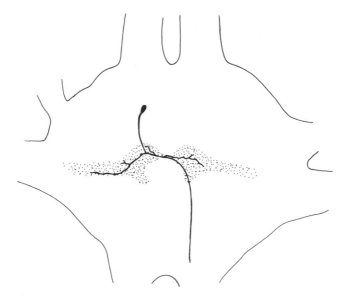

FIG. 10. *A camera lucida reconstruction of an acoustic inter-neuron that sends its axon posterior, to lower (thoracic?) ganglia. The outlines of the acoustic neuropile have been stippled in for reference; the neuron sends projections to both acoustic neuropiles.*

the "proximate" aspects. It is obvious that detailed information about neurophysiology and neuroanatomy of the acoustic nervous system will enlighten the analysis of the ontogeny of acoustic behavior. As was done in the earlier sections we can divide our study in terms of the development of sound production in the male and sound reception in the female.

Only adult males are capable of producing sounds by stridulatory movements of their forewings. Crickets develop to adulthood by going through a series of 9–12 molts; only with the adult molt are the wings differentiated for song production. Thus, the larval crickets, although capable of performing most adult behaviors (except those concerned with reproductive function) are mute. However, the absence of functional wings does not preclude the presence of the neuronal network that is responsible for generating the temporal pattern of the calling song; this question was investigated in *Teleogryllus* by Bentley and Hoy (1970). They found that by lesioning a particular part of the brain the neural network that produces song pattern could be activated in the last larval stage. Although the larval wings are so small as to be scarcely recognizable as wings, they are nonetheless set in motion by the lesion. Electrodes inserted into the wing muscles revealed that the contraction pattern

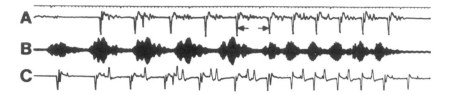

FIG. 11. *Electromyograms from stridulatory muscles of a late-stage larval (immature) cricket. Previous to the recording, a brain lesion had been made that "released" the song production system into activity. Electrodes inserted into the stridulatory muscles demonstrate that the temporal pattern of these muscles is appropriate for generation of the species calling song, in this case* T. commodus. *No acoustic output results from this immature animal because its wings are not yet differentiated with respect to sound generating components. This demonstrates that the "song generator" is already formed and capable of functioning even though the actual sound-producing organs themselves are not yet developed. In the figure, A and C are the myogram records from the immature cricket; A is a recording from a wing opener muscle, whereas C records the activity of both the opener muscles (downward-going spikes) and the closer muscles (upward going spikes). We have superimposed an oscillograph of a portion of the calling song (B) of an adult* T. commodus *to demonstrate the nearly perfect match between calling song and the motor output pattern of the immature song system. Of particular interest is the switch from the chirp to the trill (indicated by the arrow), which is diagnostic of calling song for this species. It is present in both the adult song and in the temporal pattern of activity in the immature wing muscles. (From Bentley and Hoy, 1970.)*

of the muscles matched that of the adult pattern during the production of calling song (Fig. 11). Thus, the neural network for calling song generation is already present during postembryonic development and "in place" before the actual sound-producing structures (the forewings) have developed. Normally, this network is suppressed until the final molt to adulthood.

The ontogeny of the auditory pathways is largely unknown at present although the ontogeny of the tympanal organ has been studied in detail (Ball and Young, 1974). Certain findings from our detailed studies of neuroanatomy seem worth mentioning in this context. It is clear that the cell bodies of the acoustic interneurons that we have been able to identify neuroanatomically are not distributed throughout the prothoracic ganglion at random. Although the data are certainly incomplete, we note that the somata of interneurons that send an axon anterior are clustered in a discrete part of the ganglion (anterior and lateral in the ganglion, on the dorsal surface) and the few examples of interneurons that have a descending axon have their somata in a different location (Figs. 8, 10). In addition, some neurons pass through the ganglion, that is, they have a process in both the anterior and posterior connectives. These cells do not have their somata in T1. If this observation holds up for other interneurons then we might consider the intriguing possibility that interneurons of like function (ascending or descending) are embryologically related (perhaps arising clonally from a common progenitor cell). This would be a remarkable and powerful generalization, and one which certainly justifies the value of making detailed neurophysiological and neuroanatomical maps of the nervous system. It is our opinion that understanding the ontogeny of behaviors such as acoustic communication in crickets will only be possible after making detailed studies of the adult system; for this reason our priorities lie in making a complete neuroanatomical and neurophysiological map of the auditory pathways in the cricket.

References

Alexander, R. D. 1962. Evolutionary changes in cricket acoustical communication. *Evolution* 16:443–467.

Ball, E., and D. Young. 1974. Structure and development of the auditory system in the prothoracic leg of the cricket *Teleogryllus commodus* (Walker). II. Postembryonic development. *Z. Zellforsch.* 147:313–324.

Bentley, D. R. 1971. Genetic control of an insect neuronal network. *Science* 174:1139–1141.

Bentley, D. R. 1977. Control of cricket song patterns by descending interneurons. *J. Comp. Physiol.* 116:19–38.

Bentley, D. R., and R. R. Hoy. 1970. Postembryonic development of adult motor patterns in crickets: a neural analysis. *Science* 170:1409–1411.

Bentley, D. R., and R. R. Hoy. 1972. Genetic control of the neuronal network generating cricket song patterns. *Anim. Behav.* 20:478–492.

Casaday, G., and R. Hoy. 1977. Auditory interneurons in the cricket *Teleogryllus oceanicus:* physiological and anatomical properties. *J. Comp. Physiol.* 121:1–13.

Ewing, A. W. 1969. The genetic basis of sound production in *Drosophila pseudoobscura* and *D. persimilis. Anim. Behav.* 17:555–560.

von Helversen, D., and O. von Helversen. 1975a. Verhaltengenetische untersuchungen am akustischen kommunikationssystem der feldheuschrecken (Orthoptera, Acrididae). I. Der gesang von artbastarden zwischen *Chorthippus biguttulus* und *Ch. mollis. J. Comp. Physiol.* 104:273–299.

von Helversen, D., and O. von Helversen. 1975b. Verhaltengenetische untersuchungen am akustischen kommunikationssystem der feldheuschrecken (Orthoptera, Acrididae) II. Das lautschema von artbastarden zwischen *Chorthippus biguttulus* und *Ch. Mollis. J. Comp. Physiol.* 104::301–323.

Hill, K. G. 1974. Carrier frequency as a factor in phonotactic behavior of female crickets (*Teleogryllus commodus*). *J. Comp. Physiol.* 93:7–18.

Hill, K. G. 1977. Sensitivity to frequency and direction of sound in the auditory system of crickets (Gryllidae). *J. Comp. Physiol.* 121:79–97.

Hill, K. G., J. J. Loftus-Hills, and D. F. Gartside. 1972. Pre-mating isolation between the Australian field crickets *Teleogryllus commodus* and *T. oceanicus* (Orthoptera: Gryllidae). *Aust. J. Zool.* 20:153–163.

Hoy, R. R., J. Hahn, and R. C. Paul. 1977. Hybrid cricket auditory behavior: evidence for genetic coupling in animal communication. *Science* 195:82–84.

Hoy, R. R., and R. R. Paul. 1973. Genetic control of song specificity in crickets. *Science* 180:82–83.

Huber, R. 1975. Sensory and neuronal mechanisms underlying acoustic communication in orthopteran insects. In *Sensory physiology and behavior*, eds., R. Galun, P. Hillman, I. Parnas, and R. Werman, New York: Plenum.

Liberman, A. M., F. S. Cooper, D. S. Shankweiler, and M. Studdert-Kennedy. 1967. Perception of the speech code. *Psychol. Rev.* 74:431–461.

Loftus-Hills, J. J., M. J. Littlejohn, and V. G. Hill. 1971. Auditory sensitivity of the crickets *Teleogryllus commodus* and *T. oceanicus. Nature New Biol.* 233:184–185.

Marler, P. 1973. Learning, genetics, and communication. *Social Res.* 40:293–310.

Michelsen, A., and H. Nocke. 1974. Biophysical aspects of sound communication in crickets. *Adv. Insect Physiol.* 9:247–296.

Paton, J. 1975. Ph.D. dissertation, Cornell University, Ithaca, N. Y. (Unpublished.)

Regen, J. 1913. Über die anlockung des weibchens von *Gryllus campestris* L. durch telephonisch übertragene stridulations-laute des männchens. *Pflüger's Arch. fur Physiol.* 155:193–200.

Rehbein, H. G., K. Kalmring, and H. Romer. 1974. Structure and function of acoustic neurons in the thoracic ventral nerve cord of *Locusta migratoria* (Acrididae). *J. Comp. Physiol.* 95:263–280.

Shuvalov, V. F., and A. V. Popov. 1973. Study of the significance of some parameters of calling signals of male crickets *Gryllus bimaculatus* for phonotaxis of females. *J. Evol. Biochem. Physiol.* 9:177–182.

Tyrer, N. M., and E. M. Bell. 1974. The intensification of cobalt-filled neurone profiles using a modification of Timm's sulphide-silver method. *Brain Res.* 73:151–155.

Walker, T. J. 1957. Specificity in the response of female tree crickets (Orthoptera, Gryllidae, Oecanthinae) to calling songs of the males. *Ann. Entomol. Soc. Am.* 50:626–636.

Zaretsky, M. D. 1972. Specificity of the calling song and short term changes in the phonotactic response by female crickets *Scapsipedus marginatus* (Gryllidae). *J. Comp. Physiol.* 79:153–172.

Chapter Four

The Development of Behavior in Drosophila

JOHN M. RINGO

Zoology Department
University of Maine
Orono, Maine 04473

Introduction

As a model system for studying behavioral development, the genus *Drosophila* has much to offer. First, there are over 1200 known species, distributed throughout the world and occupying a variety of ecological niches. Many of these species can be cultured in the laboratory, cheaply and conveniently. Second, the behavior of *Drosophila* in confined spaces in the laboratory is indistinguishable from their behavior in the field (Spieth, 1968). Third, there is a wealth of background information— genetic, taxonomic, ecological, and physiological—to draw upon for ethological investigations. Fourth, most species of *Drosophila* have moderate-sized behavioral repertoires of stereotyped displays, including the courtship signaling system. Finally, *Drosophila* show species-specific patterns in the development of stereotyped displays and rituals (Spieth, 1974a). Although species of *Drosophila* show a rudimentary capacity to learn (Quinn *et al.*, 1974; O'Hara *et al.*, 1976), experience generally appears to have little influence on their behavior.

The potential for developmental studies of *Drosophila* is enhanced by the discrete steps of the life cycle: egg, three successive larval instars, pupa, and adult. The species-characteristic length of the cycle, egg to egg, varies for most species between ten days and ten weeks, which allows the investigator considerable freedom to choose a species whose lag time between any two developmental events is appropriate to a given experiment. *Drosophila* displays observable behavior in two stages of the life cycle, larva and adult, since pupae apparently do not move.

LARVAL BEHAVIOR

Larvae are much simpler in behavior than adults. Ostensibly, larval functions are limited primarily to feeding and simple taxes. To date, larval behavior has not been extensively investigated, possibly owing to the undramatic quality of larval feeding and the difficulty of following the progress of a burrowing animal. However, the few studies of behavioral development in larvae have yielded interesting findings.

The most important aspect of larval behavior is feeding. Larvae feed almost continuously except for general locomotion and brief pauses between instars. Feeding consists of repeated probing movements by the mouthparts, accompanied by ingestion of food during each cycle of extension and retraction. In *D. melanogaster* the rate of probing increases to a relative maximum near the beginning of the third instar and drops off slightly during the third instar (Burnet and Connolly, 1974). In between bouts of feeding, larvae slowly move a short distance to a new feeding site; the frequency of movement from site to site is approximately constant through larval life (Burnet and Connolly, 1974).

Larvae of the cardini species group are apparently unique in *Drosophila* in that they are cannibalistic. This cannibalism probably appears first in the second instar and continues through the third instar (Futch, 1962). The victims are often pupae that have become trapped in the food.

A second major aspect of larval behavior is migration from the food to a pupation site. Most species of *Drosophila* migrate away from the food site, some pupate on the food surface, and some do both (de Souza *et al.,* 1970). Species that pupate away from the food undergo a sudden change of behavior late in the third larval instar. Some species tend to be positively geotactic until the end of the third instar, when they become negatively geotactic, facilitating movement away from the food (Burnet and Connolly, 1974). Some species, for example many Hawaiian *Drosophila,* burrow into the soil after leaving the food; burrowing reduces the opportunity for predation.

Another migratory adaptation is skipping, a behavior unique to the cardini group (Futch, 1962). To skip, a larva bends double and then suddenly propels itself through the air several centimeters. Skipping first appears late in the third instar, just before pupation. Skipping facilitates rapid migration from the food site in order for flies to pupate on or in the surrounding soil.

In *D. willistoni* there is genetic variation for a developmental switch that affects larval migration. De Souza *et al.* (1970) selected strains that consistently pupated on the food and other strains whose larvae consistently migrated away from the food before pupating. The behavioral polymorphism is controlled by two alleles of one gene and some polygenic modifiers (de Souza *et al.*, 1970).

MATURATION IN ADULTS

The flashier behavioral repertoire of adults includes feeding, courting, mating, oviposition, cleaning, antipredator behavior, and several taxes. The development of adult behavior has been studied much more intensively in females than in males, apparently because there are several abrupt developmental changes in female behavior and because these changes depend upon oogenesis and copulation. In contrast, the only major developmental change in male behavior is the switch-on of sexual behavior, which has not yet been connected to any physiological event.

The two major components of female behavioral development are changes in sexual receptivity (and accompanying changes in behaviors used to reject male suitors) and oviposition. Both are correlated with maturation of the ovaries and growth of the juvenile hormone-producing glands, the corpora allata. The chain of events connecting these correlated behavioral and physiological changes is being investigated in several laboratories.

Sexual receptivity in virgin females is switched on at a species-specific time, which in *D. melanogaster* is approximately 48 hours after emergence (Manning, 1967). Sexually immature, unreceptive females use several stereotyped actions to reject and repel males (Spieth 1952; Connolly and Cook, 1973). Apparently the switch-on of receptivity involves the lowering of the acceptance threshold to a fixed point (Cook, 1973). Receptivity declines gradually in old virgins—beginning eight days after eclosion in *D. melanogaster* (Manning, 1967).

The correlation between the maturation of the ovaries and switch-on of receptivity led to the hypothesis that distention of the ovaries causes a lowered acceptance threshold via neural feedback to the brain. This hypothesis was ruled out because females congenitally lacking ovaries or having underdeveloped ovaries become receptive at the usual time (Smith, 1956; Manning, 1967). Another hypothesis is that juvenile hormone (JH), which has gonadotrophic effects in insects generally (Schneiderman, 1971), acts on the brain directly or through an intermediate neurosecretory factor to increase receptivity. *D. melanogaster* females in which corpora allata had been implanted in the pupal stage showed precocial sexual receptivity and had larger ovaries than controls within 24 h of emergence (Manning, 1966), supporting the idea that JH or a factor triggered by JH increases receptivity. Bouletreau-Merle (1973) confirmed this hypothesis by applying JH topically to *D. melanogaster* females; she found the precocious appearance of oöcytes and early switch-on of sexual receptivity.

After copulation females again become unreceptive. The stimulus of copulation itself, the presence of sperm in the sperm receptacles, and a secretion produced in the (male) paragonial gland all contribute to the switch-off of receptivity (Manning, 1967; Burnet *et al.,* 1973). However, multiple inseminations can occur under many circumstances, including natural conditions (Spieth, 1974a). Interestingly, the relative frequencies of three female rejection signals change after fertilization or with the implantation of a paragonial gland (Burnet *et al.,* 1973).

Concomitant with the switch-off of receptivity *Drosophila* females increase oviposition. This increase is apparently caused by two or more factors. One is the expansion of the maturing ovaries, which may feed back to the central nervous system via stretch receptors (Grossfield and Sakri, 1972; Burnet and Connolly, 1974). Oviposition is also increased by paragonial gland secretions (Merle, 1969) of which the sex peptide transferred during copulation is the active component (Chen and Buhler, 1970).

The onset of male courtship displays is delayed for hours or even days after eclosion, the length of delay being correlated with the species' life span. After sexual activity is switched on, males court other males as well as females, and in many cases other species are courted as well as conspecifics. The sexual preferences that males have for conspecific females may increase with age (Spieth, 1958).

Very little work has been done on the maturation of nonsexual behaviors in *Drosophila.* Lipps (1973) found little evidence for maturation in the cleaning behavior of ten species of *Drosophila,* except for the effects of senescence. Hay (1972) observed a significant interaction between age and time of day in the general activity of *D. melanogaster* males; he suggested that further work on the ontogeny of male behaviors in *Drosophila* is needed.

FATE MAPS

Behavioral development can be studied from the standpoint of embryonic determination of adult behavior. One can ask, "Where, when, and how are cells that control particular behaviors determined?" The *where* part of the question is being answered by applying Sturtevant's fate-mapping technique (Garcia-Bellido and Merriam, 1969) to *D. melanogaster* (Hotta and Benzer, 1970).

A fate map is a two-dimensional representation of blastoderm sites that are fated to become larval, pupal, or adult tissues. Maps are constructed by using gynandromorphs with specified genotypes. The map distance between two sites is proportional to the probability that they are of different sexes (i.e., that one site has an XX karyotype and that the other site is XO). With this technique, one can map the embryological sites at which sex-linked genes act to produce particular phenotypes, including behavior patterns. The affected adult structure causing a particular behavior is called the *behavioral focus.*

Fate maps have been constructed for mutant alleles of several genes (Hotta and Benzer, 1970, 1972; Homyk, 1977) causing such diverse behaviors as abnormal electroretinograms, ether-induced leg-shaking and abnormal flight reactions. Some of the foci for these mutant behaviors map to the central nervous system. Hotta and Benzer (1976) and Nissani (1977) have also mapped the foci that determine normal sex-limited courtship and mating behaviors. The focus of male orientation and wing vibration maps inside the head, possibly in the brain. The focus of (male) attempted copulation is inside the thorax, possibly in the thoracic ganglion. Female acceptance of males is controlled by a focus inside the head, possibly in the brain. As Hotta and Benzer (1972, 1976) have pointed out, the behavioral fate map of *D. melanogaster* will show much more topological detail when more mutants, including internal markers, are available.

Present Studies on Maturation of a Lek Species

LEK DROSOPHILA

A group of species with rich potential for the study of behavioral ontogeny are the many slow-maturing, lek *Drosophila* endemic to the Hawaiian islands. Although these species are considerably less convenient to rear in the laboratory than their smaller continental relatives such as *D. melanogaster* or *D. virilis,* they offer several advantages to the ethologist. Their sexual maturation takes days or even weeks instead of hours, and males give many lek displays in addition to courtship itself. Females, too, give a number of special displays which may be lek-related. It is thought that these species are subject to intense sexual selection, which enhances behavioral and morphological dimorphism of the sexes (Spieth, 1974b; Ringo and Hodosh, 1978).

Lek species behave differently from non-lek, continental species (Carson *et al.,* 1970; Spieth, 1974a). Continental *Drosophila* court and mate at the feeding-oviposition sites; courting males dance through the population, and flies interact vigorously and conspicuously. Lek species differ sharply; they are quiet and cryptic at feeding-oviposition sites. This is probably an antipredator adaptation (Spieth, 1974b). After feeding briefly, sexually mature males leave the food and fly to a lek in the nearby vegetation. The leks of some species are subdivided into territories, in which case males vigorously defend their areas, using ritualized displays and fighting. Males also advertise for females at these lek sites with special displays that often involve the dispersion of pheromones. A few species exhibit communal displays (Ringo, 1974, 1976). Receptive females fly to the lek, where courtship and mating take place. Courtship in some species is elaborate and lengthy.

Drosophila grimshawi is one species whose repertoire, while far from the most complex of Hawaiian *Drosophila,* contains the four main types of behavior given by lek species: (1) agonistic displays, (2) phero-mone-dispersing behavior, (3) a communal display, and (4) courtship containing visual displays. All the stereotyped behaviors of *D. grimshawi* have been described elsewhere (Spieth, 1966; Ringo, 1976; Ringo and Hodosh, 1977). Briefly, the eight behaviors used in the studies reported or reviewed here were: *abdomen drag* (used to lay down odor trails), *court, joust* (a complex communal display), *curl* and *slash* (two agonistic displays), *shiver* (a wing display whose adaptive significance is not known), *clean,* and *fly.*

INSECTARIUM OBSERVATIONS

In order to determine the approximate ages at which various lek displays in *D. grimshawi* are switched on, and to let flies interact in a relatively large space, 83 males and 61 females aged 1–48 h were placed in a 55 l clear plastic box lined on the bottom with sand and containing food and water. Flies were observed briefly several times each day at irregular intervals, as well as each day between 1700 and 1715 h. The study lasted 38 days.

The aggressive display, *slash,* was first seen on day 1—i.e., the day on which flies were introduced; *curl* was first seen on day 4; *joust* was first observed on day 8. By day 13, when flies were 13–15 days old, *joust* had reached its maximum frequency of about 10–15 flies per observation. Several courtships were observed near *jousting* males, and on day 14 the termination of one copulation was seen at a *jousting* site. No other mating was seen, which is not surprising, given the brief duration of copulation in this species ($\leq$ 5 min). Flies *jousted* on the drier, lighter sides of the insectarium; flies did not *abdomen drag, court,* or *joust* on the bottom sand. All lek displays—*abdomen drag, court, joust, curl,* and *slash*—tended to occur more frequently in male aggregations. Females tended to occupy locations away from male aggregations. No territorial behavior was observed. However, the odor trails produced by *abdomen dragging* flies were not random, but tended to turn frequently so that each odor trail crisscrossed a small area (1–100 cm^2) many times.

THE ONTOGENY OF MALE DISPLAYS

Following up the insectarium observations, I measured the frequency of six stereotyped displays (*abdomen drag, court, joust, curl, slash,* and *shiver*) and of two behaviors not specific to leks (*clean* and *fly*) in *D. grimshawi* males at ages 1, 8, 15, 22, and 29 days. "*Fly*" was operationally defined as flight or jumping. It was found that males will engage in lek activities in small 100 cc cells. Their displays in this "minilek" situation are indistinguishable from those in a large container or from field behavior (Spieth, 1966). Flies were sexed at eclosion while lightly etherized. Males were housed under standard culture conditions (Spieth, 1972) until testing. To measure behavioral frequencies, five males were aspirated into a 100 cc plastic cell closed on one end with moist cellulose sponge, and were observed for 60 s periods every other minute for an hour. During each minute of observation the number of flies engaging in each behavior was tallied. Two cells were viewed during alternate minutes, each cell containing a different age group. In this way each of the ten pairwise combinations of ages was observed in three replicates. This incomplete randomized-blocks design called for 12 replicates of each age group. Data were transformed using the Freeman-Tukey square root transformation (Mosteller and Youtz, 1961).

Blocks (days) had a negligible effect on each behavior, as shown by an analysis of variance (Yates, 1940). Therefore, simple one-way analyses of variance were performed. The results were: *abdomen drag* ($F = 30.33$, $p < .001$); *court* ($F = 36.93$, $p < .001$); *joust* ($F = 11.60$, $p < .001$); *curl* ($F = 5.97$, $p < .001$); *clean* ($F = 9.13$, $p < .001$); *fly* ($F = 23.37$, $p < .001$); *shiver* ($F = 11.59$, $p < .001$); *slash* ($F = 0.89$, $.25 < p < .50$); in each case, df $= 4, 55$. The frequency of *slash* did not vary significantly with age, whereas the frequencies of the other seven behaviors changed significantly as the flies matured.

Three principal developmental patterns are apparent from the data, graphed in Figure 1. (1) Four lek behaviors, *abdomen drag, court, joust,* and *curl,* were virtually absent until 15 days when they became more frequent, rising to relative maxima at 22 or 29 days. The first three of these became very common, while *curl* remained relatively rare. (2) The two non-lek behaviors, *fly* and *clean,* were less frequent at 1 day than at the other ages. Variation after one day was irregular, particularly with *fly*. These two behaviors reflected the flies' general activity, which was signifi-

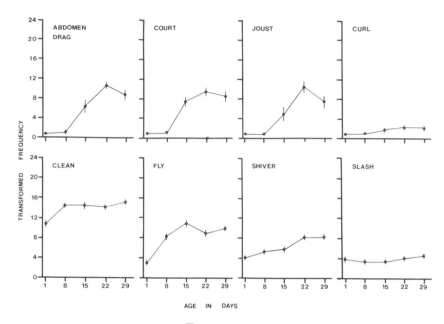

FIG. 1. *Frequencies* ($\overline{X} \pm SE$) *of eight behaviors in* D. grimshawi *males at ages of 1, 8, 15, 22, and 29 days, post-eclosion.*
The Freeman-Tukey square root transformation was applied to
the behavioral frequencies.

cantly lower on day 1. (3) The remaining displays, *slash* and *shiver,* did
not undergo any sudden week-to-week shifts. Both were relatively common at all ages. *Slash* remained approximately constant at all ages, while
shiver increased gradually with increasing age. Overall, one-day-old flies
tended to sit quietly, engaging in little observable activity. *Drosophila
grimshawi* males are altricial for the four lek behaviors and precocial for
the other behaviors. Topoff and Mirenda (Chapter 5) discuss the concepts
of precocial vs. altricial behavior.

The diversity of behavior seemed to increase with age, mainly because sexual and agonistic behaviors increased dramatically on days 15
and 22. To test this idea quantitatively, behavioral diversity of each
"minilek" was measured by the Shannon-Weaver index of normalized
mean information, H/H_{max}. The terms of the fraction are defined as
follows: mean information, $H = -\Sigma p_i \log p_i$; $H_{max} = -\log (1/n)$. This
index of information (entropy) was chosen because it is widely used in
biology and is sensitive to differences in rare events (Peet, 1974), which
makes it useful for less common behaviors such as *curl.* The plot of
H/H_{max} against age (Fig. 2) clearly shows a shift from low behavioral

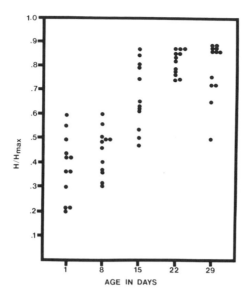

FIG. 2 *Behavioral diversity of* D. grimshawi *males as a function of age.* H/H_{max} *was plotted for ages of 1, 8, 15, 22, and 29 days, post-eclosion.*

diversity in the one-day and eight-day groups to high diversity in older flies. In contrast to Berg (Chapter 1), who measured individual variability, I measured group variability to examine age-dependent diversity of lek *Drosophila* behavior (see also M. Bekoff, Chapter 9).

The relative frequency of each behavior changed with age, so that each age had a different profile across the eight behaviors. To measure the magnitude of the changes in behavioral profiles with age, a simple multivariate distance measure, $H_{jk} = -\Sigma \log_2 s_{ijk}$, was used (Sneath and Sokal, 1973). The measure is modified from Gower's (1971) coefficient of similarity. The value of s_{ijk} is defined by $s_{ijk} = w_{ijk}(1 - |p_{ij} - p_{ik}|)$, where $w_{ijk} = 0$ if populations j and k both lack behavior i and $w_{ijk} = 1$ otherwise, and where p_{ij} and p_{ik} are the relative frequencies of behavior i in populations j and k, respectively. The distance H_{1k} between day 1 flies and other age groups increased dramatically the first three weeks, declining slightly at 29 days (Fig. 3A).

Testis size was checked in 20 males of each age used in the experiment. Mean testis widths did not vary significantly with age; for all groups, $\bar{X} \pm SE = .16 \pm .14$ mm. However, testes became pigmented with age. On day 1, there was no pigment observed; on day 8 testes were

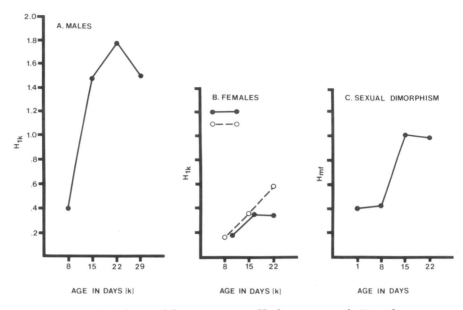

FIG. 3. *Behavioral distance measure* H_{jk} *between populations of maturing* D. grimshawi. *A. Distance was measured between one-day-old males and four older groups of males—8, 15, 22, and 29 days, post-eclosion. B. Distance was measured between one-day-old females and three older groups of females—8, 15, and 22 days, post-eclosion. Open circles represent groups measured in the presence of male pheromone; closed circles represent groups measured in the absence of male pheromone. C. Age-specific sexual dimorphism is measured from males to females exposed to male pheromone, for ages of 1, 8, 15, and 22 days, post-eclosion.*

yellow; on day 15 they were yellow with dark tinges; on days 22 and 29 they were dark brown.

THE ONTOGENY OF FEMALE DISPLAYS

The behavioral repertoires of both male and female lek *Drosophila* species are more varied and complex than those of their non-lek relatives. Normally, *Drosophila* females do not court, although females in certain mutant strains of *D. melanogaster* show more or less regular male courtship, and so have been designated as "lesbian" (Cook, 1975). Although *D. grimshawi* females do not court, they do show the typical "male" lek behaviors, *abdomen drag, joust,* and *curl* under certain conditions. However, compared with males, their behaviors are often incomplete, occurring in brief and infrequent bouts. *Shiver* and *slash* occur in roughly equal frequencies in both sexes. Hodosh and Ringo (unpublished) found that a male pheromone facilitates the behavior in mature ($\geqslant 18$ days) virgin females.

Hodosh and Ringo (unpublished thesis) measured the frequencies of seven behaviors (*abdomen drag, joust, curl, clean, fly, shiver,* and *slash*) in virgin females aged 1, 8, 15, and 22 days. Flies were treated and behaviors were measured as in the male ontogeny experiment above except that (1) ten flies per cell were used; (2) one age was observed at a time; (3) one cell contained male pheromone while the alternate cell was free of male odors; (4) five replicates per odor treatment per age were used. The behavioral frequencies were treated with the Freeman-Tukey square root transformation (Mosteller and Youtz, 1961).

Three developmental patterns emerged from the data shown in Figure 4. (1) The three "male" lek behaviors, *abdomen drag, joust,* and

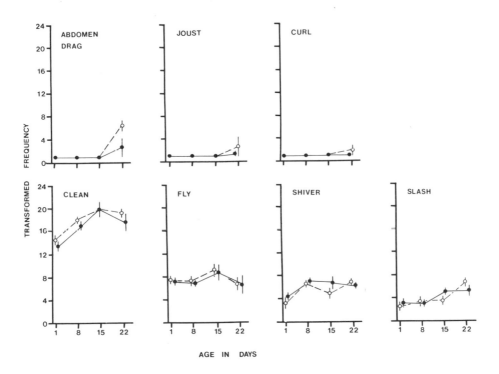

FIG. 4. *Frequencies* ($\overline{X} \pm SE$) *of seven behaviors in* D. grimshawi *females at ages of 1, 8, 15, and 22 days, post-eclosion. Open and closed circles as in Fig. 3B.*

curl, did not appear during the first two weeks, but were switched on some time before 22 days of age. Male odor appeared to facilitate all of these behaviors, but in fact the effect of odor was significant only with *abdomen drag.* The interaction of odor and maturation is similar to that found by Topoff and Mirenda (Chapter 5). The behaviors were much less frequent than in males. (2) *Clean* was less frequent on day 1, and varied irregularly thereafter. (3) *Fly, shiver,* and *slash* remained about constant with age, although *slash* increased slightly in older flies. Just as lek behaviors were switched on in males between 8 and 15 days, these same behaviors were switched on in females between 15 days and 22 days. When females were housed with males, about 50% of the females became inseminated by day 15, indicating that sexual receptivity in females is switched on a few days before the "male" lek behaviors appear.

Behavioral diversity, measured by the Shannon-Weaver formula, increased with age in females. The increase was much greater in flies exposed to male pheromone (odor condition) than in flies not exposed to male pheromone (control). This interaction between age and pheromone is undoubtedly caused by the increase in "male" displays, as the increase in diversity was slight and gradual in controls, but took a sharp rise at 22 days in the odor condition (Fig. 5).

The distance measure, H_{jk}, was computed from the day 1 group to the other ages, for both treatments. In the odor condition, females diverged steadily with age from the 1-d-old flies. Controls did not diverge after 15 days (Fig. 3B). The divergence of the profile of female behaviors from day 1 to older groups was much smaller than the maturational changes in the male profiles.

Sexual dimorphism was computed using H_{jk} for females and males of the same ages. Only odor condition females were used because male odor was present in the male groups, and would also be present in nature at leks, and because the male-female differences are thereby minimized, making for a more conservative estimate of sexual dimorphism. Males and females differed little on days 1 and 8, but suddenly diverged on day 15, because male lek displays were switched on (Fig. 3C). Interestingly, however, sexual dimorphism was never as great as the change in the male behavioral profile from day 1 to after sexual maturation. This may have been caused by the general inactivity of day 1 males compared with day 1 females.

JUVENILE HORMONE AND FEMALE BEHAVIORAL DEVELOPMENT

Juvenile hormones (JH) and their mimics are known to produce three effects on adult insect behavior: (1) to switch on female sexual receptivity;

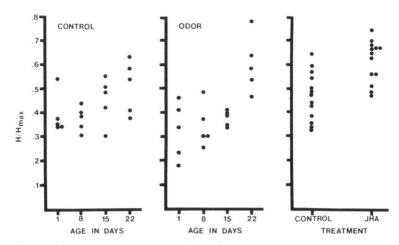

FIG. 5 *Behavioral diversity of* D. grimshawi *virgin females as a function of age or hormone treatment. Diversity is measured by the normalized Shannon-Weaver index,* H/H_{max}. *A. Control females were measured in clean cells at 1, 8, 15, and 22 days, post-eclosion. B. Odor group females were measured in cells containing male pheromone at 1, 8, 15, or 22 days, post-eclosion. C. Females were measured in the presence of male pheromone at 16–17 days, post-eclosion. JHA flies were treated at 24–48 h, post-eclosion with 6 nl of ZR515 dissolved in 2 μl of acetone. Controls were treated with 2 μl of acetone.*

(2) to switch on male sexual behavior; and (3) to accelerate the appearance of female age-specific behaviors. The first of these effects is known for grasshoppers (Engelman and Barth, 1968), house flies (Adams and Hintz, 1969), mosquitoes (Lea, 1968; Gwadz *et al.,* 1971), and fruit flies (Manning, 1966; Bouletreau-Merle, 1973; Ringo and Pratt, 1978). The second effect has been found in grasshoppers. Loher (1960) and Pener (1967) found that the removal of corpora allata (CA) from *Schistocerca* males lowered male sexual behavior; implantation of active CA into allatectomized males restored male sexual behavior. Broza and Pener (1969) found the implanted CA produced intense sexual behavior in diapausing *Oedipoda* males. The third effect has also been reported for several species. A JH analog (JHA) accelerates age polyethism of honey bees (Jaycox *et al.,* 1974). Juvenile hormone and several analogs alter the social dominance of female *Polistes,* a social wasp, by increasing the frequency of behaviors correlated with social dominance (Barth *et al.,* 1975). Ringo and Pratt (1978) found that a JHA accelerates the develop-

ment of "male" lek behaviors in *D. grimshawi* virgin females; their basic findings and a re-analysis of their data are given below.

In one experiment, designed to test for the first type of behavioral effect, either 6 nl of the JHA, ZR515, dissolved in acetone, or acetone alone, was applied topically to the abdomen of a newly eclosed *D. grimshawi* female. Groups of these flies were housed with sexually mature males, and dissected at 9, 11, 13, or 15 days of age. The JHA flies began to become inseminated on day 9, and the percentage of inseminated females was significantly higher in the JHA group until day 15, indicating a six-day acceleration of sexual receptivity.

A second experiment investigated the effects of ZR515 on the frequency of stereotyped behaviors in *D. grimshawi* virgin females. Flies were treated as above except that the hormone or acetone was applied 24–48 h after eclosion, and that females were housed without males, to insure virginity. At 16–17 days post-eclosion, groups of five females were placed in plastic cells containing male pheromone. A group of JHA flies and a group of controls were viewed during alternate minutes, and the number of flies showing each of the seven behaviors measured in the female ontogeny experiment (*abdomen drag, joust, curl, clean, fly, shiver,* and *slash*) were tallied. In all, 13 replicates were performed. The frequencies were transformed by the Freeman-Tukey square root transformation (Mosteller and Youtz, 1961); means and standard errors of the transformed data appear in Figure 6.

Abdomen drag and *slash* were increased significantly by ZR515, as shown by Mann-Whitney U tests. The Shannon-Weaver diversity index was computed for each test group. Behavioral diversity was shifted sharply higher in the ZR515 group, a reflection of an overall acceleration in behavioral development (Fig. 5).

The results suggest that JH controls the appearance of lek displays in the repertoire of *D. grimshawi* virgin females. JH may act directly on the central nervous system (Engelmann and Barth, 1968), or it may facilitate the release of neurosecretory factors that cause developmental changes in the central nervous system (Manning, 1967). Whatever the mechanism of JH action on the nervous system, it entails a slow process with a lag time of weeks between a rise in JH level and behavioral changes in *D. grimshawi*. In fact, no direct short-term effects of JH on behavior have been demonstrated (Staal, 1971). Probably growth or other slow physiological processes underlie the behavioral changes induced by JH.

It is tempting to say that JH masculinizes *D. grimshawi* females, inasmuch as the hormone accelerates the appearance of principally male lek behaviors. The most probable explanation for this seeming anomaly is that the mechanisms for producing behavioral dimorphism between the sexes is imperfect. This may be related to the fact that lek behaviors in

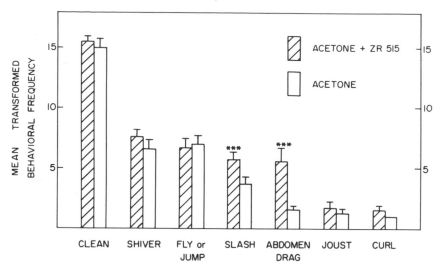

FIG. 6. *Frequencies* ($\overline{X} \pm$ SE) *of 7 behaviors in* D. grimshawi *virgin females, 16–17 days, post eclosion. The Freeman-Tukey square root transformation was applied to the behavioral frequencies.* Patterned bars: *Flies were treated at 24–48 h, post-eclosion, with 6 nl of ZR515 dissolved in 2 μl of acetone.* Clear bars: *Control flies treated with 2 μl of acetone. Statistically significant Mann-Whitney U tests are denoted with asterisks:* ***p $<$ .001. (*Data from Ringo and Pratt, 1978.*)

Drosophila are phylogenetically recent traits and are still evolving rapidly (Spieth, 1974b; Ringo, 1976, 1977).

Acknowledgments

Financial assistance was provided by University of Maine faculty research grant no. 5-2-26136. The juvenile hormone analog was donated by Zoecon Corporation of Palo Alto, California. David Leonard and Ralph Hodosh offered valuable suggestions on writing the manuscript. Special thanks go to Gregory Howick, who prepared the figures.

References

Adams, T. S., and A. M. Hintz. 1969. Relationship of age, ovarian development, and the corpus allatum to mating in the housefly, *Musca domestica. J. Insect Physiol.* 15:201–215.

Barth, R. H., L. J. Lester, P. Sroca, T. Kessler, and R. Hearn. 1975. Juvenile hormone promotes dominance behavior and ovarian development in social wasps (*Polistes annularis*). *Experientia* 31:691–692.

Bouletreau-Merle, J. 1973. Réceptivité sexuelle et vitellogenèse chez les femmes de *Drosophila melanogaster:* effets d'une application d'hormone juvénile et de deux analogues hormonaux. *C. R. Acad. Sci. Paris* 277:2045–2048.

Broza, M., and M. P. Pener. 1969. Hormonal control of the reproductive diapause in the grasshopper, *Oedipoda miniata*. *Experientia* 25:414–415.

Burnet, B., K. Connolly, M. Kearney, and R. Cook. 1973. Effects of male paragonial gland secretion on sexual receptivity and courtship behaviour of female *Drosophila melanogaster*. *J. Insect Physiol.* 19:2421–2431.

Burnet, B., and K. Connolly. 1974. Activity and sexual behavior in *Drosophila melanogaster*. *In* ed., J. H. F. van Abeelen. The genetics of behaviour. Amsterdam: North-Holland. pp. 236–243.

Carson, H. L., D. E. Hardy, H. T. Spieth, and W. S. Stone. 1970. The evolutionary biology of the Hawaiian Drosophilidae. *In* ed., M. K. Hecht and W. C. Steere, pp. 437–543. Essays in evolution and genetics in honor of Theodosius Dobzhansky. New York: Appleton-Century-Crofts.

Chen, P. S., and R. Buhler. 1970. Paragonial substance (sex peptide) and other free ninhydrin-positive components in male and female adults of *Drosophila melanogaster*. *J. Insect Physiol.* 16:615–627.

Connolly, K., and R. Cook. 1973. Rejection responses by female *Drosophila melanogaster:* their ontogeny, causality and effects upon the behaviour of the courting male. *Behaviour* 44:142–146.

Cook, R. 1973. Physiological factors in the courtship processing of *Drosophila melanogaster*. *J. Insect Physiol.* 19:397–406.

Cook, R. 1975. "Lesbian" phenotype of *Drosophila melanogaster? Nature* 254: 241–242.

Engelmann, F., and R. H. Barth. 1968. Endocrine control of female receptivity in *Leucophaea maderae* (Blatteria). *Ann. Entomol. Soc. Am.* 61:503–505.

Futch, D. G. 1962. Hybridization tests within the cardini species group of the genus *Drosophila*. Univ. Tex. Publ. 6205:539–554.

Garcia-Bellido, A., and J. R. Merriam. 1969. Cell lineage of the imaginal discs in *Drosophila* gynandromorphs. *J. Exp. Zool.* 170:61–75.

Gower, J. C. 1971. A general coefficient of similarity and some of its properties. *Biometrics* 27:857–871.

Grossfield, J., and B. Sakri. 1972. Divergence in the neural control of oviposition in *Drosophila*. *J. Insect Physiol.* 18:237–241.

Gwadz, R. W., L. P. Lounibos, and G. B. Craig, Jr. 1971. Precocious sexual receptivity induced by a juvenile hormone analog in females of the yellow fever mosquito, *Aedes aegypti*. *Gen. Comp. Endocrinol.* 16:47–51.

Hay, D. A. 1972. Behavioural rhythms in cultures of immature *Drosophila melanogaster*. *Experientia* 28:922–923.

Homyk, T. 1977. Behavioral mutants of *Drosophila* melanogaster. II. Behavioral analysis and focus mapping. *Genetics* 87:105–128.

Hotta, Y., and S. Benzer. 1970. Genetic dissection of the *Drosophila* nervous system by means of mosaics. *Proc. Natl. Acad. Sci. U.S.A.* 67:1156–1163.

Hotta, Y., and S. Benzer. 1972. Mapping of behaviour in *Drosophila* mosaics. *Nature* 240:527–535.

Hotta, Y., and S. Benzer. 1976. Courtship in *Drosophila* mosaics: sex-specific foci for sequential action patterns. *Proc. Natl. Acad. Sci. U.S.A.* 73:4154–4158.

Jaycox, E. R., W. Skowronck, and G. Guynn. 1974. Behavioral changes in worker honey bees (*Apis mellifera*) induced by injections of a juvenile hormone mimic. *Ann. Entomol. Soc. Am.* 67:529–534.

Lea, A. O. 1968. Mating of virgin *Aedes aegypti* without insemination. *J. Insect Physiol.* 14:305–308.

Lipps, K. 1973. Comparative cleaning behavior in *Drosophila*. Doctoral dissertation, Univer. of California at Davis, 166 pp.

Loher, W. 1960. The chemical acceleration of the maturation process and its hormonal control in the male of the desert locust. *Proc. Roy. Soc. London* B 153:380–397.

Manning, A. 1966. Corpus allatum and sexual receptivity in female *Drosophila melanogaster*. *Nature* 211:1321–1322.

Manning, A. 1967. The control of sexual receptivity in female *Drosophila*. *Anim. Behav.* 15:239–250.

Merle, J. 1969. Fonctionnement ovarien et réceptivité sexuelle de *Drosophila melanogaster* après implantation de fragments de l'appareil génitale male. *J. Insect Physiol.* 14:1159–1168.

Mosteller, F., and C. Youtz. 1961. Tables of the Freeman-Tukey transformations for the binomial and Poisson distributions. *Biometrika* 48:433–440.

Nissani, M. 1977. Gynandromorph analysis of some aspects of sexual behaviour of *Drosophila melanogaster Anim. Behav.* 25:555–566.

O'Hara, E., A. Pruzan, and L. Ehrman. 1976. Ethological isolation and mating experience in *Drosophila paulistorum*. *Proc. Natl. Acad. Sci. U.S.A.* 73:975–976.

Peet, R. K. 1974. The measurement of species diversity. *Annu. Rev. Ecol. Syst.* 5:285–307.

Pener, M. P. 1967. Effects of allatectomy and sectioning of the nerves of the corpora allata on oocyte growth, male sexual behaviour, and colour change in adults of *Schistocerca gregaria*. *J. Insect Physiol.* 13:665–684.

Quinn, W. G., W. A. Harris, and S. Benzer. 1974. Conditioned behavior in *Drosophila melanogaster*. *Proc. Natl. Acad. Sci. U.S.A.* 71:708–712.

Ringo, J. M. 1974. Behavioral characters distinguishing two species of Hawaiian *Drosophila, Drosophila grimshawi* and *Drosophila pullipes* (Diptera: Drosophilidae). *Ann. Entomol. Soc. Am.* 67:823.

Ringo, J. M. 1976. A communal display in Hawaiian *Drosophila* (Diptera: Drosophilidae). *Ann. Entomol. Soc. Am.* 69:209–214.

Ringo, J. M. 1977. Why 300 species of Hawaiian *Drosophila*? The sexual selection hypothesis. *Evolution* 31:694–696.

Ringo, J. M., and R. J. Hodosh. 1978. A multivariate analysis of behavioral divergence among closely related species of endemic Hawaiian *Drosophila*. Evolution. (In press.)

Ringo, J. M., and N. R. Pratt. 1978. A juvenile hormone analog induces precocial sexual behavior in *Drosophila grimshawi*. *Ann. Entomol. Soc. Am.* (In press.)

Schneiderman, H. A. 1972. Insect hormones and insect control. *In* eds., J. J. Menn and M. Beroza, pp. 3–27. Insect juvenile hormones. New York: Academic Press.

Smith, J. M. 1956. Fertility, mating behaviour and sexual selection in *Drosophila subobscura. J. Genet.* 54:261–279.

Sneath, P. H. A., and R. R. Sokal. 1973. Numerical taxonomy. San Francisco: W. H. Freeman.

de Souza, H. M. L., A. B. da Cunha, and E. P. dos Santos. 1970. Adaptive polymorphism of behavior evolved in laboratory populations of *Drosophila willistoni. Am. Naturalist* 104:175–189.

Spieth, H. T. 1952. Mating behavior within the genus *Drosophila* (Diptera). *Bull. Am. Mus. Nat. Hist.* 99:399–474.

Spieth, H. T. 1958. Behavior and isolating mechanisms. *In* eds., A. Roe and G. G. Simpson, pp. 363–389. Behavior and evolution. New Haven: Yale Univ. Press.

Spieth, H. T. 1966. Courtship behavior of endemic Hawaiian *Drosophila. Univ. Tex. Publ.* 6615:245–313.

Spieth, H. T. 1968. The evolutionary implications of sexual behavior in *Drosophila. Evol. Biol.* 2:157–191.

Spieth, H. T. 1972. Rearing techniques for the Hawaiian species of *Drosophila grimshawi* and *Drosophila crucigera. Drosophila Inf. Svc.* 48:155–157.

Spieth, H. T. 1974a. Courtship behavior in *Drosophila. Annu. Rev. Entomol.* 19:385–405.

Spieth, H. T. 1974b. Mating behavior and evolution of the Hawaiian *Drosophila. In* ed. M. J. D. White, pp. 94–101. Genetic mechanisms of speciation in insects. Sydney: Australia and New Zealand Book Co.

Staal, G. B. 1972. Biological activity and bioassay of juvenile hormone analogs. *In* eds., J. J. Menn and M. Beroza, pp. 69–94. Insect juvenile hormones. New York: Academic Press.

Yates, F. 1940. The recovery of inter-block information in balanced incomplete block designs. *Ann. Eugenics* 10:317–325.

Chapter Five

In Search of the Precocial Ant

HOWARD TOPOFF and JOHN MIRENDA

Department of Psychology
Hunter College of C.U.N.Y.
New York, New York 10021 and Department of
     Animal Behavior
The American Museum of Natural History
New York, New York 10024

Introduction

It is well known that similarities in biological characteristics among distantly related species can arise through convergent evolution in similar ecological contexts. Such similarities may occur in patterns of behavior as readily as in characteristics at other levels of biological organization. In both the arthropod and chordate phyla, for example, natural selection has produced such similar accomplishments as terrestrial life, complex sensory systems for orientation and communication, and a social organization sufficiently integrated to be called a society.

Furthermore, because natural selection can operate on any stage of the life cycle, developmental processes may also exhibit convergent properties. Many species of vertebrates, for example, have been characterized as either precocial or altricial, depending upon their state of maturation at the time of hatching or birth. Although these are relative concepts and in reality should be applied to specific sensory and motor systems (Gottlieb, 1968, 1971), they have nevertheless proved useful even when applied to organisms as a whole. In birds, for instance, precocial species hatch out covered with down, eyes opened, legs well developed, and soon able to leave the nest and feed by themselves. Altricial species, by contrast, are hatched naked, usually blind, and are fed by the parents. Between these extreme conditions, there exists a wide variety of intermediate types (Table 1).

According to Nice (1962), the precocial state in birds corresponds to the reptilian pattern and therefore represents the primitive condition. Nevertheless, there does not seem to have been any consistent evolution toward the altricial condition (which has evolved independently in different orders of birds at different times). Indeed, the relative degree of maturation in birds often correlates better with ecological requirements than with phylogenetic position. Thus ground-nesting species such as cranes, which may be more vulnerable to early predation, tend to be more precocial than treetop nesters such as storks.

Although this aspect of behavioral development has not been analyzed as thoroughly for mammals as for birds, a somewhat similar classification could be devised. Young mice, rabbits, and bats are born blind and naked, and can thus be considered altricial. Wolves, foxes, dogs, and cats are born with hair, are soon able to crawl about, but are blind for several days. These might fit into the semi-altricial category. Deer, moose, and other ungulates are semiprecocial. Among the most precocial of all mammals are the seals. Newly born fur seals, for instance, are able to stand up and vocalize from 15 to 45 s after birth (Bartholomew, 1959).

As is the case with birds, the degree of maturation at birth often cuts across phylogenetic position. Both rabbits (genus *Sylvilagus*) and hares (*Lepus*) belong to the family Laporidae. Nevertheless, most species of hares are born with their eyes open, while rabbits are typically blind at birth. Although the distributions of these genera overlap to some extent, many species of hares are adapted to habitats that are considerably more exposed than those of rabbits (Hall and Kelson, 1959). Likewise, herbivo-

TABLE 1. Relative maturity of young birds at hatching

A. Precocial. (Eyes open, down-covered, leave nest first day or two)

    1. Completely independent of parents (megapodes)

    2. Follow parents but find own food (ducks, shorebirds)

    3. Follow parents and are shown food (quail, chickens)

    4. Follow parents and are fed by them (grebes, rails)

B. Semiprecocial. (Eyes open, down covered, stay at nest although able to walk, fed by parents—gulls, terns)

C. Semi-altricial. (Down-covered, unable to leave nest, fed by parents)

    1. Eyes open (herons, hawks)

    2. Eyes closed (owls)

D. Altricial (Eyes closed, little or no down, unable to leave nest, fed by parents—passerines)

Source: After Nice (1962).

rous ungulates, which also live in open areas, are frequently more precocial than predatory species and those living in relatively concealed habitats.

Although the concepts *precocial* and *altricial* have been systematically analyzed only with reference to vertebrates, we propose that analogous differences in developmental processes have evolved in the social Hymenoptera. In ants, beginning with the period of eclosion from the pupal stage of development, species differences in the following characteristics are known to exist: (1) the degree to which mature adults are necessary for assisting the callows in removing the pupal skin (and cocoon) during eclosion (Haskins and Haskins, 1950); (2) the time when mature adult pigmentation develops (Wheeler, 1910, 1928); (3) the sequence of temporal division of labor within the nest (Wilson, 1971); and (4) the time when individuals first leave the nest to forage for food (Topoff and Mirenda, in press).

As in vertebrates, developmental processes differ in the degree to which they correlate with phylogenetic position. One developmental characteristic that is typically found in primitive ants is the ability of callows to eclose without the assistance of mature adult workers. Thus, in the primitive subfamily Ponerinae, Wheeler (1910, 1928) found that the pupal skin of *Stigmatomma pallipes* is shed while the callow is still inside the cocoon, and that eclosion can occur in the absence of adults. Species within another primitive subfamily, the Myrmeciinae, represent a somewhat transitional stage in the evolution of this process. When cocoons of *Myrmecia pilosula* were isolated, the pupal skin was also shed in the cocoon, and the callows advanced to full pigmentation. Nevertheless, approximately one-half of the individuals died when adults were not permitted to assist in tearing open the cocoons and removing the callows (Haskins and Haskins, 1950). Species within the army ant subfamily Dorylinae vary greatly with respect to the presence or absence of a cocoon (Table 2). In all cases that have been studied, however, adult participation is essential for callow eclosion. In *Neivamyrmex nigrescens,* whose

TABLE 2. Presence (+) or absence (−) of cocoons during doryline ant development

GENUS	WORKER BROOD	SEXUAL BROOD
Eciton	+	+
Labidus	+	+
Neivamyrmex	−	+
Dorylus	−	−
Aenictus	−	−

worker larvae pupate without cocoons, each mature pupa is attended by 1–5 adult ants. The adults scrape off the moist pupal skin, generally working from the head toward the abdominal region. Once removed, it is transported throughout the nest and is avidly licked by many adult workers.

The index of behavioral development that we have been studying is the posteclosion age for the transition from activities such as tending the brood, conducted inside the nest (called "*Innendienst*" in the German literature), to foraging and other activities conducted outside the nest (*Aussendienst*).

The fact that newly eclosed callow ants tend to remain in the nest has been documented for species of *Camponotus* (Buckingham, 1910; Kiil, 1934), *Oecophylla* (Ledoux, 1950), *Formica* (Otto, 1958; Dobrzańska, 1959), *Pogonomyrmex* (McCook, 1879), *Messor* (Goetsch and Eisner, 1930; Ehrhardt, 1931), *Myrmica* (Ehrhardt, 1931; Weir, 1958), and *Myrmecia* (Freeland, 1958). In the few studies of age polyethism in which the relevant quantitative data have been reported, this behavioral trait does not correlate with phylogenetic position. Thus, in the primitive species *Myrmecia forceps,* callows remain inside the nest for approximately one month (Freeland, 1958). In the formicine species *Formica polyctena* (Otto, 1958), the transition from *Innendienst* to *Aussendienst* takes place after 40 days. Finally, callows of the myrmicine genus *Messor* (Goetsch and Eisner, 1931) require from one to several months before they leave the nest.

A very different picture of behavioral development emerges from studies of predatory, nomadic army ants. Thus in *Neivamyrmex nigrescens, Eciton hamatum,* and *E. burchelli,* callows join in the predatory raids only 3–7 days after eclosion. Furthermore, unlike most ant species, which typically occupy a single nest, these army ants exhibit behavioral cycles, each consisting of a distinct nomadic and statary phase (Schneirla, 1971). Throughout each nomadic phase, the entire colony emigrates almost daily to new nesting sites, and the first emigration typically occurs within 24 h after the callows' eclosion. During each emigration, the entire callow population (together with the adult workers and queen) leaves the bivouac and follows the colony's chemical trail to the new nest. The callows' extremely early participation in both the emigrations and predatory raids suggests a relatively advanced state in the degree of their sensory and motor development at the time of eclosion.

The aims of this chapter therefore are: (1) to summarize the results of experiments on temporal changes in the behavior of callows of the nearctic army ant *Neivamyrmex nigrescens;* (2) to demonstrate that the behavioral development of the callows is best conceptualized through analyses of the social interactions between callows and mature adult ants;

and (3) to suggest possible ecological parameters that could have facilitated the evolution of rapid maturational rates.

Experiment 1

INTRODUCTION AND METHODS

As one index of behavioral development, we examined the callows' performance in a pattern of behavior which in the field is exhibited only by workers mature enough to participate in the predatory raid: the ability to form a raiding front at the start of foraging in an area around the nest containing no previously deposited chemical trail. This situation was simulated on a small scale in the laboratory by allowing groups of ants to emerge from a plastic enclosure onto a large sheet (46 × 57 cm) of filter paper on which concentric circles were drawn at 2-cm intervals from the center. The enclosure consisted of a circular holding chamber that was connected to a curved tunnel (Fig. 1). The lid of the holding chamber had a small downward-projecting plastic strip that functioned as a door

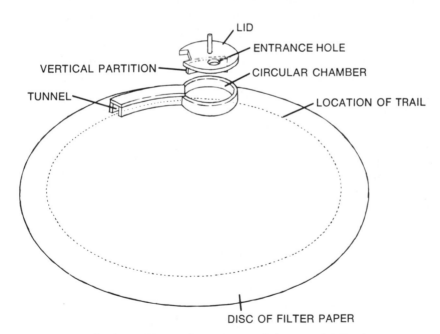

FIG. 1. *Apparatus used to compare behavior of callow and mature adult army ants. For Experiment 1, the curved tunnel exited onto the center of a large sheet of filter paper.*

between the chamber and adjoining tunnel (for additional details, see Topoff, Lawson, and Richards, 1972).

In the first test series, 20 mature adult workers (collected from an emigration on the first nomadic night) were placed in the holding chamber. After an adjustment period of 30 sec, the door between the holding chamber and tunnel was opened. Each test lasted 2 min, during which we recorded the ants' rate of exit from the tunnel, and their rate of movement across the substrate. The test was repeated ten times, with a different group of 20 ants and a clean sheet of filter paper for each test. Between tests, we washed the plastic apparatus with 70% ethyl alcohol (to remove odors), followed by distilled water. After the adult test series, three additional series of ten tests were conducted, using groups of 20 callows removed from the same colony on nomadic days 1, 2, and 5.

RESULTS

A comparison of the rates of exit of callow and adult ants is presented in Figure 2. Adult ants formed into a small "raiding front" and began leaving the tunnel as soon as the door opened. By the end of 30 s, over 50% of the ants had left the tunnel and were advancing across the filter paper. At the end of the test, on the average, more than 80% of the adults had left the enclosure.

By contrast, when groups of 1-day-old callows were placed in the apparatus, they promptly clustered tightly together and remained virtually motionless throughout the test (Fig. 3). On the average, across ten tests, only 5% of the callows left the enclosure. The few that did leave were individuals that by chance did not enter the main cluster. Callows removed from their colony on nomadic days 2 and 5 exhibited a progressively decreasing tendency to cluster. Figure 2 shows that their rates of exit are also progressively adult-like, although 5-day-old callows still did not perform as well as mature adults. The rates of exit of the ants over the 2-min test interval are significantly different for the four groups ($P < 0.001$, as determined by a repeated, one-way analysis of variance, followed by a pair-wise comparison using the Tukey test).

After leaving the tunnel, the movement of the adult ants across the filter paper was similar to their behavior in the field at the start of foraging in an area containing no previously deposited trail. Their outward progress proceeded in a step-wise fashion. A relay trail-laying process began, with each ant advancing several centimeters and then abruptly turning around and reentering the group. Ants coming up behind followed the trail deposited by the preceding ant and each, in turn, extended the trail slightly before backtracking. In this way the column

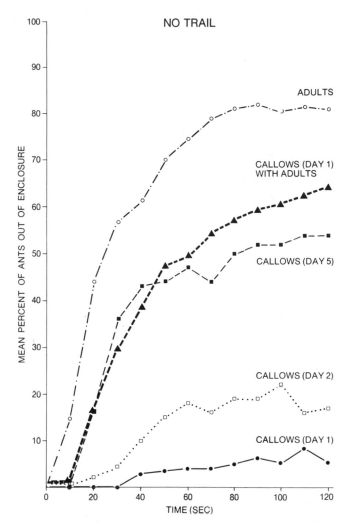

FIG. 2. *Rate of exit from enclosure of mature adult and callow ants on substrate containing no previously deposited chemical trail. Each point represents mean of ten tests. Thin-lined graphs show results of Experiment 1. Heavy curve is from Experiment 3.*

advanced steadily, with each ant taking a turn at trail laying as it reached the front.

The step-wise advance of the ants at different stages of development is shown in Figure 4. The median time for mature adults to reach the

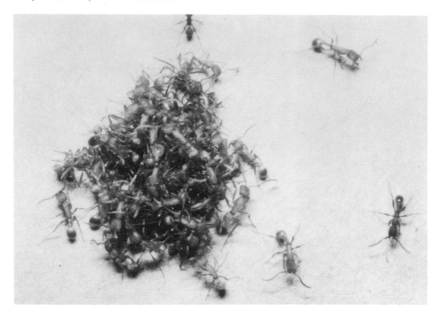

FIG. 3. *Clustering behavior of newly eclosed callow workers. The cluster illustrated here contains approximately 100 lightly pigmented callows. Several fully pigmented adult ants (one located at right of photograph) are included for comparison.*

outer boundary of the filter paper was 35 s (as shown in the upper graph). Note that the shape of the curve is a rather steep staircase. The steps are compressed laterally because many rapidly moving ants contributed to the advance of the "raiding front." Because fewer callows left the enclosure (and those that did moved more slowly), the step-wise advance of their "raiding front" is greatly exaggerated. For 1-day-old callows (lower graph in Fig. 4), the median farthest distance from the center reached by the end of the test was only 5 cm. On successive days, their pattern of advance also became more adult-like.

Experiment 2

INTRODUCTION AND METHODS

The previous experiment showed that when callows are confined in an environment in which all social stimuli arise only from other callows, they exhibit a strong approach behavior and form into motionless clusters. In

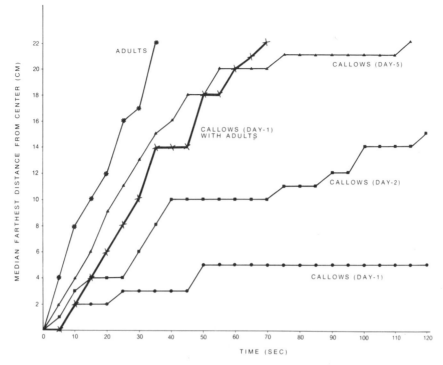

FIG. 4. *Rate of advance of "raiding front" of mature and callow ants over substrate containing no previously deposited chemical trail. Each point represents median of ten tests. Heavy curve is from Experiment 3.*

the nest, however, there exists an additional chemical information which originates from adults and is not confined to the nest. For example, after a raiding column is established, an unbroken trail is deposited from within the nest to all raiding sites. Because newly eclosed callows are extremely responsive to the odor trail (Topoff, Boshes, and Trakimas, 1972), trail following is probably an important process by which callows orient toward the new nest during an emigration. However, because the trail is present throughout the raid (when the callows are inside the nest), the trail alone should not be capable of arousing the callows to mass trail following. To verify this, a petroleum ether extract of the odor trail was deposited in a circle (8-cm diameter) on a disc of filter paper (according to procedures detailed by Topoff *et al.,* 1972). The apparatus and procedures were identical to those of Experiment 1, except that the plastic enclosure and tunnel were placed directly over the chemical trail. Thus, the adult workers and callows were constantly exposed to the stimulus.

RESULTS

As in the first experiment, the adults remained active inside the enclosure and streamed out of the tunnel at the start of each test. Because the trail was concentrated, the adults did not form into a "raiding front" with two-way traffic. Instead, each ant ran without turning, around the entire circular course. By the end of 60 s, 70% of the adults had left the enclosure (Fig. 5). The behavior of the 1-day-old callows was again considerably different in that only 15% of them were out of the enclosure by the end of 60 s. Thus, despite the presence of a strong chemical trail extending from the enclosure, through the tunnel, and around the filter paper, most callows promptly formed into an inactive cluster.

Experiments 3 and 4

INTRODUCTION AND METHODS

The previous experiments were conducted with separate groups of callow and adult ants. The following two experiments, therefore, explored the behavioral interactions between workers comprising the two age groups. Procedures were identical to those of Experiments 1 and 2, except that 20 adult workers and 20 newly eclosed callows were combined in the holding chamber. Data were recorded for the two groups separately. In addition, to determine that the observed effects were not the result of crowding, an additional series of tests was included, using 10 ants of each age group. The results were identical to those reported below.

RESULTS

In contrast to their clustering behavior in Experiments 1 and 2, the callows never formed into a cluster in the presence of adult ants. The activity of the adults kept the callows dispersed and constantly moving inside the holding chamber. In the no-trail condition of Experiment 3, callows and adults formed into a mixed "raiding front" as soon as the door to the tunnel opened (Fig. 2, heavy curve). Although the rate of exit of the callows was still lower than that of mature adults alone, their performance was significantly better than callows' of the same age that were not combined with adults ($\chi^2 = 82.7$, $p < 0.001$). After leaving the tunnel, the callows that were physically interacting with adult ants also

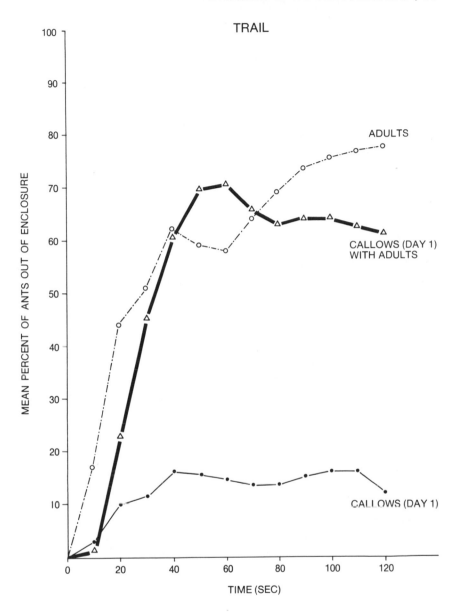

FIG. 5. *Rate of exit from enclosure of mature adult and callow ants on substrate containing an extract of the chemical trail. Each point represents mean of ten tests. Heavy curve is from Experiment 4.*

advanced across the filter paper in an adult-like manner (Fig. 4, heavy curve).

In the presence of a trail (Experiment 4), the adult ants had an even greater effect on the callows' behavior (Fig. 5, heavy curve). By the end of 60 s, approximately 70% of the callows had left the enclosure, as compared with 45% when callows and adults were combined in the no-trail condition of Experiment 3. In fact, in the presence of the chemical trail, the performance of the callows was not significantly different from that of their mature adult sisters ($\chi^2 = 0.05, p > 0.9$).

Experiment 5

INTRODUCTION AND METHODS

In Experiments 3 and 4, the adult ants prevented callows from clustering together. In laboratory nests, however, adult ants at the start of an emigration disperse callows that are already tightly clustered in a large central mass (Topoff and Mirenda, in press). The mechanism of this dispersal could be mechanical stimulation of the callows by adults, the secretion by adults of an arousing chemical, or a combination of both. To determine the nature of the arousing stimulus, a two-story trail-following experiment was designed. As in Experiment 2 and 4, a circular trail was deposited on a disc of filter paper, and 20 callows placed in the holding chamber. The chamber was positioned so that the callows could exit in a counterclockwise direction (Fig. 6). At the other end of the trail, an identical enclosure containing 20 adults was positioned with its tunnel exiting in a clockwise direction. A second disc of filter paper containing a trail of equal concentration and diameter was placed over both enclosures, and a third enclosure containing 20 callows was placed on the upper filter paper, directly over the lower callow chamber. Thus, the upper disc of porous filter paper served as both the roof of the lower callow chamber and the floor of the upper one. After the callows in the lower and upper chambers had formed into a cluster, the doors of all three enclosures were opened.

RESULTS

As expected, as soon as the door to the adult chamber was opened, the ants streamed out and followed the circular trail towards the lower callow chamber. Upon entering the callow chamber, the adults spread out and ran repeatedly over and through the callow cluster. Because the callow chamber was a dead end for the adults, many promptly turned around

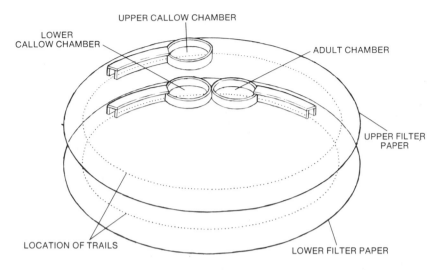

UPPER CALLOW CHAMBER

LOWER CALLOW CHAMBER

ADULT CHAMBER

UPPER FILTER PAPER

LOCATION OF TRAILS

LOWER FILTER PAPER

FIG. 6. *Two-story trail following apparatus for determining mechanism of callow arousal by adult workers. The upper filter paper sits on top of the lower chambers. It functions as roof and floor of lower and upper callow chambers respectively.*

and followed the trail back towards their "home" chamber. As the adult ants contacted the callows, the cluster dispersed and callow activity increased. Across three tests, by the end of 2 min (starting when the first adult ant entered the callow chamber), 70% of the callows were out of the enclosure, running with the adults along the chemical trail. The callows in the upper chamber, although separated from the lower one by a height of only 1 cm, remained tightly clustered, and none left the enclosure throughout each of the three tests.

Having ruled out the possibility that callow arousal was due to volatile odors secreted by the adult ants, we next tested the possibility that nonvolatile chemicals deposited by adults on the lower filter paper were involved. To do this, we repeated the above experiment but removed the callows and adults from the lower paper as soon as the adults had dispersed the callow cluster. The clustered callows from above were then immediately introduced into the lower callow enclosure. These callows again formed into a ball, and only one individual left the enclosure throughout the 2-min test. Finally, we lightly stroked the callows with fine wire, in an attempt to simulate the mechanical stimulation provided by adult ants when they dispersed the callows. Surprisingly, this artificial stimulation caused the callows to gradually disperse, become active, and leave the enclosure. Within 1 min after the start of mechanical stimulation, 50% of the callows were out of the enclosure, following the chemical trail.

Experiment 6

INTRODUCTION AND METHODS

In the previous experiment, callows were aroused to trail following by the mechanical stimulation imparted by adult ants. It is possible, however, that the mechanism of arousal could be different in a more species-typical nest environment when callow recruitment by adults occurs in the context of an emigration to a new nesting site. To ascertain this, a cylindrical enclosure (12-cm diameter $\times$ 10-cm height) with a stainless steel mesh floor was fitted into a hole in the roof of a laboratory ant nest. As in the previous experiment, the enclosure exited onto a sheet of filter paper containing a strong chemical trail. A colony of *N. nigrescens* containing newly eclosed callows was placed in the nest, which was connected to a second nest 15 m away. Three hundred and fifty callows that had been previously removed from the colony were placed in the roof-top enclosure.

RESULTS

Throughout the evening raid, the callows in the nest and upper enclosure remained tightly clustered. After 2 h of raiding, a full-scale emigration towards the second nest developed. During the next 3 h, the entire callow and adult population emigrated, but the 350 callows in the upper enclosure remained tightly clustered and none left the chamber. As soon as the emigration was completed, the cover of the laboratory nest was removed and the 350 callows placed directly on the filter paper floor of the nest. Despite this procedure, the callows again formed into a ball. But when we lightly stroked the callows with a fine wire, they slowly dispersed, left the nest, and followed the chemical trail along the runway to the main colony in the second nest.

Discussion

THE PROCESS OF RECRUITMENT

Species of army ants that exhibit regular cycles of alternating nomadic and statary phases are unique in being the only social Hymenoptera in which an entire population of newly eclosed workers is regularly recruited, along with the adults and queen, out of the colony nest. It is appropriate, therefore, to analyze the interactions between adult and callow army ants within the conceptual framework provided by recruitment studies in general.

An examination of the literature on social insect communication shows that the term "recruitment" has been used very loosely and subsumes a variety of qualitatively different social interactions that have the effect of bringing nestmates to some point in space for food retrieval, nest construction, defense, and migration. In some ant species the process of recruitment is as direct as one worker carrying another to the target area (Hölldobler and Maschwitz, 1965; Möglich and Hölldobler, 1974). More often, however, recruitment is based upon more complex forms of communication involving chemical and tactual signals. In a primitive recruitment pattern known as "tandem running" constant recruiter contact is necessary, and only one nestmate is recruited at a time (Wilson 1959; Hölldobler, Möglich and Maschwitz, 1974; Maschwitz, Hölldobler, and Möglich, 1974). In the more advanced patterns, orientation to a deposited chemical trail becomes increasingly important. For example, in a system known as "group recruitment" a chemical trail allows groups of ants to follow a single recruiter. If the leading ant is removed, however, the group promptly disperses (Urbani, 1973, in Chadab and Rettenmeyer, 1975). The myrmicine ant species *Monomorium venustum* (Szlep and Jacobi, 1967) and *Solenopsis invicta* (Wilson, 1962) utilize a "mass recruitment" system. In this pattern, the recruiting ant is important only in initially arousing nestmates, after which they orient by following the trail deposited by the recruiter. Chadab and Rettenmeyer (1975) recently demonstrated mass recruitment to a feeding site in species of the army ant genus *Eciton*. According to these investigators, recruiting workers aroused colony members to trail following by repeatedly contacting them while running back and forth in the raid column. The results of our studies on *N. nigrescens* indicate that a similar process of mass recruitment is responsible for arousing callows at the start of an emigration.

It has previously been shown that newly eclosed callows of *N. nigrescens* are as responsive as adult ants to the colony's chemical trail (Topoff, Boshes, and Trakimas, 1972). Because foraging adult ants continuously run in and out of the nest depositing a strong unbroken trail, callows undoubtedly have an abundance of cues that could be used for leaving the nest during the predatory raid. Experiments 1 and 2, however, have shown that callows are also extremely sensitive to stimuli arising from each other, and tend to cluster tightly when adult ants are not physically interspersed with them. Finally, Experiments 3 through 5 show that (1) mature adult ants can prevent callows from clustering and can disperse preexisting clusters, and (2) the process of callow arousal by adults is based predominantly upon mechanical stimulation.

Based upon the experiments reported in this paper, as well as observations of colonies of *N. nigrescens* in laboratory nests (Topoff and Mirenda, in press), mass recruitment from the nest in this species can be divided into three relatively distinct stages. The initial arousal of the various ant subpopulations for any stage seems to be similar to that for all stages, and may be explained by considering the intensity and duration of activity exhibited by recruiting workers, and the spatial location and threshold of arousal of individuals in the nest.

The adult workers occupying the most peripheral areas in the nest are loosely organized and are the first ants out of the nest at the start of the evening raid. Immediately upon leaving the bivouac, these workers become very excited, run repeatedly in and out of the nest, and arouse other adults near the nest entrance. This is the first stage of recruitment, and it is responsible for generating the raiding front and the multi-branched trail system. The importance of this initial recruitment to open spaces can be appreciated by considering the fact that the success of the army ant predatory raid depends on a critical striking force being present at or near the site where a prey species of ant is initially located. The size range of *N. nigrescens* is 2.5–5.0 mm (Topoff, 1971), and the ants often prey upon species that are larger and which possess a variety of defensive strategies. For example, nests of *Camponotus ocreatus* are guarded by major workers with large mandibles that are capable of decapitating individuals of *N. nigrescens*. Another defensive behavior of species of *Camponotus, Pheidole,* and *Trachymyrmex* is their ability to detect the presence of army ants before the raiding column reaches their nest. Under these conditions the adult workers remove their own brood from the nest and either climb up vegetation around the nest or abandon the area altogether until the raiding column departs.

Upon encountering a supply of booty, the second stage of recruitment begins, and its intensity is directly proportional to the quantity of food located. Excited ants returning to the bivouac arouse those adults that came to occupy the outermost areas of the nest as a result of the initial stage of recruitment. Our observations of laboratory colonies indicate, however, that the magnitude of recruitment from the nest at this stage may be relatively small. If food is found far from the nest, most recruitment occurs within the trail complex itself, with the result that ants running in outlying columns "drain" into the column where an active raid is in progress. During these first two stages of recruitment, the location of the callows near the brood in the center of the nest effectively buffers them from the excitation of recruiting adults. Finally, when the foraging workers establish a new nesting site, the last and most sustained stage of recruitment takes place within the old nest. As additional layers of adult ants are aroused to trail following, the inner callow population becomes exposed to the excitatory effects of the adult recruiters, and the callows disperse. They promptly enter the thick column of ants leaving the nest and follow the chemical trail to the new nesting site.

Analyses of Behavior Development Through Studies of Social Interactions

Callows of *N. nigrescens* leave the nest in two distinct contexts, emigration and raiding, with the former initially occurring at an earlier stage of the callows' development. Because their first exit from the nest takes place only 24 h after their eclosion (during the first nomadic emigration), their movements are still quite sluggish. The success of their participation in the emigration is probably due in large part to the relative simplicity of the emigration as a behavioral process. Thus, from the callows' perspective, emigrating involves following a single, unbranched chemical trail from the old nest to the new one. Because the density of ants in the emigrating column is considerably higher than during raiding, the callows are also aided by tactile cues along the entire route.

Raiding, by contrast, appears to be a behavioral process of far greater complexity. At the height of the nocturnal raid, the trail system typically consists of an anastomosing network of columns that may extend up to 60 m from the nest. Each booty-laden worker must therefore select the appropriate trail at each of the many junctions encountered on its return trip. Preliminary results from ongoing field studies suggest that newly eclosed callows are not as efficient as mature adult ants in orienting along the trail complex. This, combined with the fact that newly eclosed callow army ants are capable only of relatively slow rates of travel on the trail (Topoff, Lawson, and Richards, 1972), suggests that their absence from the predatory raids during the first few nomadic days may be adaptive to the colony.

The detailed discussion of recruitment behavior at the beginning of this section was included to illustrate how a full appreciation of behavioral development in army ants is possible only when analyses of social interactions are combined with studies of individual maturation. In studies of embryonic motility in vertebrates, considerable attention has been paid to the relative influence of external sources of stimulation (Gottlieb, 1971; Hamburger, 1963). Although the behavioral patterns of concern in our studies obviously represent a very different level of organization, the role of external sensory input is crucial for the adaptive integration of the callows' behavior into colony life. At the start of the nomadic phase, the callows have clearly achieved a degree of sensory and motor maturation necessary for joining in the emigration. But the behavioral consequences of their maturational state are exhibited only when the callows' tendency to approach each other and form into motionless clusters is overcome by stimulation from recruiting mature ants at the appropriate time. As the callows mature within the nest during the first few days of the nomadic phase, they become less responsive to each other and assume more peripheral positions relative to the new generation of developing larvae. As a result, the callows are more easily aroused by nestmates and begin participating in the predatory raids. Thus, the callows temporal progression from emigrating to raiding is based upon an interaction between the callows' changing maturational state and the sensory processes underlying mass recruitment by their mature adult sisters.

Ecological Correlates of Precocial Behavior

In the case of army ants in particular, the correlation between the advanced state of callow maturation and emigratory behavior is relatively clear-cut. An additional factor that may have provided a selection pressure for rapid development is the group predatory behavior of army ants. Although quantitative data are lacking, our field observations, as well as those of Schneirla (1971), show that many adult workers are routinely killed during raids on other ant species. The early participation of the callows in group foraging would contribute to a rapid replacement of the adult population, and therefore ensure the maintenance of a critical striking force. This ecological requirement is heightened by the strict brood periodicity of the army ant queen, who lays a single batch of eggs approximately every six weeks. This means that when one callow population is eclosing at the start of a nomadic phase, the only brood in the nest is a population of recently hatched larvae. Under these circumstances, if callow army ants (like those of *Myrmecia, Messor,* or *Formica*) required an additional month or more before being able to participate in raiding, it is probable that the predatory population would be sorely depleted.

The hypothesis that rapid callow development has evolved as an adaptation for. group predatory behavior is also supported by studies in our laboratory by E. Kwait (personal communication) on the slave-raiding ant *Polyergus lucidus*. Callows of this species also participate in the slave raid within three to five days after their eclosion. Of additional significance is the fact that the *Polyergus* queen also lays eggs in relatively distinct periods. When the slave raiding season begins in early June, only larvae are present in the nest, and an additional three to four weeks are necessary for them to complete their larval and pupal development. This means that the first population of new adults will not appear until well into the raiding season, and by this time the colony may be in desperate need of worker replacements.

Acknowledgments

Research supported by NIMH grant MH 28557-01 and by NSF grant BNS 76-17366.

References

Bartholomew, G. A. 1959. Mother-young relations and the maturation of pup behaviour in the Alaskan fur seal. *Anim. Behav.* 7:163–171.

Buckingham, E. 1910. Division of labor among ants. *Proc. Am. Acad. Arts Sci.* 46:425–507.

Chadab, R., and C. W. Rettenmeyer. 1975. Mass recruitment by army ants. *Science* 188:1124–1125.

Dobrzańska, J. 1959. Studies on the division of labour in ants of the genus *Formica*. *Acta Biol. Exp. Warsaw* 19:57–81.

Ehrhardt, S. 1931. Über Arbeitsteilung bei *Myrmica*- und *Messor*-Arten. *Z. Morphol. Ökol. Tiere* 20:755–812.

Freeland, J. 1958. Biological and social patterns in the Australian bulldog ants of the genus *Myrmecia*. *Aust. J. Zool.* 6:1–18.

Goetsch, W., and H. Eisner. 1930. Beiträge zur biologie körnersammelnder ameisen. *Z. Morphol. Ökol. Tiere* 16:371–452.

Gottlieb, G. 1968. Prenatal behavior of birds. *Q. Rev. Biol.* 43:148–174.

Gottlieb, G. 1971. Ontogenesis of sensory function in birds and mammals. *In* eds., E. Tobach, L. Aronson, and E. Shaw, pp. 67–128. The biopsychology of development. New York: Academic Press.

Hall, E. R., and K. R. Kelson. 1959. The mammals of North America. New York: Ronald Press.

Hamburger, V. 1963. Some aspects of the embryology of behavior. *Q. Rev. Biol.* 38:342–365.

Haskins, C. P., and E. F. Haskins. 1950. Notes on the biology and social behavior of the archaic ponerine ants of the genera *Myrmecia* and *Promyrmecia*. *Ann. Entomol. Soc. Am.* 43:461–491.

Hölldobler, B., and U. Maschwitz. 1965. Der hochzeitsschwarm der rossameise *Camponotus herculeanus* L. (Hym. Formicidae). *Z. Vergl. Physiol.* 66:215–250.

Hölldobler, B., M. Möglich, and U. Maschwitz. 1974. Communication by tandem running in the ant *Camponotus sericeus*. *J. Comp. Physiol.* 90:105–127.

Kiil, V. 1934. Untersuchungen über arbeitsteilung bei ameisen (*Formica rufa* L., *Camponotus herculeanus* L., and *C. ligniperda* Latr.). *Biol. Zentralbl.* 54:114–146.

Ledoux, A. 1950. Recherche sur la biologie de la fourmi fileuse (*Oecophylla longinoda* Latr.). *Ann. Sci. Nat. Zool.* 12:313–461.

Maschwitz, U., B. Hölldobler, and M. Möglich. 1974. Tandemlaufen als rekrui-tierungsverhalten bei *Bothroponera tesserinoda* Forel (Formicidae: Pone-rinae). *Z. Tierpsychol.* 35:113–123.

McCook, H. 1879. The natural history of the agricultural ant of Texas, *Pogono-myrmex*. London: Trubner.

Möglich, M., and B. Hölldobler. 1974. Social carrying behavior and division of labor during nest movement in ants. *Psyche* 81:219–236.

Nice, M. 1962. Development of behavior in precocial birds. *Trans. Linn. Soc. N.Y.* 8:1–211.

Otto, D. 1958. Über die arbeitsteilung im staate von *Formica rufa rufo-praten-sis* Gössw. und ihre verhaltensphysiologisches grundlagen: ein beitrag zur biologie der roten waldameise. *Wiss. Abhand. Dtsch. Akad. Landwirts-schaft. Berlin* 30:1–169.

Schneirla, T. C. 1971. Army ants: a study in social organization. San Francisco: W. H. Freeman.

Szlep, R., and T. Jacobi. 1967. The mechanisms of recruitment to mass foraging in colonies of *Monomorium venustum* Smith, *M. subopacum* Em., *Tapi-noma israelis* For. and *T. simothi* Em. *Insectes Soc.* 14:25–40.

Topoff, H. 1971. Polymorphism in army ants related to division of labor and cyclic colony behavior. *Am. Naturalist* 105:529–548.

Topoff, H., M. Boshes, and W. Trakimas. 1972. A comparison of trail following between callow and adult workers of the army ant *Neivamyrmex nigre-scens*. *Anim. Behav.* 20:361–366.

Topoff, H., K. Lawson, and P. Richards. 1972. The development of trail-follow-ing behavior in the tropical army ant genus *Eciton*. *Psyche* 79:357–364.

Topoff, H., and J. Mirenda. Precocial behaviour of callow workers of the army ant *Neivamyrmex nigrescens:* importance of stimulation by adults during mass recruitment. *Anim. Behav.* (In press.)

Wheeler, W. M. 1910. Ants: their structure, development, and behavior. New York: Columbia Univ. Press.

Wheeler, W. M. 1928. The social insects: their origin and evolution. New York: Harcourt, Brace.

Weir, J. 1958. Polyethism in workers of the ant *Myrmica*. *Insectes Soc.* 5:97–128.

Wilson, E. O. 1959. Communication by tandem running in the ant *Cardiocondyla*. *Psyche* 66:29–34.

Wilson, E. O. 1962. Chemical communication among workers of the fire ant *Solenopsis saevissima* (Fr. Smith). 1. The organization of mass foraging. 2. An information analysis of the odour trail. 3. The experimental induction of social responses. *Anim. Behav.* 10:134–164.

Wilson, E. O. 1971. The insect societies. Cambridge: Harvard Univ. Press.

Section Three

IN COLD BLOOD

The typically exothermic, or cold-blooded, vertebrates have been largely ignored in ethology, particularly in comparison with studies of behavioral development in birds and mammals. Yet each of the three authors here gives convincing reasons why the so-called lower vertebrates can help illuminate problems of development. Although many classic studies of animal behavior involved work with these groups (e.g., the research of Whitman, Yerkes, Carmichael, Coghill, Lorenz, Tinbergen, Baerends, and particularly G. K. Noble), relatively few theoretical ideas or methodologies have stemmed from this work, the most notable exception being the extensive theoretical arguments based on the behavior of the stickleback fish. Yet the larval stages of fish and amphibians offer wonderful opportunities for detailed studies on physiological and sensory factors in development. Reptiles, on the other hand, being highly precocial, give opportunity for evolutionary studies on the process of neotenization and delayed development, which seems to have reappeared in most mammals and many birds (see also the chapter by Topoff and Mirenda). David Noakes contrasts the development of two widespread and important groups of fish, cichlids and salmonids. Roy McDiarmid, while not studying larvae directly, details the exquisite differences in parental behavior between two sympatric and closely related tropical frogs, opening up many questions about the consequences for development. Gordon Burghardt primarily reviews the scattered and largely unsystematic work on early behavior in diverse groups of reptiles, adding some unpublished findings from his laboratory that show that changes in behavior can occur rapidly and dramatically in reptiles, but that phylogenetic factors are more difficult to ignore than with the warm-blooded descendants of reptiles.

Chapter Six

Ontogeny of Behavior in Fishes: A Survey and Suggestions

DAVID L. G. NOAKES

Department of Zoology
College of Biological Science
University of Guelph
Guelph, Ontario, Canada N1G 2W1

Fishes have great potential for studies of behavioral ontogeny. They are tremendously diverse, comprising almost half (about 20,000 species) of the recent vertebrates. Their reproductive habits, ranging from external fertilization and scattering of thousands of eggs, to internal fertilization and viviparity of a few embryos, offer advantages for evolutionary considerations (e.g., Dawkins and Carlisle, 1976; Maynard Smith, 1977) at least comparable to the situation known for amphibians (see Chapter 7). Unlike most vertebrates, fishes can typically be aged accurately, often to days, even as wild individuals (Brothers *et al.,* 1976; Struhsaker and Uchiyama, 1976; Taubert and Coble, 1977). Many species are of great economic importance, and studies of their early life histories, including behavior, would likely be of some significance to fish management (e.g., Ware, 1975; Mason, 1976). In those species with external fertilization, embryos are readily available for observation or experimental manipulation. Despite this potential, studies of behavioral ontogeny of fishes have been less frequent, and generally more descriptive than for birds or mammals (see Chapters 9, 11, and 16). Convenience or opportunity has often dictated studies, so that we have an uneven coverage of the group. Perhaps the most common misconception, which should be dispelled at the outset, is of fishes as representative of a more primitive, ancestral condition, as compared with tetrapods. Any consideration of fishes must recognize teleosts (the vast majority of living fishes) as highly evolved, specialized products of evolution with no direct relationship, in an ancestral sense, to tetrapods. Unfortunately, study of behavioral ontogeny in amphibians and reptiles has also been infrequent (Chapters 7 and 8), so my consideration of evolutionary aspects will of necessity be mainly within fishes, and can be extrapolated only in a general sense to other groups of vertebrates.

I have chosen to concentrate on development of social behavior, in part as a personal bias, but also, I think, as a fair reflection of the greater emphasis on this area. There are studies of the development of such behavior as habitat selection, and salinity preferences (e.g., Sale, 1971; Otto and McInerney, 1970; Reynolds and Thompson, 1974) but these are generally not as numerous as studies of social responses. I will review the behavior of two families of teleosts, the salmonids and the cichlids. Not coincidentally, these fishes have been the subject of my research for the past several years, so I am most familiar with them. They also represent a convenient range of differences in their zoogeography, ecology, reproductive behavior, and taxonomic relations to illustrate the points I wish to make in that regard. Consequently, this should be taken as a selective and highly subjective treatment of this class of vertebrates. I have, however, attempted to include sufficient references to indicate my lack of detailed coverage of particular areas of research. My concentration on salmonids and cichlids, for example, largely ignores the significant body of information on the development of schooling behavior (Shaw, 1970).

Teleost fishes are a diverse evolutionary assemblage, but most current proposals of classification (e.g., Nelson, 1976; Lagler et al., 1977) agree that salmonids are typical of a relatively unspecialized basal euteleostean (higher teleostean) condition, and that cichlids are typical of perciforms, a more highly specialized and relatively advanced group. Fish have a good fossil record and an abundance of reliable features used in classification, so their evolutionary relationships are known with greater assurance than for most vertebrates.

Salmonids can be fairly readily characterized as to life history, behavior, and habitat (Hoar, 1976). They are north temperate, freshwater (or anadromous) fish (about 70 species) of moderate size. They show no parental care other than burying their eggs in coarse gravel nests (redds) or depositing them over open gravel substrates in freshwater streams or lakes. Young hatch after some weeks, and spend several additional weeks beneath the gravel as eleutheroembryos, completing their embryonic development, using their large yolk reserve (Fig. 1). At about the time the yolk is depleted the young move up through the gravel and emerge into the open water. The gas bladder is filled, the fish become neutrally buoyant, and begin feeding. Depending on the species the young (now alevins) may either become territorial or may join others of the same age

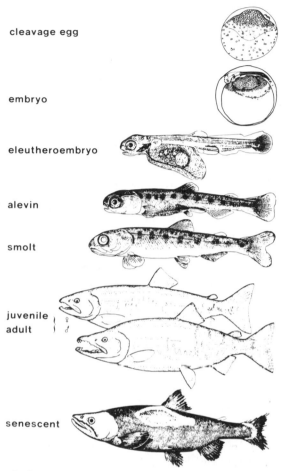

cleavage egg

embryo

eleutheroembryo

alevin

smolt

juvenile
adult

senescent

FIG. 1. *Successive developmental intervals in a representative salmonid* (Oncorhynchus *sp.*). (*From Balon, 1975a.*)

in mobile schools. They may or may not eventually move downstream as smolts (depending on the species) into a lake or ocean. After at least a year, juveniles mature and migrate back to the place of their birth to spawn. In some species adults die after spawning.

Cichlids, on the other hand, are a more advanced (evolutionary) group of about 700 species (Nelson, 1976) of small to moderate-sized fishes. They are typically specialized in their reproductive habits. All have well-developed parental care, ranging from prolonged, monogamous pairs with family units, to lek breeding systems and parental mouth brooding of embryos. Social interactions are relatively complex, and stable relationships may develop and persist over weeks or months. They are freshwater, tropical species of lakes or streams in Africa, South America, and (one genus) Asia (Lowe-McConnel, 1975). They are usually sedentary compared to the migrations of salmonids. Some species are territorial, both for reproductive and nonreproductive fish, while others may be schooling or more or less dispersed when not breeding. They have, in some cases, undergone probably the most explosive speciation known for vertebrates (Fryer and Iles, 1972). Their highly specialized feeding and reproductive habits have been suggested as important in this adaptive radiation.

Studies on salmonids have often been motivated by economic considerations. Many species are of major importance (e.g., Pacific salmon, rainbow trout) as commercial or sport fisheries. Studies on their biology are often undertaken with a view to enhancing survival or producing fish with a particular background to enhance economic harvest. On the other hand, studies of cichlids have usually been undertaken as basic research. Until recently (e.g., Turner, 1977) there has been little commercial importance attached to cichlids for harvest or angling, notwithstanding the local importance of many species to native harvest as food. Consequently, studies of courtship and spawning, and responses of parents and young to each other are often examples of "classical" ethological studies. Unfortunately, field studies of cichlids have not kept pace (e.g., Baylis, 1974; Barlow, 1974; Lowe-McConnel, 1975; Ward and Wyman, 1977), so it is more difficult to assess the significance of laboratory behavioral studies. In salmonids, the balance has, if anything, tended to favor field studies, with less emphasis on the ethological approach.

I will review in turn a summation of behavioral ontogeny in each of these families, and then relate that information to a proposal for classification of fishes, based on their reproductive habits, to demonstrate the potential utility of this latter scheme in understanding behavioral ontogeny.

The behavioral development of salmonids may be conveniently divided into the time before and the time after emergence from the gravel redd. The behavior of embryos after hatching, but before emergence, has been described in some detail by Bams (1969), Dill (1969), Dill and Northcote (1970), Peterson (1975), Mason (1976), Dill (1977), and others. The concurrent changes in sensory, and to some extent motor, capabilities have also been documented (e.g., Ali, 1959; Sharma, 1975; Verraes, 1977). There is a progressive development of visual capability such that at the time of emergence young are photopositive (changed from photonegative) (Mason, 1976), show marked photochemical changes in the retina, and possess full visual acuity, as evidenced by feeding experiments (Ali, 1959). Ahlbert (1976) found good agreement between the development of cone cells in the retina and feeding habits in salmon (*Salmo salar*) and trout (*S. trutta trutta*), from embryos through to the adult stage. Prior to emergence, embryos are capable of significant sensory and motor responses (Bams, 1969; Dill, 1969; Mason, 1976; Dill, 1977). They are markedly photonegative, positively geotactic, and thigmokinetic, all of which keep them beneath the surface of the gravel. There is also evidence that they may respond to physical factors such as oxygen and water flow, to disperse beneath the gravel (Dill and Northcote, 1970). Until they emerge from the gravel, the young still have a number of embryonic features and would be taken readily to be predators if they emerged prematurely. Mason (1976) has documented the progressive, but almost abrupt, shift which takes place in photobehavior of embryos at about the time of emergence from the gravel (Fig. 2). Dill (1977) has described in detail the motor patterns of embryos before and after emer-

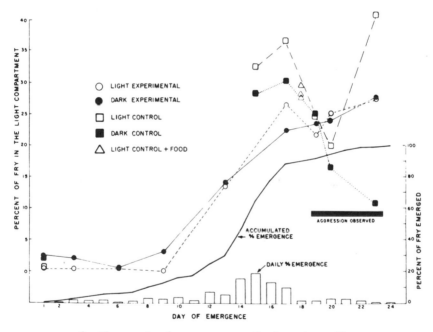

FIG. 2. *Changes in photoresponses of coho salmon* (Oncorhynchus kisutch) *fry. The curves indicate the increasing proportion of fry found in the light* (vs. *dark) half of the test chamber. The histogram shows the concurrent rate of emergence of fry from simulated redds. The black horizontal bar indicates the period during which aggressive behavior was observed.* (*From Mason, 1976.*)

gence. Motor patterns associated with regulation and/or maintenance of body position or orientation appear first in development, followed by patterns associated with the upward movement through the gravel (Fig. 3). All this leads to a progressive, integrated development of adaptive patterns, and changes in behavior. At about the time of emergence there is a marked alteration in behavior including the aforementioned photoresponses. Most of the important behavioral patterns associated with the embryonic period (and correlated with basic embryonic anatomy), are either terminated at the time of transition (e.g., thigmokinesis), or "change in sign" (e.g., phototaxis), again in concert with adaptive developmental changes in anatomy and morphology. Social interactions (agonistic behavior) and ingestive behavior are initiated. The apparently concurrent onset of these two behaviors (Cole, 1976; Mason, 1976; Dill, 1977) is of particular interest. Mason (1976) concluded that the change in photobehavior is indicative of an ontogenetic change unaffected by previous feeding history or other manipulations. Dill (1977) has suggested that the concurrent onset of feeding and agonistic behavior reflects a common underlying motivation. Certainly the motor patterns for the two types of behavior are, at least initially, very similar, and a reasonable case could be made for the development of agonistic patterns from ingestive patterns. This hypothesis remains to be tested, but is a suggestion to which I will return in my discussion of cichlids. However, Hinde (1970) has noted the problems resulting from such a functional classification of behavior during ontogeny. In the orange chromide (*Etroplus maculatus*) for example, the study which he cites, the same motor pattern may be part of different functional and causal systems in young and adult (Ward and Barlow, 1967). Others (Albrecht, 1966; Heiligenberg, 1965) have found that motor patterns of feeding and fighting are homologous in some cichlids, and share motivational relations. Stimuli which elicit feeding also promote fighting, and vice versa.

In species such as rainbow trout (*Salmo gairdneri*), which are non-schooling after emergence (Hoar, 1976), the social behavior following emergence is initially simple, consisting essentially of straightforward attacks, chases and/or bites (Cole, 1976). These overt aggressive acts are shown first after emergence, while threat postures (head down, dorsal fin

FIG. 3. *The development of behavior in* (a) *rainbow trout* (Salmo gairdneri) *and* (b) *Atlantic salmon* (S. salar). *Open bars indicate the age over which each behavior pattern was performed by most fish, the single line represents the total range of occurrence for that act. The mean age of emergence from the redd is indicated by* E(T). *Hover did not occur in all groups, so has no open bar.* (*From Dill, 1977.*)

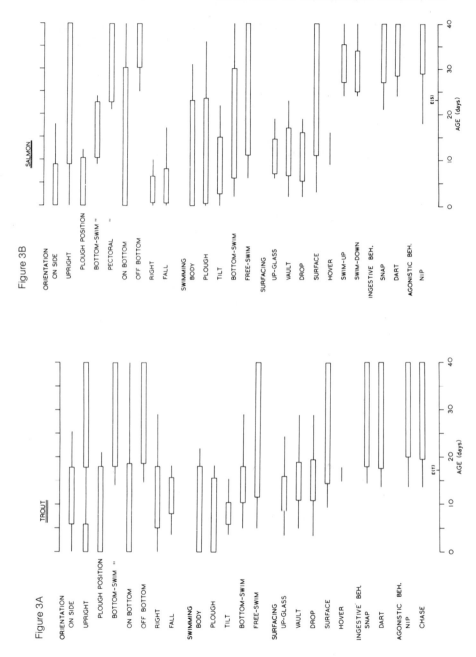

Figure 3A

TROUT

Figure 3B

SALMON

erection) do not appear until about one week later (Fig. 4). With time more elements are added to the behavioral repertoire and individual encounters become increasingly complex. Threat replaces overt attack as predominant in agonistic encounters (Fig. 5). How much of this change is the result of ontogenetic change and how much is a consequence of experience remains to be determined (Cole and Noakes, ms).

Prior to emergence salmonids feed little, if at all (Twongo and MacCrimmon, 1976), although there is evidence of feeding in some species (Dill, 1967). Shortly after emergence, when the final remnants of the yolk are absorbed, ingestive responses are initiated (Twongo and MacCrimmon, 1976; Dill, 1977). At first these appear poorly coordinated, and are directed to a variety of objects, including inappropriate items. With time and experience feeding responses become more narrowly directed to appropriate food objects, and become more efficient in terms of successful completion of feeding sequences. Some of the improvement, as measured by the success of feeding attempts, is a result of developmental change and not feeding experience itself (Twongo and Mac-

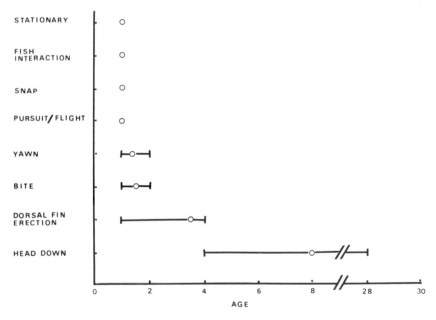

FIG. 4. *Age of onset (from time 50% of alevins free-swimming) of behavior patterns in rainbow trout* (Salmo gairdneri) *alevins. Open circle depicts mean, while left and right brackets indicate range of days over which each pattern was first observed in different observation tanks. (From Cole, 1976.)*

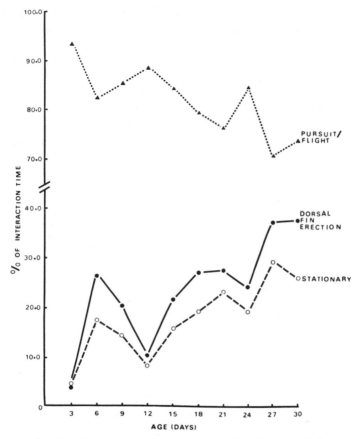

FIG. 5. *Grand means, pooled over successive 3-day periods, showing changes in the percentage of interaction time engaged in aggressive (pursuit/flight), and threat (dorsal fin erection, stationary) by rainbow trout* (Salmo gairdneri) *alevins. (From Cole, 1976.)*

Crimmon, 1976), but feeding experience does enhance subsequent feeding success. Up to a point continued food deprivation has little apparent effect on subsequent development of feeding behavior. After some time, however, designated the "point of no return" (Blaxter, 1969), fish cannot be successfully introduced to feeding. They show feeding responses, but apparently their energetic reserves are depleted past the point for maintenance and performance of behavior to prolong their metabolism. This phenomenon is well established for marine fishes, and the values for prey density, prey size, time to feeding, and other parameters have been established for a number of species (Blaxter, 1974).

As mentioned, field studies of ecology and life history, including to some extent behavior of salmonids, have often been carried out (e.g., Jenkins, 1969). One area that is an excellent illustration of the importance of behavioral ontogeny to the life of fishes is the homing behavior shown by many salmonids. Largely as a result of detailed studies by Hasler and his co-workers (Hasler, 1966), we have a good understanding of many aspects of this behavior. It is the early experience of the young, specifically their exposure to chemicals (natural or experimental), that is instrumental in determining their subsequent homing response. Evidence from tagging and recovery, transplanting fish at different ages, experimental imprinting of fish to artificial chemical cues, and neurophysiological evidence of responses to imprinting chemicals clearly demonstrate the crucial role of early experience for the specificity of homing in the subsequent spawning migration (Scholz et al., 1976). In fact, the phenomenon is generally referred to as imprinting, in the same sense as the term is used in classical studies of precocial birds (Chapter 10). This is not to deny the effects of other factors, including genetically based predispositions for more successful homing by local stocks to particular streams (Bams, 1976), nor does it necessarily account for the significant proportion of time and travel in the open waters of lakes or oceans. However, the critical role of early experience in the chemically based stream homing is clearly established.

The generalized summary of behavioral development in cichlids requires an appreciation of their reproductive habits. Within the cichlids, which are a unitary taxon by conventional classification, there are at least two distinct reproductive guilds (Balon, 1975a), mouth brooders and nest-spawning guarders. In terms of behavioral development, it is necessary to make this distinction between mouth brooders and substrate brooders, as proposed by Baerends and Baerends-van Roon (1950) (see also: Barlow, 1974; Balon, 1975a). Mouth brooding is clearly more specialized and more highly evolved (Balon, 1975a). Embryos of such species develop within the oral cavity of either the male or female parent (Oppenheimer, 1970) and lack a larval period. Instead, like salmonids, they have a special alevin phase in the juvenile. On the other hand, substrate brooding species have a larval period, but no alevin phase in the juvenile (Balon, 1975a). This makes comparisons of behavioral development somewhat complicated between species differing in reproductive habits. This is particularly true since many investigators mistakenly use the time of hatching as indicative of the developmental stage for comparisons in these fishes. This point stresses the need for uniformity and conformity in the definition and naming of developmental intervals in fish development (Balon, 1975b).

For example, in a comparison of the young of various *Tilapia* species, Brestowsky (1968) found that young mouth brooders and substrate brooders undergo essentially similar behavioral development for attachment (social cohesion) to the parents. (Note: these mouth brooding *Tilapia* species have since been re-assigned to the genus *Sarotherodon* by Trewavas [1973]). The major difference is that young mouth brooders pass through the critical period for the development of this response while still inside the parent's mouth, so the subsequent attachment behavior is not normally expressed. Under experimental conditions, embryos can be removed from the parent's mouth and tested under comparable circumstances as young substrate brooders. If the mouth brooder young are induced to swim earlier (by puncturing the yolk sac), they show the same critical period and subsequent attachment behavior as substrate brooder young.

In young of substrate brooding species, behavior becomes increasingly complex with age, as new patterns are added to the behavioral repertoire (Wyman and Ward, 1973). Behavioral patterns which originate as ingestive responses (at least in a functional sense), become incorporated into the repertoire as social responses. These are first directed towards parents and siblings, but come to be employed in sexual or territorial contexts as well. Based on a number of studies of the orange chromide, *Etroplus maculatus* (e.g., Cole and Ward, 1970; Quertermus and Ward, 1969; Wyman and Ward, 1973), Ward and co-workers have proposed one of the few theoretical models for the development of behavior in fishes (Fig. 6). To simplify somewhat, they propose that the entire behavioral repertoire of this species develops from two initial motor patterns, glancing and micronipping, under the influence of maturational changes in the young and environmental factor. Data currently available for both cichlids and salmonids (as mentioned previously) are in at least general agreement with this model. Clearly, a major task for those of us working in this field is to rigorously test this model and to refine it, or alternatives, as both an explanatory and a descriptive tool.

A number of studies have investigated the features of primary importance in the social attachment between parents and young cichlids. Kuhme (1963, 1964), Myrberg (1966, 1975), and others have demonstrated the importance of chemical cues in the communication system between parents and young. Sound may also be involved (Myrberg *et al.,* 1965), but I will discuss the visual modality, as there are more studies of that aspect. Using artificial dummies of parents it has been found that the initial responsiveness of young in substrate brooding species is to the dominant colors of the parents (Noble and Curtis, 1939; Baerends and Baerends-van Roon, 1950; Peters, 1937, 1941; Kuhme, 1962; Kuenzer, 1964).

In one of the few studies combining behavioral and anatomical

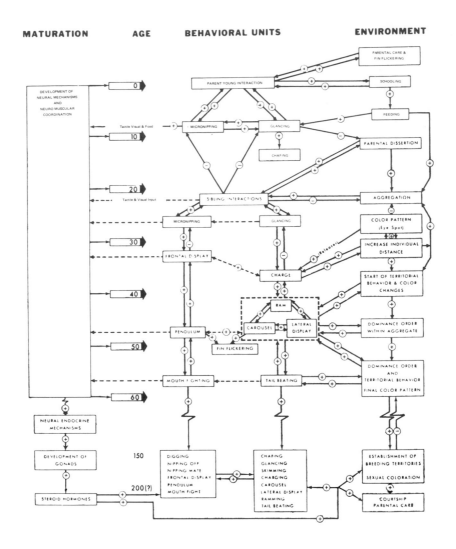

FIG. 6. *A model representing the ontogeny of behavior in*
Etroplus maculatus. *Solid arrows marked "+" indicate facili-*
tation, solid arrows marked "−" indicate inhibition. Dotted
arrows represent environmental feedback to the organism. En-
vironmental factors implied by the dotted arrows are those
indicated feeding into the model from the right. Dotted arrows
marked "+" are proposed routes of facilitation between be-
havioral units. (*From Wyman and Ward, 1973.*)

measures, Baerends *et al.* (1960) found an increase in visual acuity with increasing age in young *Aequidens portalagrensis,* a substrate brooding species. Ahlbert (1975) has summarized findings of different studies on the development of receptor cells in the retina, and/or early behavior in various species of fishes, including some cichlids. In general, there is a fairly clear and predictable correlation between the development of the photoreceptors and the requirements for the visual system at that time. Detection of movement, visual acuity, and sensitivity are some obvious properties which develop in synchrony with behavioral changes. These correlations are by no means perfect, however, and in some cases there is still a lack of information, usually behavioral, to clarify the significance of the observed anatomical developments (see references cited by Ahlbert, 1975). Peters (1965) and Kuenzer (1975), among others, have done similar studies with mouth brooding species. Features of color, color patterns, and movement are all important in releasing parent-oriented responses in these young. Peters (1963) investigated the responsiveness in young of two *Tilapia* species to parental dummies. One was a substrate brooding species, the other a mouth brooding species (but see Trewavas, 1973). Young mouth brooders normally approach and make contact with the parent (in this species) as a prelude to reentering the adult's mouth. Young of the substrate brooding species would approach, but not closely, and would not contact the parent (in this species). The responses of naive young (raised in isolation from parents) conformed to the species-typical behavior. Interestingly, hybrid young from these two species behaved more like the mouth brooding species.

In my work with cichlids I carried out some experiments directed to similar questions. In *Cichlasoma citrinellum,* a substrate brooding species, young normally contact the parents quite frequently as a trophic response (Noakes and Barlow, 1973a). With time and experience they come to show marked preferences in this contacting, depending on the color and sex of the parents [this species is markedly polymorphic (Fig. 7)]. Normal colored fish and males are contacted more than gold colored or female fish (Fig. 8). Cross-fostered young show significant effects of their prior contacting experience (Noakes and Barlow, 1973b) (Table 1), but young raised with only the sight of their siblings and parents do not develop the preferences. On their initial exposure young contact either parent about equally, regardless of color or sex (Table 2). The basis for the development of these preferences is as yet unknown, but there are histological correlates of the preferences in the skin of the adults (Noakes, 1973). This and similar responses in other cichlids has led to the suggestion that imprinting may occur in cichlids (Noakes and Barlow, 1973a; Myrberg, 1975), comparable to the much-studied phenomenon in birds

FIG. 7. *Young* Cichlasoma citrinellum, *with normal (grey) father and gold mother (partly hidden), on their first day of free swimming. (Photo by G. W. Barlow.)*

(Chapter 10). Particularly since *C. citrinellum* is a polymorphic species the question is of some interest, as the possible evolutionary consequences of imprinting in such a polymorphic species have been outlined by Kalmus and Smith (1966), Seiger (1967), Seiger and Dixon (1970), Colgan *et al.* (1974), and Cooke *et al.* (1975). Even partial imprinting can establish a balanced polymorphism in the population. Ferno and Sjolander (1976), and Weber and Weber (1976) investigated experimentally the possibility of lasting effects of early social experience on subsequent mating preferences in the convict cichlid, *Cichlasoma nigrofasciatum,* a substrate brooding species. Their results disagree but, in part at least, the discrepancy might reflect procedural differences. There is some suggestion that fish do have a significant sexual preference for the previously experienced color morph. Whether this occurs in naturally polymorphic

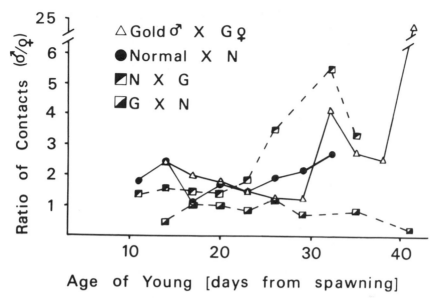

FIG. 8. *The ratio of contacts (male/female) for young* C. cit-rinellum *of different ages, according to parental colors. (From Noakes and Barlow, 1973a.)*

species (the color morph they used is only a domestic variety) remains to be seen. There are naturally occurring polymorphic cichlids (e.g., Barlow, 1973, 1976; Barlow and Munsey, 1976; Barlow, Rogers and Cappeto, 1977) in which the evolutionary and/or ecological consequences of such a mechanism could be quite significant. Larson (1976), for example, has documented a situation in three-spined sticklebacks (*Gasterosteus aculeatus*) of extensive spatial and trophic segregation of two phenotypes within a lake, maintained at least in part by inherent behavioral differences.

By way of a summary, I will draw attention to a proposal (Balon, 1975a) for the classification of fishes by other than conventional systematics. A detailed consideration of behavioral development in fishes shows that distantly related species may have rather similar features, while more closely related forms may be quite dissimilar. The cichlids, for example, are a unitary taxon (family) but belong to at least two distinct reproductive guilds. Conversely, the same reproductive habit has evolved independently in unrelated groups, e.g., livebearing in poeciliids and in the coelacanth, *Latimeria chalumnae* (Balon, 1977). A system which classifies fishes according to their reproductive habits (including ethological, ecological, and embryological features) appears to have utility in this regard. I will not recount extensively here the details of the proposal I

TABLE 1. Contacting behavior of cross-fostered young *Cichlasoma citrinel-lum*. Young raised with their own parents show consistent contacting prefer-ences. Young contact foster parents similarly, but the effect of prior experience may alter their behavior (Family No. 3).

FAMILY NO.	CONTACTING PREFERENCE OF YOUNG TO			
	OWN PARENTS		FOSTER PARENTS	
	MALE	FEMALE	MALE	FEMALE
1	Normal >	Normal	Gold <	Normal
2	Normal >	Gold	Gold >	Gold
3	Gold >	Gold	Normal =	Gold
4	Gold <	Normal	Normal >	Normal

Direction of arrow indicates significant preference; "=" indicates no preference.
Source: From data in Noakes and Barlow, 1973a.

TABLE 2. Contacting behavior of young *Cichlasoma citrinel-lum* raised in isolation from their parents

PARENTS		CONTACTING PREFERENCES	
MALE	FEMALE	CONTROLS	ISOLATES
Normal ×	Normal	$\male > \female$	$\male = \female$
Normal ×	Gold	$\male > \female$	$\male = \female$
Gold ×	Gold	$\male > \female$	$\male = \female$
Gold ×	Normal	$\male < \female$	$\male = \female$

Controls were equal numbers of siblings raised with parents. Contacting measured in both groups in successive trials on same day (preferences indicated as in Table 1). From Noakes and Barlow, 1973a.

wish to follow (Table 3) (Balon, 1975a). The important selective force in the evolution of reproductive habits in fishes appears to be predation on embryos and early developmental stages (Balon, 1975a; Ware, 1975). Adaptations for reproduction, whether the production of extremely large numbers (several million) of pelagic eggs, burying eggs in gravel redds, guarding of nest and young, mouth brooding of embryos, or internal development (the ultimate in parental protection) are clearly directed towards the reduction of losses of young to predators. Which of these particular adaptations (Table 3) is shown by a particular fish will, of course, be a consequence of its ecology and phylogeny.

An understanding of this phenomenon and an appreciation of the impact of these reproductive specializations on, (a) developmental features of the young, and (b) the reproductive (and therefore, social) behavior of

TABLE 3. Ecoethological guilds of fishes

SUBSECTION		GUILD	
A. Nonguarders			
A.1	Open substratum spawners	A.1.1	Pelagophils
		A.1.2	Litho-pelagophils
		A.1.3	Lithophils
		A.1.4	Phyto-lithophils
		A.1.5	Phytophils
		A.1.6	Psammophils
A.2	Brood hiders	A.2.1	Lithophils
		A.2.2	Speleophils
		A.2.3	Ostracophils
		A.2.4	Aero-psammophils
		A.2.5	Xerophils
B. Guarders			
B.1	Substratum choosers	B.1.1	Lithophils
		B.1.2	Phytophils
		B.1.3	Aerophils
		B.1.4	Pelagophils
B.2	Nest spawners	B.2.1	Lithophils
		B.2.2	Phytophils
		B.2.3	Psammophils
		B.2.4	Aphrophils
		B.2.5	Speleophils
		B.2.6	Polyphils
		B.2.7	Ariadnophils
		B.2.8	Actinariophils
C. Bearers			
C.1	External	C.1.1	Transfer brooders
		C.1.2	Forehead brooders
		C.1.3	Mouth brooders
		C.1.4	Gill-chamber brooders
		C.1.5	Skin brooders
		C.1.6	Pouch brooders
C.2	Internal	C.2.1	Ovi-ovoviviparous
		C.2.2	Ovoviviparous
		C.2.3	Viviparous

Source: From Balon, 1975a.

the adults, should not be underestimated. Depending on their reproductive habit fishes have corresponding adaptation in amount and density of yolk, embryonic respiratory structures, and so on, as adaptations for respiration are a major feature of the developmental stages of fishes

(Balon, 1975a). These depend on the nature and availability of the oxygen supply in the places where parents deposit the eggs (e.g., under gravel, in open water, inside the oral cavity, etc.). Given these overriding constraints, some questions concerning behavioral development become clearer, in that the general scheme of behavioral development in a species is a necessary consequence of its reproductive habit.

In birds and mammals we are familiar with the concepts of altricial and precocial young. Those animals, however, show only a limited range of reproductive habits. Birds incubate eggs, usually by one or both parents brooding them directly. Mammals have internal development, and give birth to living young. Hatching or birth may occur at different stages in development, giving relatively altricial or precocial young. In fishes, the moment of birth or hatching is also a relatively artificial boundary as far as development is concerned. It may or may not correspond to developmental changes of more basic significance (e.g., initiation of exogenous feeding) as far as comparisons across species are concerned. The example from the scheme of classification of developmental intervals proposed by Balon (1975b) will illustrate this point (Fig. 1), as well as my previous comments comparing mouth brooding and substrate brooding species of cichlids. One need not look far into the literature to appreciate the confusion resulting from the use of terms such as eyed eggs, wrigglers, and newly hatched larvae. Any comparison of behavioral data between species, or between studies, depends on a biologically meaningful designation of developmental intervals. Whatever the terminology used, the important point to bear in mind is that behavior is one facet of a fish as a "survival machine" (Dawkins, 1976), and as such must be closely coordinated and adapted to the environment. During their ontogeny, some fishes (e.g., salmonids) must adapt to very different environments (e.g., beneath gravel, in the open water of a stream), so we can expect to find corresponding behavioral adaptations at those times. Unless the behavioral ontogeny of these species is viewed in the context of their ecology (particularly their reproductive habits), it will be difficult to interpret the data in a biologically meaningful way.

I hope this review, though not exhaustive, may encourage further studies of behavioral ontogeny in fishes. Quite clearly much more descriptive information is required even to qualify the generalizations I have attempted or to make specific tests of proposed models. Unfortunately, fish lack much of the readily quantified, ongoing maintenance behavior typical of birds and mammals which has served so well for the precise, analytical studies of behavioral development in those animals (Chapters 9 and 16). Nonetheless, the great diversity of reproductive habits and, hence, developmental differences found in fishes offer obvious advantages for studies directed along those lines.

Acknowledgments

Supported by grant A6981 from the National Research Council of Canada. Many of the figures are reproduced from other sources, and I thank the authors and publishers concerned for their kind permission to use them here. A number of colleagues have contributed helpful comments and constructive criticism of my thinking on this topic. My debt to Eugene Balon is obvious. I wish also to thank George Barlow, Jeff Baylis, Kassi Cole, Pat Colgan, Larry Dill, and Peter Dill in this regard. Of course I remain responsible for any errors and all interpretations in this paper.

References

Ahlbert, I.-B. 1975. Organization of the cone cells in the retinae of some teleosts in relation to their feeding habits. Thesis, Department of Zoology, Univer. of Stockholm, Sweden.

Ahlbert, I.-B. 1976. Organization of the cone cells in the retinae of salmon (*Salmo salar*) and trout (*Salmo trutta trutta*) in relation to their feeding habits. *Acta Zool.* 57:13–35.

Albrecht, H. 1966. Zur Stammesgeschichte einigen bewegungsweisen bei fischen, untersucht am verhalten von *Haplochromis* (Pisces, Cichlidae). *Z. Tierpsychol.* 23:270–302.

Ali, M. A. 1959. The ocular structure, retinomotor, and photobehavioral responses of juvenile Pacific salmon. *Can. J. Zool.* 37:965–996.

Baerends, G. P., and J. M. Baerends–van Roon. 1950. An introduction to the study of the ethology of cichlid fishes. *Behavior Suppl.* 1:1–242.

Baerends, G. P., B. E. Bennema, and A. A. Vogelzang. 1960. Uber die anderung der sehscharfe mit dem wachstum bei *Aequidens portalagrensis* (Hensel) (Pisces, Cichlidae). *Zool. Jahr. Syst. Board* 88:67–78.

Balon, E. K. 1975a. Reproductive guilds of fishes: a proposal and definition. *J. Fish. Res. Board Can.* 32:821–864.

Balon, E. K. 1975b. Terminology of intervals in fish development. *J. Fish. Res. Board Can.* 32:1663–1670.

Balon, E. K. 1977. Early ontogeny of *Labeotropheus* (Ahl 1927) (Mbuna, Cichlidae, Lake Malawi), with a discussion on advanced protective styles in fish reproduction and development. *Environ. Biol. Fish.* 2:147–176.

Bams, R. A. 1969. Adaptations of sockeye salmon associated with incubation in stream gravels. *In* ed. T. Northcote, pp. 71–88. Symposium on salmon and trout in streams. Institute of Fisheries, University of British Columbia.

Bams, R. A. 1976. Survival and propensity for homing as affected by presence or absence of locally adapted paternal genes in two transplanted populations of pink salmon (*Oncorhynchus gorbuscha*). *J. Fish. Res. Board Can.* 33:2716–2725.

Barlow, G. W. 1973. Competition between color morphs of the polychromatic Midas cichlid *Cichlasoma citrinellum*. *Science* 179:806–807.

Barlow, G. W. 1974. Contrasts in social behavior between Central American cichlid fishes and coral-reef surgeon fishes. *Am. Zool.* 14:9–34.

Barlow, G. W. 1976. The Midas cichlid in Nicaragua. *In:* ed., T. B. Thorson, pp. 333–358. Investigations of the Ichthyofauna of Nicaraguan Lakes. University of Nebraska Press, Lincoln.

Barlow, G. W., and J. W. Munsey. 1976. The Red-Devil-Midas-Arrow cichlid species complex in Nicaragua. *In* ed., T. B. Thorson, pp. 359–369. Investigations of the Ichthyofauna of Nicaraguan Lakes. Lincoln: University of Nebraska Press.

Barlow, G. W., W. Rogers, and R. V. Cappeto. 1977. Incompatibility and assortative mating in the Midas cichlid, *Behav. Ecol. Sociobiol.* 2:49–59.

Baylis, J. R. 1974. The behavior and ecology of *Herotilapia multispinosa* (Pisces, Cichlidae). *Z. Tierpsychol.* 34:115–146.

Blaxter, J. H. S. 1969. Development: eggs and larvae. *In* eds., W. S. Hoar and D. J. Randall, pp. 177–252. Fish physiology. New York: Academic Press.

Blaxter, J. H. S., ed. 1974. The early life history of fish. New York: Springer-Verlag.

Brestowsky, M. 1968. Vergleichende untersuchungen zur elternbindung von *Tilapia*-Jungfischen (Cichlidae, Pisces). *Z. Tierpsychol.* 25:761–828.

Brothers, E. B., C. P. Mathews, and R. Lasker. 1976. Daily growth increments in otoliths from larval and adult fishes. *Fish. Bull.* 74:1–8.

Cole, J. E., and J. A. Ward. 1970. An analysis of parental recognition by the young of the cichlid fish, *Etroplus maculatus* (Bloch). *Z. Tierpsychol.* 27:156–276.

Cole, K. S. 1976. Social behaviour and social organization of young rainbow trout, *Salmo gairdneri*, of hatchery origin. M.Sc. thesis. Department of Zoology, Univ. of Guelph. 100 pp.

Cole, K. S., and D. L. G. Noakes. 1977. Development of social behaviour in young rainbow trout, *Salmo gairdneri* (Pisces: Salmonidae). (Manuscript submitted for publication.)

Colgan, P., F. Cooke, and J. T. Smith. 1974. An analysis of group composition in assortatively mating populations. *Biometrics* 30:693–696.

Cooke, F., C. D. MacInnes, and J. P. Prevett. 1975. Gene flow between breeding populations of Lesser Snow Geese. *Auk* 92:493–510.

Dawkins, R. 1976. The selfish gene. New York: Oxford University Press.

Dawkins, R., and T. R. Carlisle. 1976. Parental investment, mate desertion and a fallacy. *Nature* 262:131–133.

Dill, L. M. 1967. Studies on the early feeding of sockeye salmon alevins. *Can. Fish. Cult.* 39:23–24.

Dill, L. M. 1969. The sub-gravel behavior of Pacific salmon larvae. *In* Symposium on salmon and trout in streams. ed., T. Northcote, pp. 89–99. Institute of Fisheries, Univ. of British Columbia.

Dill, L. M., and T. G. Northcote. 1970. Effects of gravel size, egg depth, and egg density on intragravel movement and emergence of coho salmon

(*Oncorhynchus kisutch*) alevins. *J. Fish. Res. Board Can.* 27:1191–1199.

Dill, P. A. 1977. Development of behaviour in alevins of Atlantic salmon, *Salmo salar,* and rainbow trout, *S. gairdneri. Anim. Behav.* 25:116–121.

Ferno, A., and S. Sjolander. 1976. Influence of previous experience on the mate selection of two colour morphs of the convict cichlid, *Cichlasoma nigrofasciatum* (Pisces, Cichlidae). *Behav. Proc.* 1:3–14.

Fryer, G., and T. D. Iles. 1972. The cichlid fishes of the Great Lakes of Africa. Edinburgh: Oliver and Boyd.

Hasler, A. D. 1966. Underwater guideposts. Univ. of Wisconsin, Madison.

Heiligenberg, W. 1965. A quantitative analysis of digging movements and their relationship to aggressive behaviour in cichlids. *Anim. Behav.* 13:163–170.

Hinds, R. A. 1970. *Animal behaviour,* 2nd ed. A synthesis of ethology and comparative psychology. New York: McGraw-Hill.

Hoar, W. S. 1976. Smolt transformation: evolution, behavior, and physiology. *J. Fish. Res. Board Can.* 33:1233–1252.

Jenkins, T. M., Jr. 1969. Social structure, position choice and microdistribution of two trout species (*Salmo trutta* and *Salmo gairdneri*) resident in mountain streams. *Anim. Behav. Monogr.* 2:57–123.

Kalmus, H., and S. M. Smith. 1966. Some evolutionary consequences of pegmatypic mating systems (imprinting). *Am. Naturalist* 100:619–635.

Kuenzer, P. 1964. Weitere Versuche zur Auslosung der Nachfolgreaktion bei Jungfischen von *Nanacara anomala* (Cichlidae). *Naturwissenschaften* 17:419–420.

Kuenzer, P. 1975. Analyse der auslosenden Reizsituationen fur die Anschwimm-, Eindring-, und Fluchtreaktion junger *Hemihaplochromis multicolor* (Cichlidae). *Z. Tierpsychol.* 38:505–545.

Kuhme, W. 1962. Das schwarmverhalten elterngefuhrter Jungcichliden (Pisces). *Z. Tierpsychol.* 19:513–538.

Kuhme, W. 1963. Chemisch ausgeloste Brutpflege und Schwarmreaktionen bei *Hemichromis bimaculatus* (Pisces). *Z. Tierpsychol.* 20:688–704.

Kuhme, W. 1964. Eine chemisch ausgeloste Schwarmreaktion bei jungen Cichliden (Pisces). *Naturwissenschaften* 51:120–121.

Lagler, K. F., J. E. Bardach, R. R. Miller, and D. R. M. Passino. 1977. *Ichthyology,* 2nd ed. New York: Wiley.

Larson, G. L. 1976. Social behavior and feeding ability of two phenotypes of *Gasterosteus aculeatus* in relation to their spatial and trophic segregation in a temperate lake. *Can. J. Zool.* 54:107–121.

Lowe-McConnell, R. H. 1975. Fish communities in tropical freshwaters: Their distribution, ecology, and evolution. London: Longman.

Mason, J. C. 1976. Some features of coho salmon, *Oncorhynchus kisutch,* fry emerging from simulated redds and concurrent changes in photobehavior. *Fish. Bull.* 74:167–175.

Maynard Smith, J. 1977. Parental investment: A prospective analysis. *Anim. Behav.* 25:1–9.

Myrberg, A. A. 1966. Parental recognition of young in cichlid fishes. *Anim. Behav.* 14:565–571.

Myrberg, A. A., Jr. 1975. The role of chemical and visual stimuli in the preferential discrimination of young by the cichlid fish *Cichlasoma nigrofasciatum* (Gunther). *Z. Tierpsychol.* 37:274–297.

Myrberg, A. A., Jr., E. Kramer, and P. Heinecke. 1965. Sound production by cichlid fishes. *Science* 149:555–558.

Nelson, J. S. 1976. *Fishes of the world.* New York: Wiley.

Noakes, D. L. G. 1973. Parental behavior and some histological features of scales in *Cichlasoma citrinellum* (Pisces, Cichlidae). *Can. J. Zool.* 51:619–622.

Noakes, D. L. G., and G. W. Barlow. 1973a. Ontogeny of parent-contacting in young *Cichlasoma citrinellum* (Pisces, Cichlidae). *Behaviour* 46:221–225.

Noakes, D. L. G., and G. W. Barlow. 1973b. Cross-fostering and parent-offspring responses in *Cichlasoma citrinellum* (Pisces, Cichlidae). *Z. Tierpsychol.* 33:147–152.

Noble, G. K., and B. Curtis. 1939. The social behavior of the jewel fish, *Hemichromis bimaculatus* Gill. *Bull. Am. Mus. Nat. Hist.* 75:1–46.

Oppenheimer, J. R. 1970. Mouthbreeding in fishes. *Anim. Behav.* 18:493–503.

Otto, R. G., and J. E. McInerney. 1970. Development of salinity preference in pre-smolt coho salmon, *Oncorhynchus kisutch. J. Fish. Res. Board Can.* 27:793–800.

Peters, H. M. 1937. Experimentelle untersuchungen über die Brutpflege von *Haplochromis multicolor,* einem maulbrutenden Knochenfisch. *Z. Tierpsychol.* 1:201–218.

Peters, H. M. 1941. Fortpflanzungsbiologische und tiersoziologische studien an fischen. 1. *Hemichromis bimaculatus. Z. Morphol. Ökol. Tiere* 37:387–425.

Peters, H. M. 1963. Untersuchungen zum problem des angeborenen vehaltens. *Naturwissenschaften* 50:677–686.

Peters, H. M. 1965. Angeborenes verhalten bei buntbarschen. II. Das problem der erblichen grundlage des kontaktberhaltens. *Umschau* 22:711–716.

Peterson, R. H. 1975. Pectoral fin and opercular movements of Atlantic salmon (*Salmo salar*) alevins. *J. Fish. Res. Board Can.* 32:643–647.

Quertermus, C. J., and J. A. Ward. 1969. Development and significance of two motor patterns used in contacting parents by young orange chromides (*Etroplus maculatus*). *Anim. Behav.* 17:624–635.

Reynolds, W. W., and D. A. Thomson. 1974. Ontogenetic change in the response of the Gulf of California grunion, *Leuresthes sardina* (Jenkins and Evermann), to a salinity gradient. *J. Exp. Mar. Biol. Ecol.* 14:211–216.

Sale, P. F. 1971. Apparent effect of prior experience on a habitat preference exhibited by the reef fish, *Dascyllus aruanus* (Pisces: Pomacentridae). *Anim. Behav.* 19:251–256.

Scholz, A. T., R. M. Horrall, J. C. Cooper, and A. D. Hasler. 1976. Imprinting to chemical cues: The basis for home stream selection in salmon. *Science* 192:1247–1249.

Seiger, M. B. 1967. A computer simulation study of the influence of imprinting on population structure. *Am. Naturalist* 101:47–57.

Seiger, M. B., and R. D. Dixon. 1970. A computer simulation study of the effects of two behavioral traits on the genetic structure of semi-isolated populations. *Evolution* 24:90–97.

Sharma, S. C. 1975. Development of the optic tectum in brown trout. New approaches in research. *In Vision in fishes,* ed., M. A. Ali, pp. 411–417. New York: Plenum.

Shaw, E. 1970. Schooling in fishes: critique and review. *In The development and evolution of behavior,* eds., L. R. Tobach, J. S. Rosenblatt, and D. S. Lehrman, pp. 452–480. San Francisco: Freeman.

Struhsaker, P., and J. H. Uchiyama. 1976. Age and growth of the nehu, *Stolephorus purpureus* (Pisces: Engraulidae), from the Hawaiian Islands as indicated by daily growth increments of sagiattae. *Fish. Bull.* 74:9–17.

Taubert, B. D., and D. W. Coble. 1977. Daily rings in otoliths of three species of *Lepomis* and *Tilapia mossambica. J. Fish. Res. Board Can.* 34:332–340.

Trewavas, E. 1973. On the cichlid fishes of the genus *Pelmatochromis* with the proposal of a new genus for *P. congicus*; on the relationship between *Pelmatochromis* and *Tilapia* and the recognition of *Sarotherodon* as a distinct genus. *Bull. Br. Mus. Nat. Hist. Zool.* 25:1–26.

Turner, J. L. 1977. Changes in the size structure of cichlid populations of Lake Malawi resulting from bottom trawling. *J. Fish. Res. Board Can.* 34:232–238.

Twongo, K. T., and H. R. MacCrimmon. 1976. Significance of the timing of initial feeding in hatchery rainbow trout, *Salmo gairdneri. J. Fish. Res. Board Can.* 33:1914–1921.

Verraes, W. 1977. Postembryonic ontogeny and functional anatomy of the ligamentum manibulo-Hyoideum and the ligamentum interoperculo-Mandibulare, with notes on the opercular bones and some other cranial elements in *Salmo gairdneri* Richardson, 1836 (Teleostei: Salmonidae). *J. Morphol.* 151:111–120.

Ward, J. A., and G. W. Barlow. 1967. The maturation and regulation of glancing off the parents by young orange chromides (*Etroplus maculatus;* Pisces, Cichlidae). *Behaviour* 28:1–56.

Ward, J. A., and R. L. Wyman. 1977. Ethology and ecology of cichlid fishes of the genus *Etroplus* in Sri Lanka: preliminary findings. *Environ. Biol. Fish.* 2:137–145.

Ware, D. M. 1975. Relation between egg size, growth, and natural mortality of larval fish. *J. Fish. Res. Board Can.* 32:2503–2512.

Weber, P. G., and S. P. Weber. 1976. The effects of female colour, size, dominance and early experience upon mate selection in male convict cichlids, *Cichlasoma nigrofasciatum* Gunther (Pisces, Cichlidae). *Behaviour* 54:116–135.

Wyman, R. L. and J. A. Ward. 1973. The development of behavior in the cichlid fish *Etroplus maculatus-*(Bloch). *Z. Tierpsychol.* 33:461–491.

Chapter Seven

Evolution of Parental Care in Frogs

ROY W. MC DIARMID

Department of Biology
University of South Florida
Tampa, Florida 33620

The implications of increased reproductive effort and parental investment
and the manner in which such activities influence the evolution of life
history strategies have received considerable attention in recent literature
(Williams, 1966; Trivers, 1972; Wilson, 1975; Dawkins and Carlisle, 1976;
Stearns, 1976; Maynard Smith, 1977). Most of the supportive data for
hypotheses regarding these phenomena come from detailed studies of the
reproductive ecology and behavior of social insects, birds, and mammals
(reviewed in Wilson, 1976; Thornhill, 1976), although some studies have
focused on other groups (e.g., Wilbur, 1977). Salthe and Mecham (1974)
summarized the available literature on amphibian reproductive patterns
and commented on the complexity of the situation in frogs as compared
to that of salamanders and caecilians. While some of the complexity of
frog reproductive patterns certainly is attributable to the overall diversity
of the group, it reflects in part the inadequacy of available information.
With the exception of Crump's work (1974), no detailed studies of the
reproductive ecology of complex frog communities (those which are most
likely to contain species with some form of parental care, e.g., tropical
forest communities) from which general patterns can be drawn exist.
Instead, we must deal with a woefully inadequate literature of anecdotal
and often conflicting data about a group of organisms that probably
exhibits the greatest array of reproductive modes found in any vertebrate
class. Detailed studies on about ten species of frogs are available. Unfortunately, all these species exhibit the typical aquatic pattern of repro-

duction in which the eggs, deposited in water, hatch into free-swimming larvae (tadpoles), which, after a period of growth, metamorphose and usually leave the water to grow into adults. In this pattern there is no parental investment in the offspring after the eggs are laid. With the notable exception of work on *Leptodactylus ocellatus* (Vaz-Ferreira and Gehrau, 1975) detailed studies of other species that have a derived reproductive pattern including some form of post–egg-laying parental investment generally are lacking. In fact, Maynard Smith (1977) recently has commented on the need for, and value of, a comparative study of anuran parental care.

In this chapter I review the available information on parental care, one form of parental investment, in frogs. I include pertinent data from a detailed study of the reproductive ecology and behavior of some species of tropical frogs, family Centrolenidae, to illustrate several factors I consider important in the evolution of parental care in frogs. Finally, I pose some questions and make suggestions that I believe will be important for shaping relevant future research in this area of behavioral ecology.

Parental Care in Frogs

Parental care may be defined as any behavior that enables an individual to increase the survivorship of its offspring. For purposes of this review, I restrict my consideration to parental care in post–egg-laying situations, i.e., care for eggs and/or larvae by either parent acting singly or in concert with the other. I assume that this behavior decreases the ability of the investing parent to invest in additional offspring (Trivers, 1972) and thus may be considered as a subset of parental investment. Energy or effort devoted to gamete production, though part of parental investment, is not parental care. Likewise, I consider energy devoted to territorial defense or courtship as a component of parental care only if it is clear that this effort directly benefits the offspring rather than (or in addition to) functioning as a prerequisite for successful mating. For example, if a defended territory provides protection for young rather than or besides providing the appropriate resources needed for egg formation, mating, etc., then I include it in the discussion of parental care.

A review of the literature on frog reproductive behavior and life history strategies indicates that parental care is widely distributed among frog families (Table 1). Some form of such care has been reported from 14 of 20 families from temperate and tropical areas. Even so, parental care is relatively rare among frogs as it is reported from less than 10% of all

TABLE 1. The distribution of parental care among frog families

FAMILY	PARENTAL CARE	PERCENTAGE OF SPECIES WITH PARENTAL CARE
Leiopelmatidae	♂	100%
Pipidae	♀	30%
Rhinophrynidae	—	0%
Discoglossidae	♂	25%
Pelobatidae	—	0%
Pelodytidae	—	0%
Myobatrachidae	♂, ♀, both	5%
Leptodactylidae	♂, ♀, both	5%
Bufonidae	♂, ♀	1%
Brachycephalidae	—	0%
Rhinodermatidae	♂	100%
Dendrobatidae	♂, ♀, both	100%
Pseudidae	—	0%
Hylidae	♂, ♀	5%
Centrolenidae	♂	50%
Microhylidae	♂, ♀	30%
Sooglossidae	♂	100%
Ranidae	♂?, ♀	1%
Hyperoliidae	—	0%
Rhacophoridae	♀?	1%

species. As more data on reproductive biology of some groups (e.g., Microhylidae) become available, these figures may change, but I believe that 10% is a reasonable estimate. Parental care apparently is absent from six families. It occurs in fewer than 30% of the species included in nine families, and in six of these, species reported to exhibit parental care account for less than 5% of the family. Usually in these six families (all of which include large numbers of species), one group of species (e.g., species of pouch–brooding or egg–carrying hylid frogs, species of Australian *Pseudophryne,* some species of *Eleutherodactylus,* etc.) exhibits most or all the parental care recorded for that family.

Among five other families in which more than 50% of the included species have some form of parental care, three families are small. The Leiopelmatidae and the Sooglossidae have three species each; the Rhinodermatidae has two species. In the two remaining families parental care has been reported for about half of the nearly sixty species in each. In the Dendrobatidae it probably characterizes the entire family (Silverstone, 1976; also see comments by Wells, 1977). In the Centrolenidae, in contrast to the Dendrobatidae, an ethocline from species lacking parental care to those with parental care exists.

Salthe and Mecham (1974) presented three general explanations to account for the evolution of parental care in amphibians, arguing that more than a single explanation is necessary because of the parallel development of the behavior as a component of several different reproductive strategies. The three explanations are that parental care evolved in amphibians (1) as a mechanism to increase reproductive success in the absence of high fecundity by reducing predation on early life stages through guarding and active defense or through spatial dispersion of larvae; (2) as a mechanism to decrease developmental abnormalities caused by yolk layering or insufficient oxygen, through constant manipulation or jostling of the eggs; (3) as a mechanism to provide a more suitable microhabitat for the developing offspring in a generally unfavorable environment (i.e., out of water), by covering the eggs to decrease desiccation, or carrying the eggs and/or larvae around in the terrestrial environment or moving them to aquatic environments at different stages in their development.

Each of these mechanisms is important in and could account for the evolution and development of parental care in amphibians. However, Salthe and Mecham (1974) also indicated that parental care in amphibians almost always is associated with terrestrial reproductive patterns (i.e., reproduction in unfavorable habitats) and pointed to a trend of relatively larger ova and smaller clutches in those species with such care. There are some notable exceptions to this generalization concerning terrestrial reproduction, particularly among more primitive salamanders (e.g., Cryptobranchidae, Sirenidae, Necturidae) which tend to be aquatic. In addition, in aquatic frogs of the genus *Pipa,* the female parent carries the developing eggs in pockets on her back.

Likewise, there are several potential selective situations that act on both adult and egg-larval stages that could account for increased ovum size and decreased clutch size. Ovum volume and clutch size are negatively correlated so that larger eggs are found in smaller clutches. Salthe and Duellmen (1973) reported increased ovum size in lotic breeders as well as in species with terrestrial development. They also found a positive correlation between ovum size and female snout-vent length and between ovum size and hatchling size. These findings suggest that lower fecundity (i.e., smaller clutches) can develop without the added investment of parental care. For example, within the hylid genus *Smilisca,* clutch size decreases and ovum diameter increases in a comparison of stream-breeding species to pond-breeding species. None has any form of parental care (Duellman and Trueb, 1966). Heyer (1969) documented a similar trend in the genus *Leptodactylus.* Available data for frogs suggest, however, that parental care evolves in concert with reduced clutch size. All species of frogs for which parental care has been documented (Tables 1 and 2) have smaller clutches than their closest relatives which lack parental care. Thus, lower fecundity initially may be a preadaptation for parental care. Once the early stages in the evolution of parental care appear, selection should favor a continued gradual reduction in clutch size and a concomitant increase in parental care. It seems, then, that the mechanisms proposed by Salthe and Mecham (1974) must be viewed in a more synthetic frame-work, which also incorporates the apparent aquatic exceptions discussed previously. Wilson (1975) outlined a theory of parental care that involves a set of environmental conditions which acting singly or in concert will

TABLE 2. Classification of major patterns of parental care in frogs with representative examples

---

I. Investment at fixed site (philopatry)

    A. Nest or burrow

        1. eggs and larvae aquatic; male or female attendance (e.g., *Hyla rosenbergi, Leptodactylus ocellatus*).

        2. terrestrial eggs; male and/or female attendance; larvae aquatic (e.g., *Pseudophryne* spp., *Hemisus* spp.)

        3. terrestrial eggs; direct development; male or female attendance (e.g., *Breviceps* spp., *Phrynomantis* spp., *Leiopelma* spp.)

    B. No nest or burrow

        1. eggs and larvae aquatic; male attendance (e.g., *Nectophryne afra*)

        2. terrestrial or arboreal eggs; male attendance; larvae aquatic (e.g., *Centrolenella* spp.)

        3. terrestrial or arboreal eggs; direct development; male or female attendance (e.g., *Eleutherodactylus* spp., *Hylactophryne* spp.)

II. Investment at mobile site (i.e., parent)

    A. Aquatic environment

        1. eggs on back of female; larvae aquatic (e.g., *Pipa carvalhoi*)

        2. eggs on back or in stomach of female; direct development (e.g., *Pipa pipa, Rheobatrachus silus*)

    B. Terrestrial environment

        1. terrestrial eggs; larval transport in vocal sac of male or on back of male or female; larvae aquatic (e.g., *Rhinoderma rufum, Dendrobates* spp.)

        2. terrestrial eggs; direct development

            a. on back of male (e.g., *Sooglossus* spp.)

            b. in vocal sac or inguinal pouches of male (e.g., *Rhinoderma darwini, Assa darlingtoni*)

        3. eggs on legs of male, or back or pouch of female; larvae aquatic (e.g., *Alytes obstetricans, Gastrotheca* spp.)

        4. eggs on back or pouch of female; direct development (e.g., *Stefania* spp., *Gastrotheca* spp.)

---

favor modification of life history parameters and eventually result in parental care. Trivers (1972) and more recently Dawkins and Carlisle (1976) and Maynard Smith (1977) discussed some interesting ideas concerning the relative contribution of each parent to its offspring and made some predictions as to which parent was most likely to provide post–egg-laying investment in the form of parental care. By combining some of these ideas with what is known about frog reproduction and the environmental factors associated with the adaptive responses expected in organisms with parental care, it is possible to generate a classification of anuran parental care (Table 2) and to interpret the evolution of sexual strategies with regard to parental care in frogs. The patterns defined in Table 2 are illustrated in Figure 1 (e.g., *Nectophryne afra* has male parental care at the egg site in the aquatic environment and is listed as item IB_1 in Table 2 and shown in the upper right corner of Figure 1 using the IB_1 designation).

Classification of Parental Care

There is little doubt that the primitive reproductive mode of frogs is aquatic and involves external fertilization of eggs, free–swimming larvae, metamorphosis, and terrestrial existence of juveniles and adults. The major hazards affecting survivorship of frogs in the aquatic environment act primarily on the eggs and early larval stages. While data are sparse, the studies that have been done report mortality on egg and larval stages in excess of 90%. If this is representative for frogs, which have the primitive aquatic mode of reproduction, then selection favoring mechanisms to reduce aquatic mortality should be strong. I think the major mortality in the aquatic environment is attributable to predation on the eggs and/or early larvae (see also Heyer *et al.,* 1975). Other mortality factors include changes in the physical parameters of water quality (changes in temperature, O_2 availability, etc.) and probability of drying up, especially in temporary ponds and puddles. In these relatively unpredictable situations (e.g., small puddles), competition may play a secondary role. A physical factor of importance in lotic habitats is associated with a sudden rise in water level and the increased velocity of rain-swollen streams. These conditions are common in tropical areas during the rainy season and generally have favored a temporal shift in aquatic stream-breeding species to dry season reproduction (McDiarmid, personal observation). Another possible solution to these unpredictable fluctuations in water level is to get the early stages out of the water. This response will require favorable conditions on land. Such conditions are

AQUATIC ENVIRONMENT

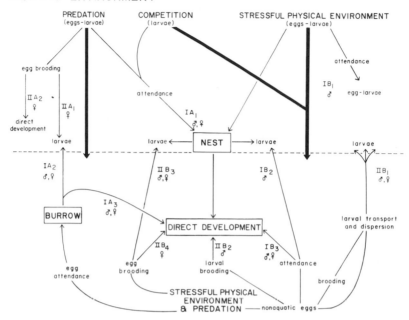

TERRESTRIAL ENVIRONMENT

FIG. 1. *A diagrammatic representation of the ecological and evolutionary distribution of parental care in frogs. The dashed line represents the interface between the aquatic and terrestrial environments. The major selective pressures of each environment and the stages upon which they operate are shown at the top and bottom of the diagram. Large arrows indicate a movement out of the aquatic environment as a solution. The narrow arrows represent evolutionary solutions to the selective pressures in each environment. The pattern of parental care is indicated to the side of each arrow and refers to the classification in Table 2. Arrows without associated patterns represent solutions without parental care.*

best in tropical environments during the wet season when humidity is highest and problems of egg and larval desiccation at a minimum.

Adaptive responses that represent shifts toward increased parental care in the aquatic environment are not uncommon in aquatic salamanders (Salthe and Mecham 1974) but relatively rare in frogs. In part this is due to the terrestrial habits of most adult frogs. Indeed, the only examples (Table 2) of increased parental care in the aquatic environment are found in the species of aquatic frogs of the genus *Pipa,* in *Rheo-*

batrachus silus, and in the small bufonid *Nectophryne afra.* Females of *Pipa* transport eggs in small pockets on their backs where the eggs either hatch into larvae (IIA_1) which become free-swimming (e.g., *Pipa carvalhoi*) or have a modified form of direct development (IIA_2) with no free-swimming larvae (e.g., *Pipa pipa*) (Dunn, 1948). In the Australian myobatrachid *Rheobatrachus silus* the embryos and developing larvae are brooded in the stomach (Corben *et al.,* 1974). In *Nectophryne afra* (IB_1) the male spends considerable time swimming in place, stirring up water currents directed toward the eggs and tadpoles (Scheel, 1970). This behavior may function to reduce any developmental abnormalities which result from low aeration or poor water circulation.

The major adaptive responses to the hazards of the aquatic environment in frogs are associated with a tendency to get the egg and early larval stages out of the water. This trend includes a wide variety of adaptive modifications of the basic pattern (see recent review by Lamotte and Lescure, 1977) and is indicated by large arrows in Figure 1. Initial stages include species with nonaquatic eggs, aquatic larvae, and no parental care. The evolutionary sequence continues with increased terrestriality, first with the eggs, and then with eggs and larvae, and eventually culminates in some form of direct development, a reproductive mode which has developed independently in many unrelated families of frogs (Salthe and Mecham, 1974; Lamotte and Lescure, 1977). Parental care may have been associated with any of these sequential stages. As the selective advantages of increasing parental care manifested themselves, initially eggs, and then eggs and larvae were tended by one of the parents. In some situations selection favored male attendance; in others it favored that of the female parent. As selection for maintaining or providing the larvae with a suitable "aquatic" habitat out of the water became stronger, parental care became more and more important. As a result, the nature of parental care expanded from an initial one of egg attendance only, to include attendance of later stages, larval transport, and brooding. In this framework, attendance is viewed as an association of either parent with the eggs and/or tadpoles at a fixed site. Attendance is a form of philopatry and probably evolved with the increasing use of nests or burrows. These initial evolutionary efforts to get out of the aquatic environment because of predation and/or physical stress were enhanced by selection favoring philopatric behavior (e.g., construction of a foam nest, burrow, etc.) that provided protection for eggs or eggs and early larval stages. In some species foam nests obviously are structures to prevent egg and/or early larval desiccation. Whether they also protect early developmental stages from predation is unknown but worthy of consideration. Initially, these nests and burrows must have been in close proximity to water and, thereby, allowed easy return to the aquatic habitat by larger, more mobile larvae that presumably were better able to deal with the rigors of aquatic existence. I view these initial evolutionary experiments as efforts to maintain the aquatic nature of the organisms rather than efforts to become more terrestrial. In so doing, these behavioral modifications led to increased terrestriality. As reproductive modes became more terrestrial, the rigors of the terrestrial environment began to play a greater and greater role in molding reproduction. Probably, the physiological and developmental requirements of the developing egg, particularly with respect to desiccation, favored increasing attendance by one of the parents. It is likely that terrestrial predation also played a leading role in selecting for attendance. In some, this involved transport of the larvae back to the water. As the rigors of transport increased (e.g., increased susceptibility of the transporting parent to predation), selection favored reduction in the free-swimming larval stage, culminating in a form of direct development. If parental care in the form of egg attendance were important in the initial stages of increasing terrestriality, the investment of the attending parent already would be considerable, and the amount needed to get the offspring to near hatching would be less. It follows that the evolution of additional investment would more likely occur than a reduction or abandonment of parental care. The culmination of this evolutionary progression is the actual brooding by one parent, usually the female. As used here, brooding is the form of parental care in which the eggs and/or larvae are maintained and carried on the parent or in specialized parental structures suited for their maintenance. This trend ultimately results in direct development of the young in or on the parent. Direct development in its several forms is wide spread among frogs and reflects the evolutionary experimentation in reproductive modes that so characterizes this vertebrate group (Lutz, 1947; Orton, 1949; Jameson, 1957; Goin and Goin, 1962; Lamotte and Lescure, 1977).

The evolutionary scheme outlined above finds considerable support among frogs that have various kinds of parental care (Table 2) and is schematically diagrammed in Figure 1. The initial stages of attendance are found in the mud-nest-building species of the *boans* group of hylid frogs and the foam–nest–building frog *Leptodactylus ocellatus* (IA₁). Males of *Hyla rosenbergi, H. faber,* and others are territorial and actively defend mud-depression nests constructed at the edge of streams or ponds. The nests are used as sites for egg deposition and early larval development and effectively isolate these stages from aquatic predators (Breder, 1946). Here the male is territorial, larger than the female (Duellman, 1970), and is involved in parental care (guarding) at the nest which contains his eggs and/or larvae. In these species, the male apparently continues to attract females to the nest, and the added cost of parental care seems to be out-weighed by the benefit of increased survivorship of the young resulting in an increase in his reproductive success. This system has considerable poten-tial in elucidating the relationship between male territoriality and parental care and currently is being studied (Kluge, personal communication).

In *Leptodactylus ocellatus* either the female (usually) or the male (occasionally) or both guard the foam nest and the included eggs. On hatching the tadpoles form a school which moves around the pond feeding. The female follows the aggregated tadpoles around the pond and will attack potential predators attempting to feed on the larvae (Vaz-Ferreira and Gehrau, 1975). Unfortunately, data are lacking as to the preamplectic behavior of either parent. In this case it is clear that at-tendance by the parent usually the female, serves to protect the eggs and larvae from predation particularly by birds. This is the only documented case of protection of free–swimming larvae by a parent. Reports (e.g., Rose, 1962; Poynton, 1964) of the defense of larvae of the African ranid *Pyxicephalus adspersus* by adults generally have been discounted (Wager, 1965) but may deserve additional study considering the findings for *L. ocellatus.*

The next step in this evolutionary progression involves the develop-ment of more terrestrial reproduction but with retention of aquatic larvae (IA$_2$). In the African ranid *Hemisus,* the female sits on the eggs in a burrow she constructs near an aquatic site. After the eggs hatch in the burrow, she tunnels laterally to the water, and the larvae follow (Wager, 1965). Similar patterns involving either parent are found in the Australian myobatrachid *Pseudophryne* (Salthe and Mecham, 1975). In some species of *Pseudophryne,* the males actively defend burrows which also serve as nest sites (Pengilley, 1971). Presumably, desertion of the burrow and contained clutch by the male would decrease his likelihood of attracting another female as quickly as would be possible if he remained. As burrows apparently are contested resources in short supply and are required for successful reproduction, his continued attendance at the egg site, rather than the female's attendance, is not unexpected. This will be true only if he is potentially able to attract other females to the original site. These ideas find some support in a recent study of *Pseudophryne* by Woodruff (1977). With this sequence established, the role of the male or the female at a nest site in those species (*Leiopelma* spp., Stephenson, 1951; *Phrynomantis* spp., Zweifel, 1972; *Breviceps* spp., Wagner, 1965; etc.) which have direct development (IA$_3$) is more easily interpreted. These same patterns are indicated in species which have terrestrial eggs but no nest site (IB$_3$); for example, males of *Hylactophryne* spp. are territorial and attend eggs (Jameson, 1950). The males of certain species of *Centrolenella* are territorial and involved in varying degrees of egg attendance (BI_2). These forms will be discussed in more detail below. Either males or females guard eggs in some species of *Eleutherodactylus* (Myers, 1969; Drewry and Jones, 1976). Little is known about their preamplectic behavior and further comment is unjustified at this time.

The situation in the Malagasy microhylid frogs of the subfamily Cophylinae is somewhat intermediate between modes IA_2 and IA_3. Blommers-Schlösser (1975) reviewed the breeding habits of these frogs and reported direct development with parental care as characteristic of the subfamily. Eggs are deposited in burrows on the ground and attended by the female in the fossorial species *Plethodontohyla tuberata.* In the arboreal species *Platyhyla grandis, Plethodontohyla notosticta,* and *Anodontohyla boulengeri* the eggs are deposited in phytotelmes or other water filled holes in bamboo or tree trunks and attended by the male. In the arboreal species the tadpoles are free-swimming but do not feed. Interestingly the males are larger than females in *P. grandis* pairs. Other microhylid frogs in the subfamilies Asterophryinae and Sphenophryinae (Tyler, 1963; Zweifel, 1972) are more easily assigned to pattern IA_3. Many more data on all three subfamilies are needed before their arrangement in the classification can be made with assurance.

The transition from fixed to mobile (I to II) sites in my classification (Table 2) is accomplished through species with terrestrial eggs and larval transport (Fig. 1). This pattern is found among the Dendrobatidae (IIB_1) where males (occasionally females) transport the larvae on their backs from the place of egg deposition in forest leaf litter to an aquatic site and in *Rhinoderma rufum* (Formas et al., 1975). Again the territorial nature of most male dendrobatids is well known (Silverstone, 1975, 1976) and probably is responsible for the paternal care in this species. Males of the sooglossid frogs have the same mode, but the tadpoles undergo direct development (IIB_{2a}) on the males' backs rather than becoming free-swimming (Salthe and Mecham, 1974). I interpret male brooding (IIB_{2b}) in the vocal pouch in *Rhinoderma darwini* (Cei, 1962; Busse, 1970) and *R. rufum* (Formas et al., 1975) and in inguinal pouches in *Assa darlingtoni* (Ingram et al., 1975), as the expected evolutionary results of initial efforts at transport of the larvae back to an aquatic site. The fact that *Rhinoderma rufum* has free-swimming larvae and that the larvae of *Rhinoderma darwini* will develop in aquatic situations (Cei, 1962) but at a much slower rate and occassionally in moist terrestrial situations (Busse, 1970), supports this view. With one exception, all other forms of brooding, either with a free-swimming larval stage (IIB_3) or with direct development (IIB_4), involve parental care by the female. I suspect two factors are important here: (1) females often are larger than males and, therefore, more likely to successfully carry the fertilized eggs which they have produced, and (2) the males in these species probably are capable of multiple fertilizations. Brooding by the male in these species potentially could reduce his ability to successfully mate with another female. On the other hand, the female probably has expended her entire egg complement and, therefore, will not be ready to breed again for some time. Considering this aspect of the female reproductive cycle, her larger initial investment, and the relatively higher costs to her of getting another clutch to the same stage as her current one, selection would favor increased parental care through her brooding rather than the male's. The same pattern can be used to explain the brooding role of the female in the aquatic systems with *Pipa* spp. and *Rheobatrachus silus* (IIA_1, IIA_2).

The single exception to female egg-brooding is found in the discoglossid frog, *Alytes obstetricans*. After an elaborate mating, the male carries the eggs entwined on his hind limbs. When the eggs are near hatching, he moves to a suitable aquatic site. There the eggs hatch, releasing typical free-swimming tadpoles. During drier nights, the male may immerse the eggs in water (Boulenger, 1897). This form of brooding, which allows such activity, apparently has been favored by the relatively harsh terrestrial environment. In view of what is known about other forms of paternal care, I predict that males carrying eggs are not appreciably hindered in the successful completion of additional courtship and have more to gain by brooding a clutch and continuing to attract other females than by abandoning it. Boulenger (1897) remarked that the male is "little impeded in his movements" on land and occasionally successfully fertilizes a second female, adding another clutch of eggs.

In summary, parental care is relatively widespread but uncommon in frogs. The selective pressures (i.e., predation, habitat unpredictability) of the aquatic environment operating on eggs and early larval stages have favored mechanisms which remove these stages of the life cycle from the water. In some instances problems associated with the terrestrial environment, particularly egg desiccation, selected for parental care in the form of attendance at the eggs in these initial efforts at getting away from the hazards of an aquatic existence. As terrestrial pressures (e.g., predation) came into play, selection favored moving egg sites farther away from the water's edge. The greater the distance from an aquatic site, the greater the problems of larval return to the water. Parental care in the form of larval transport became crucial. With the increasing terrestriality, problems of desiccation were added to those of predation and transport and eventually resulted in direct development. Increased parental care coupled with direct development accounts for the incredible diversity of brooding behavior known in frogs.

Parental Care: Case Histories

The scarcity of data on parental care in frogs became obvious to me in reviewing the available literature for this paper. There are very few studies on the reproductive ecology and behavior of any species of frog. None has been published that specifically examines the role of parental care and its effect on species reproductive behavior or individual fitness. The work on *Leptodactylus ocellatus* (Vaz-Ferreira and Gehrau, 1975) is a start, but even in this study many questions remain unanswered. In part the paucity of data is a reflection of the secretive nature of individuals of many of the species that have tending behavior (especially at burrows) and of females of species with brooding behavior. Rarely is there the opportunity to compare the relative selective advantage of different forms of parental care in two similar species. It is appropriate, therefore, to describe the essentials of just such a system in two species of *Centrolenella*. These data are part of a long term study on the reproductive ecology and behavior of centrolenid frogs (McDiarmid and Adler, 1974; McDiarmid, 1975, and unpublished ms).

Species of glass frogs, genus *Centrolenella*, breed on vegetation along small streams in lowland and mid-elevation forests in Central and South America. In many species the males are territorial (McDiarmid and Adler, 1974; Duellman and Savitzky, 1976), and in some they have parental care in the form of attendance at the egg site (pattern IB_2, Fig. 1). I have looked at two species of *Centrolenella, C. colymbiphyllum,* and *C. valerioi,* over three field seasons along a small, forest stream near Rincón de Osa, Puntarenas Province, Costa Rica. Selected data from this study are presented here to (1) document the nature of parental care in these frogs, (2) illustrate the relative advantages of differential investment between two closely related species, and (3) examine the evolutionary consequences of parental care with respect to differential survivorship. To my knowledge, this study is the first documenting the relative advantages of differential parental care between two closely related species that are ecologically sympatric and essentially identical in their reproductive behavior. Males of both species are nocturnal and establish territories on appropriate leaves above the stream at least from April through September. Males advertise their reproductive readiness from the undersides of leaves and attract females from the surrounding forest. Eggs are deposited on the underside of the leaves and attended by the male. In both species males continue to advertise and attract females after egg clutches are deposited. Individuals with up to eight clutches (= females) on one leaf have been recorded for male *colymbiphyllum* and those with up to seven for male *valerioi*. The mean number of clutches per male in 1973 was 2.6 for *C. valerioi* and 1.8 for *C. colymbiphyllum*. The major difference between males of the two species is in the amount of care each devotes to his clutches. Male *colymbiphyllum* spend each night at a leaf site calling and attending any clutches present (Fig. 2). As dawn breaks, *colymbiphyllum* males move off the defended calling and egg site and hide in the vegetation back from the stream. The few males that have been located in the day time are adpressed against the underside of a leaf in a concealed site some distance from the egg clutch. They spend the entire day in this "sleeping" position and return to the calling site each night to resume vocalization and attendance. In contrast, male *C. valerioi* also spend the daylight hours in attendance at the egg site. Instead of assuming the adpressed "sleep" posture, they occupy a position next to their clutch or clutches (Fig. 3) often with their head or a front foot resting on the edge of one of them. In this position they are alert and able to tend the eggs during the day as well as at night. Colored photographs of the sleep posture of male *C. colymbiphyllum* and diurnal guarding behavior of male *C. valerioi* are available elsewhere (McDiarmid, 1975). What we have, then, are two species with very similar ecologies but with different amounts of parental care.

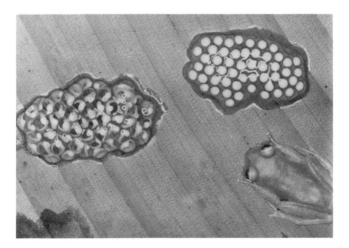

FIG. 2. *A male* Centrolenella colymbiphyllum *attending two egg clutches at night. The uniformly green male retreats to the surrounding vegetation to spend each day but returns to the egg site each night to continue calling.*

FIG. 3. *A male* Centrolenella valerioi *attending two egg clutches during the day. The male is reticulate patterned and alert and resembles his egg clutches especially during the day. He resumes calling at the same site each night.*

In Table 3, selected aspects of the reproductive ecology of the two species are compared. Of particular interest are three points. (1) The proportions of male *colymbiphyllum* to *valerioi* for each year, which are remarkably consistent: I interpret this as an indication of the relative stability of numbers of individuals both within and between the species. (2) The higher percentage predation of *colymbiphyllum* eggs as compared to those of *valerioi* in 1971 and 1973: The high mortality of *colymbiphyllum* eggs in 1971 was primarily the result of diurnal predation by a small wasp during drier periods of the study. The wasp searches the under–leaf surfaces for suitable prey. When it finds an egg mass, it lands on the jelly, withdraws a developing egg in its mandibles, and flies off. I suspect the eggs are used to provision a nest. Apparently wasps return to the same site until all eggs in a clutch are gone. The similarly low values in 1969 may have been due to the shorter, wetter sampling period. Similar two-week periods in 1973 had lower predation than is reflected in the 1973 totals over the eight weeks. (3) The larger clutch size of *colymbiphyllum*. In addition, yolk diameters of eggs averaged 1.5 mm in *colymbiphyllum* and 2 mm in *valerioi* within 12 hours of deposition on the leaf.

An analysis of the 1973 data is presented in Table 4. The presence of an egg clutch at a male's call site was used as a measure of individual success. About 73.5% of the 83 male *colymbiphyllum* were successful in

TABLE 3. Selected aspects of the reproductive ecology of two sympatric species of *Centrolenella* studied along a forest stream near Rincón de Osa, Puntarenas Province, Costa Rica.

	C. COLYMBIPHYLLUM			*C. VALERIOI*		
	1969	1971	1973	1969	1971	1973
Time (wk)	2	2	8	2	2	8
Stream length (m)	260	100	260	260	100	260
Number of males	40	27	83	24	14	43
Number of clutches	30	25	152	29	21	112
Success (%)	50	63	73	71	71	67
Predation (%)	3.3	44.0	26.3	3.4	9.5	14.3
Clutch size						
N	21	4	53	20	11	52
range	33–64	50–56	37–70	19–35	27–33	19–45
$\bar{x}$	49.1	53.7	50.4	29.3	29.2	28.9

TABLE 4. Comparative aspects of the reproductive behavior and ecology of two sympatric species of *Centrolenella* illustrating the role of parental care in their reproductive success

	COLYMBIPHYLLUM	*VALERIOI*
Number ♂ ♂ (a)	83	43
Number of clutches (b)	152	112
$\bar{x}$ Clutch size (c)	50.4	28.9
♂ ♂ with Egg clutch(es)		
N (d)	61	29
% (d/a)	73.5	67.4
♂ ♂ Producing larvae		
N (e)	47	27
% (e/a)	56.6	62.8
% ♂ ♂ with Eggs producing larvae (e/d)	77.0	93.1
% Clutches predated (f)	26.3	14.3
$\bar{x}$ Larval production/ ♂ (c × b/a × 1 − f)	68.0	64.5

attracting a female that laid eggs. The value was slightly lower (67.4% of 43) for male *valerioi*. In contrast, an average male *valerioi* was slightly more successful in producing larvae (62.8% of the male *valerioi* to 56.6% of the male *colymbiphyllum*). When we examine the percent of males with eggs that produced larvae (i.e., had eggs which were not predated), the values are 93.1 for *valerioi* and 77.0 for *colymbiphyllum*. The increased success of *valerioi* clearly is the result of the evolution of increased parental care in the form of diurnal attendance at the egg site. Interestingly, the mean numbers of larvae produced by a male of each species do not differ appreciably. The similar values between the two species, 68.0 in *colymbiphyllum* and 64.5 in *valerioi,* in spite of the added investment of *valerioi,* result from the greater mean clutch size of *colymbiphyllum.* The fact that the two species produce nearly equal numbers of larvae per male indicates the trade-off in fitness traits (i.e., clutch size and parental care) between the two species, i.e., that parental care is not the only successful strategy.

The dorsal color patterns of the two species also underscore the difference between them with regard to diurnal guarding behavior. *C. colymbiphyllum* are essentially uniform green with small, scattered yellow dots (Fig. 2). *C. valerioi,* on the other hand, have a reticulate green pattern on a yellowish to pale-gold background, and an attending male is strikingly similar in appearance to the egg clutch(es) which he guards, especially during the day (Fig. 3). I suggest that the dorsal coloration of *valerioi* has evolved in response to its daytime presence at the egg site through the activities of visual-hunting diurnal predators. If the predator searching for an egg clutch is small enough to be repulsed by the male (e.g., small wasp), selection should favor a strong resemblance between the guarding male and his clutch. This will increase the probability of successful defense when the predator mistakes the male for the clutch. If the predator is a frog-eating species, it may mistake the frog for another egg clutch, which is not suitable prey, and continue its search. This would not be the case if a *colymbiphyllum* were encountered and would account for the more typical diurnal behavior of this species.

Future Research

The classification proposed earlier in this chapter is based on available data. While the major dichotomies and general trends appear to me to accurately reflect the patterns of parental care in frogs as currently understood, the picture is far from complete. With the exception of my work on frogs of the genus *Centrolenella* and current work on *Hyla rosenbergi* (Kluge, personal communication) there are no detailed studies on the relative importance of parental care in frogs. The relative importance of the various forms of parental care can only be evaluated against a background of detailed ecological information concerning the population dynamics of a particular species. Data relating to specific aspects of the population biology of several species with varying reproductive modes but occurring in the same habitat are needed. These data must be generated over relatively long periods of time (an entire reproductive season) and from marked individuals in a population. Specifically, studies must consider the reproductive capabilities of individuals in a population with reference to age at first reproduction, number of reproductive efforts per season, number of eggs produced per clutch, number of clutches per lifetime, relationship between age, size, and fecundity, and the sex ratio in the breeding population at the time. In addition workers must address the question of the role of behavior in molding reproductive modes in frogs. An understanding of the preamplectic behavior of males and females in the population is crucial to unraveling the relative trade-off between precopulatory and postcopulatory investment. One correlation uncovered in this review is that between marked territorial behavior and male parental care in some species. Does the scarcity of call and/or egg deposition sites result in territorial defense by males and thus favor an additional investment in the form of guarding and attendance by the male rather than female care? If this is true, as suggested by my data for *Centrolenella,* what determines the success of one male compared to another? Is it the site (i.e., a well hidden leaf; a strategically located burrow) that is important or is it the male or some combination of both? Do females evaluate the relative quality of a potential partner by the site that he holds, or do they somehow "count" his previous successes (i.e., egg clutches) and use that information in the selection process? Of particular importance to understanding the role of parental care is a knowledge of the reproductive success of individuals in a population. Crucial questions are: How often does a female breed in a reproductive season? Do quantitative or qualitative differences exist between eggs in her first clutch and eggs in her last clutch? Does she return to the same site or the same mate each time? Does attendance or brooding preclude additional breeding at the same time? What is the probability of an individual breeding successfully a second time if it deserts its first clutch? Are multiple breedings without parental care more likely to produce successful offspring than fewer breedings with parental care? What are the major mortality factors in a population, and how do these relate to the different kinds of parental investment?

Finally, we must have a better understanding of the role of each sex in reproductively related activities to allow for a more meaningful interpretation of the observed patterns. We must know more than the sex of the attending or brooding parent to unravel those factors favoring the evolution of different sexual strategies in terms of parental investment. These are some of the questions which must be considered in selecting a suitable group for study. In addition, the ease with which natural and artificial manipulations can be done should be weighed heavily in selecting an appropriate system. It is obvious that frogs have a great deal to offer to our understanding of the ecological role and evolutionary impact of parental care. What is needed are more data.

Acknowledgments

The field work on *Centrolenella* reproductive ecology was facilitated by a Faculty Release Time Award from the University of South Florida in 1971. Additional field work in 1969 and 1973 was made possible by logistic support from the Organization for Tropical Studies and assistance from the officials of OSA Productos Forestales and Rincón Resorts. Mercedes Foster and Kathy and Andy Shumaker helped with the field work. Early drafts of the manuscript were criticized by Mercedes Foster, Harry Greene, Ron Heyer, Stan Salthe and Kentwood Wells; their comments sharpened my thinking on several issues. Steve Godley also provided ideas, technical assistance, and helped in the final preparation of the manuscript. Mercedes Foster contributed immensely to the preparation of the original drafts. Without her help the deadlines would not have been met. To all of these people, I extend my appreciation and thanks for their help.

References

Blommers-Schlösser, R. M. A. 1975. Observations on the larval development of some Malagasy frogs, with notes on their ecology and biology (Anura: Dyscophinae, Scaphiophryninae and Cophylinae). *Beaufortia* 24(309):7–26.

Boulenger, G. A. 1897. *The tailless batrachians of Europe*. Part 1. London: Ray Society. 210 pp.

Breder, C. M., Jr. 1946. Amphibians and reptiles of the Rio Chucunaque Drainage, Darien, Panama, with notes on their life histories and habits. *Bull. Am. Mus. Nat. Hist.* 86(8):375–436.

Busse, K. 1970. Care of the young by male *Rhinoderma darwini. Copeia* 1970:395.

Cei, J. M. 1962. *Batracios de Chile*. Santiago: Universidad de Chile. 128 pp.

Corben, C. J., G. J. Ingram, and M. J. Tyler. 1974. Gastric brooding: unique form of parental care in an Australian frog. *Science* 186:946–947.

Crump, M. L. 1974. Reproductive strategies in a tropical anuran community. *Univ. Kansas Mus. Nat. Hist. Misc. Publ.* 61:1–68.

Dawkins, R., and T. R. Carlisle. 1976. Parental investment and mate desertion: a fallacy. *Nature* 262:131–133.

Drewry, G. E., and K. L. Jones. 1976. A new ovoviviparous frog, *Eleutherodactylus jasperi* (Amphibia, Anura, Leptodactylidae), from Puerto Rico. *J. Herpetol.* 10(3):161–165.

Duellman, W. E. 1970. The hylid frogs of Middle America. *Monogr. Mus. Nat. Hist. Univ. Kansas* 1:1–753.

Duellman, W. E., and L. Trueb. 1966. Neotropical hylid frogs, genus *Smilisca. Univ. Kansas Publ. Mus. Nat. Hist.* 17(7):281–375.

Duellman, W. E., and A. H. Savitzky. 1976. Aggressive behavior in a centrolenid frog, with comments on territoriality in anurans. *Herpetologica* 32:401–404.

Dunn, E. R. 1948. American frogs of the family Pipidae. *Am. Mus. Novit.* 1384:1–13.

Formas, R., E. Pugin, and B. Jorquera. 1975. La identidad del batracio Chileno *Heminectes rufus* Philippi, 1902. *Physis* 34(89):147–157.

Goin, O. B., and C. J. Goin. 1962. Amphibian eggs and the montane environment. *Evolution* 16:364–371.

Heyer, W. R. 1969. The adaptive ecology of the species groups of the genus *Leptodactylus* (Amphibia, Leptodactylidae). *Evolution* 23:421–428.

Heyer, W. R., R. W. McDiarmid, and D. L. Weigmann. 1975. Tadpoles, predation and pond habitats in the Tropics. *Biotropica* 7(2):100–111.

Ingram, G. J., M. Anstis, and C. J. Corben. 1975. Observations on the Australian leptodactylid frog, *Assa darlingtoni*. *Herpetologica* 31:425–429.

Jameson, D. L. 1950. Development of *Eleutherodactylus latrans*. *Copeia* 1950:44–46.

Jameson, D. L. 1957. Life history and phylogeny in the Salientians. *Syst. Zool.* 6(2):75-78.

Lamotte, M., and J. Lescure. 1977. Tendances adaptatives a l'affranchissement du milieu aquatique chez les amphibiens anoures. *La Terre et la Vie.* 31(2):225–311.

Lutz, B. 1947. Trends towards non-aquatic and direct development in frogs. *Copeia* 194(7):242–252.

Maynard Smith, J. 1977. Parental investment: a prospective analysis. *Anim. Behav.* 25:1–9.

McDiarmid, R. W. 1975. Glass frog romance along a tropical stream. *Terra, Los Angeles Co. Mus.* 13(4):14–18.

McDiarmid, R. W., and K. Adler. 1974. Notes on territorial and vocal behavior of neotropical frogs of the genus *Centrolenella*. *Herpetologica* 30(1):75–78.

Myers, C. W. 1969. The ecological geography of cloud forest in Panama. *Am. Mus. Novit.* 2396:1–52.

Orton, G. L. 1949. Larval development of *Nectophrynoides tornieri* (Roux), with comments on direct development in frogs. *Ann. Carnegie Mus.* 31:257–277.

Pengilley, R. K. 1971. Calling and associated behaviour of some species of *Pseudophryne* (Anura, Leptodactylidae). *J. Zool.* 163:73–92.

Poynton, J. C. 1964. The Amphibia of Southern Africa: a faunal study. *Ann. Natal Mus.* 17:1–334.

Rose, W. 1962. *The reptiles and amphibians of Southern Africa.* Cape Town, South Africa: Maskew Miller Ltd. 494 pp.

Salthe, S. N., and W. E. Duellman. 1973. Quantitative constraints associated with reproductive mode in anurans. In *Evolutionary biology of the anurans,* ed., J. L. Vial, pp. 229–249. Columbia: Univ. Missouri Press.

Salthe, S. N., and J. S. Mecham. 1974. Reproductive and courtship patterns. In *Physiology of the Amphibia,* ed., B. Lofts, pp. 309–521. New York and London: Academic Press.

Scheel, J. J. 1970. Notes on the biology of the African Tree-toad, *Nectophryne afra* Buchholz and Peters, 1875, (Bufonidae, Anura) from Fernando Poo. *Rev. Zool. Bot. Afr.* 81(3,4):225–236.

Silverstone, P. A. 1975. A revision of the poison-arrow frogs of the genus *Dendrobates* Wagler. *Nat. Hist. Mus. Los Angeles Co. Sci. Bull.* 21:1–55.

Silverstone, P. A. 1976. A revision of the poison-arrow frogs of the genus *Phyllobates* Bibron in Sagra (Family Dendrobatidae). *Nat. Hist. Mus. Los Angeles Co. Sci. Bull.* 27:1–53.

Stearns, S. C. 1976. Life history tactics: a review of the ideas. *Q. Rev. Biol.* 51:3–47.

Stephenson, N. G. 1951. Observations on the development of the amphicoelous frogs, *Leiopelma* and *Ascaphus. Linn. Soc. J. Zool.* 42:18–28.

Thornhill, R. 1976. Sexual selection and paternal investment in insects. *Am. Naturalist* 110(971):153–163.

Trivers, R. L. 1972. Parental investment and sexual selection. In *Sexual selection and the descent of man 1871–1971,* ed., B. Campbell, pp. 136–179. Chicago: Aldine.

Tyler, M. J. 1963. A taxonomic study of amphibians and reptiles of the Central Highlands of New Guinea, with notes on their ecology and biology. 1. Anura: Microhylidae. *Trans. Roy. Soc. South Aust.* 86:11–29.

Vaz-Ferreira, R., and A. Gehrau. 1975. Comportamiento epimeletico de la rana comun, *Leptodactylus ocellatus* (L.) (Amphibia, Leptodactylidae). I. Atencion de la cria y actividades alimentarias y agresivas relacionadas. *Physis* 34(88):1–14.

Wager, V. A. 1965. *The frogs of South Africa.* Cape Town and Johannesburg, Africa: Purnell & Sons Pty. Ltd. 242 pp.

Wells, K. D. 1977. The social behaviour of anuran amphibians. *Anim. Behav.* 25(3):666–693.

Wilbur, H. M. 1977. Propagule size, number, and dispersion pattern in *Ambystoma* and *Asclepias. Am. Naturalist* 111(977):43–68.

Williams, G. C. 1966. Natural selection, the costs of reproduction, and a refinement of Lack's principle. *Am. Naturalist* 100(916):687–690.

Wilson, E. O. 1975. *Sociobiology the new synthesis.* Cambridge, Mass.: Belknap Press. 697 pp.

Woodruff, D. S. 1977. Male postmating brooding behavior in three Australian *Pseudophryne* (Anura: Leptodactylidae). *Herpetologica* 33:296–303.

Zweifel, R. G. 1972. Results of the Archbold expeditions. No. 97. A revision of the frogs of the subfamily Asterophryinae family Microhylidae. *Bull. Am. Mus. Nat. Hist.* 148(3):411–546.

Chapter Eight

Behavioral Ontogeny in Reptiles: Whence, Whither, and Why?

GORDON M. BURGHARDT

University of Tennessee
Knoxville, Tennessee 37916

The title of this chapter deliberately implies two different messages. One is that I will be talking about "whence, whither, and why" in relation to *studies* of reptile ontogeny. The other is that I will address myself to *processes* of reptile ontogeny. But I will, in fact, do a little of both, and a complete job of neither. What I do hope is that by bringing together some scattered information and pointing out some problems this effort contributes to the ontogeny of a neglected area.

Reptiles occupy a unique and central position in the evolution of vertebrates, being the ancestors of both birds and mammals. They were the first vertebrate class to be completely liberated from the aquatic environment, developing a tough dehydration resistant egg and, secondarily, viviparity. The important behavioral correlates, not to say causes, of the reptile-mammal and reptile-bird transitions are unknown, although speculation is common and controversy frequent (e.g., Desmond, 1976; Jerison, 1973). As might be expected, the fossil evidence indicates intermediary mixes of amphibian-reptile, reptile-mammal, and reptile-bird characteristics. In addition, living species show variations in physical and behavioral characters that presage what would come or belie what has been. Not that the retention of apparently primitive characters in "advanced" classes or the presence of advanced characters in "lower" classes means that they are homologous across vertebrate classes. But such diversity is important and I, at least, have always been fascinated by monotremes, direct-developing amphibians, and parenting crocodilians. We need to assess fully the range of accomplishments within a group before making generalizations about vertebrate classes that are typological and premature (Burghardt, 1977a, 1977b). Indeed, the second sentence in this paragraph is itself suspect when amphibians living and breeding away from water (e.g. the red–backed salamander, *Plethodon cinereus*) and the completely aquatic sea snakes (Hydrophiidae) are recalled.

Detailed and comparative study of sequences of behavioral development in the life history of organisms are infrequent (Chapter 9, Burghardt, 1977c) but especially so in reptiles. The ubiquity of parental care in birds and mammals and its seeming rarity among reptiles is one of several factors leading to the neglect of reptile behavior even by those interested in the evolution and ontogeny of behavior patterns in "higher" vertebrates.

But is parental care the most salient feature in the development of behavior in birds and mammals? Skeptics, of course, can choose from among other apparently critical improvements over the dull and torpid reptile such as endothermism, increased metabolic needs, more efficient physiology, higher activity levels, greatly developed auditory system, a long and "helpless" infancy, larger and more complex brains, intellectual advances, and so on. These other differences also give apparent justification for ignoring nonsocial behavior in reptiles. It is rarely remembered that most of the above differences will not stand up in an absolute sense; as soon as only relative, average, or probabilistic differences are found, detailed analysis must replace superficial, but often plausible and ingenious, speculation of both the armchair and modeling type.

In an important 1963 symposium on "Social Organization of Animal Communities" held in London (but typically ignoring amphibians and reptiles) we have Kalmus's representative statement (1965; p. 5) concerning the evolution of social behavior, altruism, mutual recognition, and cooperation: "Most often these changes probably occurred in family groups and originated from parental behaviour." He goes on to state that it is controversial "whether natural animal communities ever started in any other way." It must be added that workers trying to deal with varieties of social organization in birds and mammals rarely invoke differences in parental care, as for example, the ecological approaches of Crook (1965) and Hall (1965) presented at the same conference. On the other hand all aspects of the evolution of life history differences should be tied to environmental variables (see Tinkle, 1972, for a stimulating discussion *re* lizards).

One reason for the above discussion is to emphasize that, while ontogenetic processes cannot be effectively studied without evolutionary considerations, our evolutionary views concerning reptile behavior have combined with other factors to hinder our accurate understanding and appreciation of vertebrate behavioral evolution. Reptiles are often viewed as being born as miniature versions of the adult (e.g., Bellairs, 1970) that do almost all the same things, behaving in a simple "instinctive" manner, in which parental care is quite irrelevant even where faint sidesteps toward it appear.

It is quite easy to develop from such a superficial conception that the important ontogenetic aspects of bird and mammal behavior arose independently and were somehow associated with parental care and prolonged immaturity. Reptiles are thus irrelevant backwater relics, somewhat like Desmond Morris' (1967) "primitive" human tribes. The rudimentary mechanisms reptiles show may have antiquarian interest, but certainly not much importance for those studying their "really successful" descendants. In addition an attitude assuming that experience and learning act in rather gross and direct ways to shape behavior connects nicely with the importance ascribed to parental care. The controversy over the deprivation experiment (e.g., Lorenz, 1965) was fought almost entirely in this arena. We could not really parcel out the factors in deprivation to everyone's satisfaction, but we could assume their importance. Reptiles did not seem to show much change, postnatally, in their behavior patterns nor did differential rearing seem to affect them. Thus they were left out of this "most important game in town." While traditional views of reptile behavior may turn out, on balance, to be true, the points I want to emphasize are that (a) such a conclusion is not now warranted, and (b) the diversity of structure, habits, and habitat among reptiles and their convergence with other vertebrate classes may illuminate topics of current interest in ontogeny, social organization, and learning (Burghardt, 1977a; Greenberg and Crews, 1977).

Deal!

Reptiles are not underdeveloped birds or mammals; they represent a highly successful and diversified group of vertebrates solving common problems in many different ways. The problem of ontogeny is essentially to determine the precursors to a behavior performed by an animal at a given time in a given context. There are many avenues for developmental processes to take (see, e.g. Burghardt, 1977c). A convenient way of looking at the problem is to consider a large shuffled deck of cards, each card representing some physical, behavioral, or ecological characteristic of reptiles. [Inherited characteristics have been compared to cards in popular writings on evolution (e.g., Thomson, 1926). The present analogy has a different emphasis.] While some of these are widely represented in vertebrates, few will be exclusively reptilian. Most will be limited to certain species. Mammals, birds, amphibians, and fish each have a different deck. But when each species is dealt its individual hand the class distinctions often become less relevant. Compare the probabilities of being dealt five cards from a pinochle deck versus a bridge deck. [A pinochle card deck omits the deuce through the eight but has two each of the remaining cards (nine through ace).] The size of the decks are almost the same and it is possible to be dealt four kings from either deck. Now it's more likely that you'll get four kings from the pinochle deck, but not at all assured, while one will never get a pair of sevens from it. The chance of getting a flush is the same with both decks. Also to be considered is the game involved. Playing lowball poker with cards from a pinochle deck would be a disaster, but again not a loser all the time against the bridge deck. Thus, although different decks evolved in somewhat different contexts, the individual hands may differ radically and be capable of rather different development, as any seasoned card player knows.

Next, consider that we walk into a gambling hall knowing neither the game that is being played nor the composition of the deck that hands are being dealt from. This, I submit, is the problem we are faced with in comparative studies of development. We have been reluctant to do the careful descriptive work that is the essential background for uncovering these two separate but related unknowns, perhaps being conned by some dramatic plays and impressive hands. Looking at what happens when the same game is played with different decks may help us understand relations between both games and decks. Needless to say, with reptiles the shuffled deck is virtually unknown, as so few hands have been studied, and even those incompletely.

I here review data for a variety of behavior patterns that illustrate some of the diversity and complexity (hands) that neonate reptiles present us with and the types of short-term experiential effects that have been demonstrated. Then I will argue that the obvious conclusions should *not* be drawn, at least without consideration of some data about the later life history of reptiles. Since I have recently discussed our work on social behavior in reptiles (Burghardt, 1977b), I will concentrate on other patterns here.

Most studies of developmental processes in reptiles begin, and often end, with neonates or juveniles. This is certainly true of most of the work from our laboratory. Cross-sectional studies comparing different age classes are rare in laboratory studies but are typical of field studies. The latter are becoming more behavioral and age, sex, and size comparisons are entering the literature. Longitudinal studies in which individual animals are followed through their life are even less common except for largely anecdotal histories of individual or pet animals, although the work on Sylvia, *Python molurus,* by Pope (1961) indicates the wealth of material to be obtained from intelligent and careful record keeping.

Maintenance Behavior

Reptiles must escape from the egg or the membrane in which they are born. Hatching from an egg shell is the more complex and has recently been reviewed (Oppenheim, 1973). But there is a paucity of information and some of the best observations go back many years, as in the excellent series of copper engravings of crocodilian emergence by Seba (1734). Hatching obviously is one behavior shown by neonates that is not found in the adults. Not only is well-coordinated behavior involved but temporary morphological changes also occur. Lizards and snakes develop a true egg tooth while crocodilians and turtles, like birds, develop an epidermal egg caruncle. While some hatching crocodilians vocalize when ready to emerge from the nest, which leads to adults opening the nest and even the eggs (references in Garrick and Lang, 1977), most reptiles that hatch from eggs must dig out of sand, dirt, leaf litter, or rotting logs. Social facilitation of digging and emergence behavior from nests occurs in green turtles, *Chelonia mydus,* via their activity (Carr and Hirth, 1961) and a similar mechanism may operate in other species. Green turtles normally emerge at night when temperatures are cool; exposure to hot sand and bright sun can lead to rapid death (Bustard, 1967). But during the day some hatchlings are found with their heads sticking out of the sand. They are, however, completely inactive and unresponsive even to prodding. Bustard (1967; p. 317) thus suggests "that the comatose condition of the few turtles which reach the surface has a dampening effect on the activity of those below. Furthermore, those at the surface act like corks in a bottle, occupying the opening of the nest shaft and making exit difficult if not impossible for those below."

Temperature dependence of activity is an important mechanism insuring survival on the trip across the hot sand. For example, the frenzied activity of hatching turtles is inhibited above 28.5°C (Mrosovsky, 1968). But one-day-old green and hawksbill, *Eretmochelys imbricata,* turtles were largely (83%) inactive at *all* temperatures. Thus the dramatic photic behavior of turtles underlying their migration from nest to sea is quickly altered. This must involve some component other than age as turtles may remain in the nest for varying periods of time. It should be noted that behavior in these turtles can be far from lethargic. A turtle but a few cm long may travel several meters in 60 s.

Locomotory behavior is similarly precocious in all reptiles, although here too detailed quantitative studies are called for. Scratching, licking, and body nipping behaviors are also present and have been described and filmed in our studies of hatching green iguanas, *Iguana iguana* (e.g., Burghardt, *et al.,* 1977).

Circadian rhythms and other activity cycles have been studied in reptiles (e.g., Heckrotte, 1962, for garter snakes) but little seems to have been done with neonates. The question of activity cycles in newly born and newly hatched snakes was studied by Jones (1974) in our laboratory. The basic method was to photograph, at 60-second intervals, the behavior of individually housed snakes. All aquaria were in an environmental chamber under constant light and temperature. The snakes were placed in them on the date of birth prior to any experience with a day–night cycle.

When three newborn ribbon snakes, *Thamnophis sauritus sackeni,* were in the box for 18 days, no evidence of a rhythm of any kind was found (the data were scored as activity occurring in any 15-min block). This was true for individual and grouped data. However, eight newly hatched corn snakes, *Elaphe guttata,* tested over about the same interval showed a clear cycle with an activity peak about 0700 and a larger one at 1900 h (Fig. 1). Six of the snakes showed the early evening peak. As no

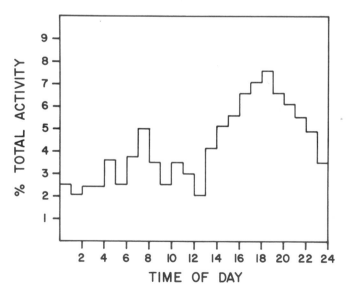

FIG. 1. *Activity profile of eight hatchling corn snakes for their first 18 days (17 days for two late hatchers) of life when maintained under constant light.*

feeding or other cues could be located as extrinsic factors, an intrinsic interpretation seems most likely, particularly since ribbon snakes showed no rhythm when studied under similar circumstances. But just as adult ribbon snakes showed a clear activity cycle that could be manipulated by the light cycle, so too did we find that the corn snakes, after eight months under typical lab conditions with a light cycle, showed an even more marked rhythm, with the peak appearing shortly after the lights were turned out (Fig. 2).

These data support the existence of a biologically appropriate activity cycle (as determined by sparse field data) either endogenous or "self-taught" in several days. This then is further evidence for the precocity of reptile behavior and individual adaptability with changing conditions.

Feeding

Although newly born reptiles can go from a few days to many weeks before their first meal, eventually they must find and ingest food. Not surprisingly, feeding has been the most thoroughly studied behavior in neonate reptiles. The role of parents or adults in shaping food choice or feeding topography is minimal, as far as we know now, even in species, such as crocodilians, where parental care is highly developed. There is some indication adult crocodilians may raise clumps of vegetation to the surface where young can feed on associated invertebrates. Herzog (personal communication) once noted, in Florida, a large adult alligator grab a submerged decomposing pig's head and hold it in its jaws at the surface while hatchlings approached and tore off pieces of flesh. Skinks, *Eumeces,* common North American lizards, are well known for the female's diligent incubation of her eggs (Noble and Mason, 1933) and, at least in one species observed in the laboratory, the young stay with the mother for up to 16 d after hatching (Evans, 1959, 1961). She not only grooms them but will apparently inhibit her usual voracious feeding upon small animals to exclude not only her own young but small insects the young may be about to attack (cf. Desmond, 1976). As I have seen neonate skinks die trying to swallow oversize prey (Burghardt, 1964), it is possible that the female plays some regulatory role here. Appleby (1971) reports that newborn vipers, *Vipera berus,* stay with the mother for a few days after hatching but it is not even known if the neonates eat during this period. In general it can be concluded that protection of the young is the main function of parental behavior wherever it occurs in reptiles.

Newly hatched turtles seem predisposed to stalk and attack small fish and insects shortly after hatching. Naive snapping turtles, *Chelydra ser-*

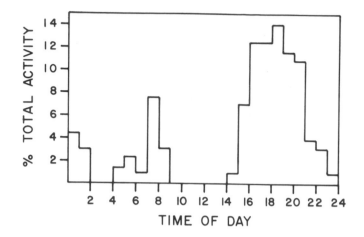

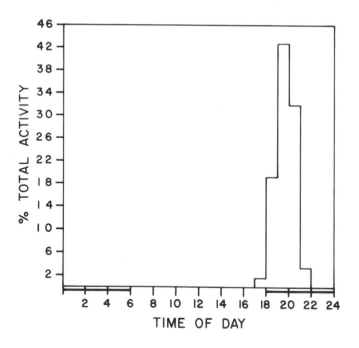

FIG. 2. Top: *Activity profile of corn snake subject 1 under constant light for first 18 days after hatching.* Bottom: *The same subject during seven days under a 12-hour light cycle.*

pentina, engaged in the characteristic neck stretching, snapping, and clawing with the first meals (Froese and Burghardt, 1974). However, the behavior did not include the rapid forward thrusting movement, typical of adults, until the turtles were a year old, at which time feeding seemed completely adult-like. Live food was not used and this may have altered the pace of development, which nonetheless occured. As for snapping, I have observed it in late embryos, especially to tactile stimulation. A rare opportunity to observe the first feeding responses of two hatchling South American mata mata turtles, *Chelus fimbratus,* was taken advantage of by Dona Drake at the Knoxville Zoological Park (personal communication). This species feeds by approaching prey (or allowing the latter to approach the turtle's snout) and then suddenly gaping wide the jaws, sucking in water and food. The topography of these movements seemed adult–like from the first time the turtles ate, however, the accuracy (and success) of the gapes improved over the first few feedings. In addition, movement of dead food (e.g., pieces of beef heart) was needed at first but later was unnecessary. Rapid, but short-term, changes in food preferences can be instilled in turtles by differential feeding in *Chelydra* (Burghardt and Hess, 1966; Burghardt, 1967) and *Chrysemys* (Mahmoud and Lavenda, 1969).

All snakes are predators and most must overcome relatively large prey. Indeed, a distinctive feature of the snake is the modified jaw apparatus that allows it to swallow prey much larger than the entire head (Gans, 1961). Venomous, constricting, and "seize and swallow" species all seem to be quite proficient in finding and ingesting their first meals. As a given individual may use, on the same or different occasions, components of two, or even all three of the above methods, detailed observations of neonate snakes are needed for understanding how various context dependent and experiential factors interact with initial perceptual, morphological, and behavioral characteristics.

We do know that naive young of many harmless species of the largest serpent family, Colubridae, recognize the chemical cues associated with their species-characteristic prey without any prior experience with that prey.[1] The vomeronasal organ, which opens into the roof of the mouth in snakes and is stimulated by chemicals picked up by the flicking tongue, is the primary chemosensory channel (Burghardt and Pruitt, 1975). Similar, but less dramatic, neonatal results have been obtained with two species of skinks (Burghardt, 1973; Loop and Scoville, 1972) and an alligator lizard (*Gerrhonotus,* Greene in Burghardt, 1977c). Geographic and intralitter (Arnold, 1977; Burghardt, 1975) differences exist. While the extract prey differences are refractory to certain experiences such as maternal diet (Burghardt, 1971, 1977c) and deprivation, they can be modified within limits in neonates by prior experience with effective extracts, differential feeding, illness–induced aversions (Burghardt and Czaplicki, unpublished), and habituation (Czaplicki, 1975, and, for new-borns, Burghardt, 1977a, and unpublished).

While many species respond to the chemical cues at birth in the sequence (when in close proximity of prey): tongue-flick, reception of necessary odor cues, attack, poison or constrict (optional), and swallow, others do not. Newborn rattlesnakes (Chiszar and Radcliffe, 1977) do not show increased arousal or tongue–flicking to the species–characteristic prey extracts that are effective in other species eating the same prey. Striking at prey is apparently critical to release "searching" even in the naive young. This supports work on adult vipers and rattlesnakes which strike at prey in response to visual (movement) cues (also thermal cues in pit vipers) and ignore the chemical cues until *after* the strike, at which time they will follow the chemical trail left by the envenomed animal (Dullemeijer, 1961; Naulleau, 1966).

Even newborn eastern garter snakes, *Thamnophis s. sirtalis,* respond with approach to small moving objects, although prey attack is not released without appropriate chemical stimuli. But facilitation of response to prey with repeated feedings does occur. In experiments performed with Doris Gove, the first feedings at several-day intervals of 11 neonate snakes upon fish were studied. The snakes were housed in our standard individual glass aquaria and a live or dead 2-cm guppy, *Poecilia reticulata,* added to the water dish. Live and dead fish were alternated for the first four feedings (Series I) using balanced orders. Four latencies were recorded during the tests. These were the time of onset of *search,* a clearcut increase in locomotor behavior and tongue flicking; *orient,* the momentary cessation of locomotion with the snake's snout pointed at the prey; *first attack,* an open-mouthed lunge at the fish; and *final attack,* the successful strike leading to ingestion. Responses to live and dead fish often differed and, for simplicity, only the final attack results will be presented here. It can be seen from Figure 3 that none of the snakes responded to the dead fish within the 30-min test period (although most eventually found and ate the fish as it was left in the dish overnight if not eaten). In Series II the snakes were pre-fed a fish with a forceps immediately prior to introduction of the test guppy. For the first two trials in the series a dead guppy only was offered and now some of the snakes began to respond. For the next two trials live and dead fish were alternated and, while the decreased latency to the dead fish is dramatic, the live fish was still more effective. Nonetheless the pre-feeding did have an apparent "arousal" effect. Series III was a longer replication of Series I, and it can be seen that response times to both stimuli continued their rapid decline (a few snakes did die during the tests). Thus the attractant

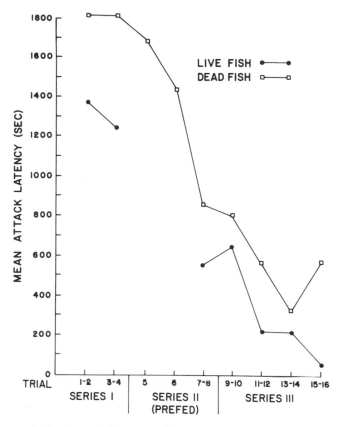

FIG. 3. *Final attack latencies of young garter snakes to live and dead fish (female guppies). See text for details.*

effect of movement can apparently be capitalized on by the snake and, however mediated, rapid learning of some kind is going on.

The propensity of many snakes to swallow prey head first has been frequently noted and is evident in many snake illustrations in natural history compendia from the eighteenth and nineteenth centuries. Although not completely predictable, a large part of the variability is accounted for by considering the relative size of the prey (Loop and Bailey, 1972). Small mice may often be swallowed other than head first but the proportion of head first ingestion increases with relative prey size. This was also established in predatory lizards such as *Varanus*. Observations on the first feeding of 22 newly hatched rat snakes, *Elaphe obsoleta,* and king snakes, *Lampropeltis calligaster,* with relatively large mice showed head-first ingestion in 19 cases (Klein and Loop, 1975).

While the authors interpret the results as evidence for innate head recognition, the possibility of rapid learning during even the first feeding due to size and shape constraints or other factors needs to be considered (Greene, 1976). Filmed first encounters with prey would be useful. A two-headed black rat snake, *Elaphe obsoleta,* that I have been observing also often swallows mice head first, complications arising when both heads attack the prey, relative head dominance being a factor in ingestion direction.

The striking and swallowing of prey in snakes seems normal in the neonate, as does the ubiquitous tongue flicking. The problem is that because these behaviors appear so normal virtually no descriptive, quantitative data on newborns has been published, in spite of all the breeding success of zoos, amateurs, and herpetologists. The coordination of jaw and swallowing movements as a limbless creature swallows an object bigger around than its own head is impressive.

Several years ago I measured the length of time taken by 13 newborn garter snakes, *Thamnophis sirtalis parietalis,* to swallow their first eight (at 4–5 d intervals) meals of hotwater-killed, 6-cm redworms, *Eisenia foetida.* As can be seen in Figure 4, a rapid and relatively linear decline in latency was found (by the fourth feeding, 11 of the 13 snakes swallowed faster than they did the first meal). But just as the head-first study could not be taken as proof of innate recognition on first encounter, so this result should not be interpreted too quickly as evidence of experience. The snakes were growing, maturing, and perhaps getting stronger while the size of the worm remained constant. Nonetheless, the similarity to the attack latency graph of the same species in Figure 1 is remarkable.

A more interesting prey-capture technique used by snakes is constriction, which has evolved in two large families (plus some smaller ones), all the Boidae (boas and pythons) and perhaps 10% of the Colubridae (containing about 70% of all snakes, mostly harmless). Constricting methods vary among colubrids but are consistent across at least four families of primitive snakes. Careful quantitative observation indicates that, in general, neonate, naive boids show an apparent adult expertise from the first prey encounter, while neonate colubrids often constrict in a more variable manner than the adult (Burghardt, 1977d; Greene, 1977; Greene and Burghardt, 1978). Thus specialized, similar motor patterns subserving the same function may have different developmental histories in different taxa. The possibility of such differences within the same family or genus needs to be studied.

There are also intriguing reports of possible ontogenetic changes in specialized and uncommon predatory tactics. Several species of vipers and pit vipers use caudal luring to attract prey (see reviews in Greene and Campbell, 1972; Heatwole and Davison, 1976). For example, young of

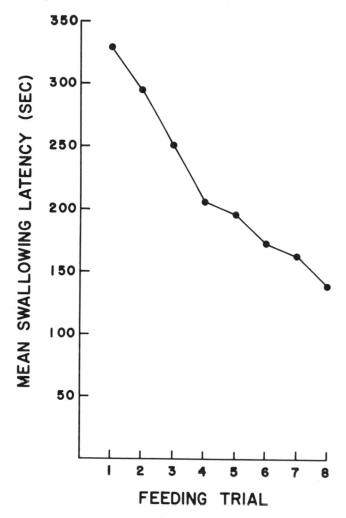

FIG. 4. *Swallowing latencies of 13 newborn garter snakes to 6-cm earthworms for the first eight feedings.*

the hump-nosed viper, *Agkistrodon hypnale,* respond to the presence of a lizard by protruding their whitish tails from their coils and wriggling them like active earthworms or grubs. The lizards are lured closer and actually seize the fake worm. Then they are struck and envenomed. Subsequently the snakes began to strike before the tail was bitten (report by G. M. Henry quoted in Smith, 1943). In only two species is it known that the behavior persists in the adult (*Bothrops bilineatus* and *Cersastes vipera*), and these are unusual in that the distinctively marked tail is retained and the diet does not shift. A good opportunity exists here for the study of ontogenetic changes in coloration, behavior, and diet.

Antipredator Responses

Many reptiles shortly after birth engage in such typical defensive behaviors as flight or, when seized: biting, tail lashing, defecation, and general writhing. Many lizards including neonates will autotomize part of the tail. Neonate snakes sometimes discharge the contents of the vile-smelling cloacal sacs. Some species, such as some pit vipers or colubrids will stand their ground and strike. Young crocodilians will often vocalize a distinctive call that presumably alerts and attracts nearby adults (Herzog and Burghardt, 1977), yet it is rarely if ever given by adults.

Tail displays are dramatic defensive maneuvers for many snakes and lizards. Their diversity and relationships have been worked out by Greene (1973). I have noted rapid tail vibration in many newly hatched rat snakes, *Elaphe*. Tail vibration in material such as leaf litter produces a remarkably rattlesnake–like sound that presumably has warning significance. I have a vivid recollection of fox snakes, *Elaphe vulpina*, engaging in this behavior, even when not completely emerged from the egg. My startle response to the sound when I approached the container with the hatching eggs was extreme, and I established that the egg shells themselves could effectively interact with the vibrating tail.

I have noted extreme differences in the defensive responses of various species of garter snakes. Whereas the eastern garter snake, *Thamnophis sirtalis,* when first captured, will often strike and bite vigorously, newborn individuals typically do not. However, after several days or a few weeks, such behavior does appear, along with body flattening and neck expansion, suggesting some type of maturation. Young of the western aquatic garter snake, *T. couchi aquaticus,* engage in such behavior immediately upon birth, whereas Butler's garter snake, *T. butleri,* whether newborn or adult, virtually never strike or bite. Such differences within the same genus would be a fruitful topic for careful comparative, ontogenetic, and physiological study.

One of the most dramatic defensive behaviors found in any snake is the bluffing and then death-feigning (letisimulation) behavior of the hog-nose snakes, *Heterodon*. Both have been described many times in several species, especially for the eastern hognose, *H. platyrhinos,* though not quantitatively (see review by Gehlbach, 1970). We have studied this behavior intensively in both adult and hatchling snakes. Its occurrence in hatchlings while partially in the egg has been noted (Raun, 1962), but we found that all the components found in the adult sequence can be found in neonates and that the bluff aspect of the behavior habituates more quickly than the death feign, albeit with great individual variability (Burghardt, Greene, and Klinghammer, in preparation).

Using our cotton swab technique to present surface substance extracts to naive Mormon racers, *Coluber constrictor mormon,* we found a jerky, spasmodic avoidance response to scorpion extract. This is the only clear evidence of a possible antipredator response to extracts in our testing situation. Admittedly, we usually pick out small potential prey items for our tests. Such behavior, however, in contrast to mere indifference should be studied. A prime possibility for study in neonates would be the chemically elicited avoidance and bridging behavior shown by pit vipers in response to king snakes, *Lampropeltis,* which are ophiophagous snakes immune to pit viper venom (e.g., Bogert, 1941). Preliminary results in our laboratory indicate this often variable and transient response can occasionally be induced to naive neonate pit vipers, (Weldon and Burghardt, in preparation).

In hatchling iguanas a whole constellation of antipredator behaviors has been identified (Greene *et al.,* 1978). These include flight (quadrapedal and bipedal running on land; climbing up and jumping from trees and shrubs; swimming, diving, and bipedal running on water), cryptic coloration and movements, group activities, vigilance, writhing, and tail autotonomy when grabbed, and perhaps tonic immobility, described in laboratory studies by Prestrude and Crawford (1970) and seen in our field studies with an adult.

As reptiles often grow to be hundreds of times larger (in weight) than newborns and live in different environments as they mature, it seems likely that over a period of time they will be preyed on by different species. While some general tactics should be useful against many kinds of predators (e.g., flight, writhing), others should be specific to certain types for the animal at a given age. As an extreme example, hatchling crocodilians are at the mercy of many birds and mammals, whereas adults are preyed on by virtually nothing other than humans in recent times. Behavior patterns associated with changing "dangers" should be studied. Comparing widely distributed species living in contrasting environments with different kinds and proportions of predators would also be of interest. Little has in fact been done in this area. Interestingly, however, the widespread ringneck snake, *Diadophis punctatus,* shows geographic variation both in the use of the tail display and its colorful tail markings (Fitch, 1975). The report by Smith (1974) on a population of this species shows that ontogenetic thinking may be important. About two thirds of small ringneck snakes (less than 25 cm) showed tail displays while only

30% of larger animals gave the response. Now the ecological and developmental factors involved are not known; many could be postulated, and captive-rearing studies would be useful. But a simple, natural selection hypothesis is not tenable unless it is postulated that tail displays, while common, are disadvantageous to younger individuals.

Gregariousness and Spacing

Spacing patterns have not been extensively studied in neonate reptiles; observations in the field are especially rare.

All turtles hatch from eggs buried in the ground. Observations of emerging, and especially migrating, freshwater and marine species are common. As clutch size can be large, groups of animals are often found together during migration and for some time thereafter. It is usually concluded that the young are following orientation cues and are not together for any social reason.

Newly hatched iguanas frequently emerge from the nest and migrate from the hatch site in small groups. Further, iguanas are often found sleeping and foraging together for at least several months post hatch (Burghardt, 1977b). Yearlings are found singly more frequently and adults observed by Dugan (personal communication) maintain greater individual distances, with little of the physical contact seen in juveniles. But even so, numerous authors have reported that several adults tend to be found in the same tree. Grouping of neonates does not seem the case in the related ground iguanas (*Cyclura,* T. Wiewandt, personal communication). But evidence that this gregariousness is not limited to one species was found during observations made in early evening in the Canal Zone, Panama, on May 11, 1976. Of eight sleeping neonate basilisks, *Basiliscus basiliscus,* in one locale, four were in close proximity on a dead branch jutting off a shallow pond, three were together on one of the many dead branches of a fallen tree, and one was apparently alone in reeds adjacent to the pond. Along a stream five hatchlings were spotted in reeds, none of the lizards were more than 3 m apart; they were found in two groups of two and a singleton. Thus 11 of 13 basilisks were found within a few centimeters of another individual in a large area (several hectares) carefully scouted by six field workers.

Observations of the aggregative tendencies of many colubrid snakes are plentiful and could be attested to by any field herpetologist in North America (Dundee and Miller, 1968; Fitch, 1975; Noble and Clausen, 1936). We have collected extensive data on newborn animals indicating the occurrence of grouping and aggregation (physical contact) under cover objects in the laboratory (Burghardt, 1977b; Burghardt and Spottswood, in preparation). We find that newborn young in the lab aggregate and selectively choose conspecifics, although not exclusively. Noble and Clausen, however, found greater exclusivity, and even repulsion between sympatric species. This may be an ontogenetic change or merely an artifact of the different methods employed.

Territory and Display

Many reptiles are territorial as adults, or at least have relatively non-overlapping home ranges. This is especially true of iguanid lizards, which have been most studied in this regard. Simon and Middendorf (in press) have observations on neonate *Sceloporus jarrovi* that indicate the establishment of territories within a few weeks after hatching, and extensive spacing even earlier. Aggressive displays were noted by 13 d in field populations and they quote other papers to support the notion of elemental "territorial behavior at birth" (see also Stamps, in press, for *Anolis aeneus*). The head-bobbing movements of several iguanid lizard species at birth or hatching have been noted (e.g., Cooper, 1971) and while these may be marker behaviors for territorial establishment, the inference should currently be resisted. In iguanas in Panama, where territoriality is not a factor, we have filmed complex head bobs along with several other displays occurring seconds after emerging from the nest hole. Greenberg (unpublished) has quantified changes in the bobbing of newborn *Sceloporus cyanogenys* until the adult form was noted at 72 h of age. During two weeks of extensive observations of 14 laboratory-born chameleons, *Chameleo bitaeniatus,* I saw chasing, head butting, bobbing, color changing, holding each other's feet or tails, and crawling over each other and "wrestling." These behavior patterns seemed most associated with movement and displacing siblings from preferred positions.

The Future

The preceding material has amply established the precocial nature of much reptile behavior, a precociousness that makes reptiles suitable for many studies beginning immediately after birth or hatching. But evidence was also presented that some fairly dramatic changes occur in the species characteristic behavior of these animals.

Some of these changes are rapid, as in waning of migratory behavior in hatchling turtles. Others occur over years, as in food habits of many species or vocalization in crocodilians. Symptomatic of our ignorance, however, is that we have little information about the time course of the more long-term processes. The development of courtship and sexual behavior has been little followed with the main exception of some of the shorter-lived lizards. And even here reproductive tracts rather than behavior are usually observed, a more than trivial example of the Heisenberg indeterminacy principle, since courtship is difficult for a lizard in formalin. Let me now add some areas deserving close study.

EMBRYOLOGY

The embryology and prenatal behavior of several species have been touched on (Bellairs, 1970). In one approach effects on prenatal and postnatal development of various conditions such as temperature, maternal diet, hormones, and so forth are examined. The effects of incubating eggs or maintaining gravid females at different temperatures have been the primary focus of the few studies available (e.g., Vinegar, 1974) and usually mortality and structural (scalation and vertebral) consequences noted. The dependence on a narrow range of temperature for proper development is shown by many birds and is probably true of most mammals as the embryos are kept at a highly constant temperature by virtue of maternal physiology. Yet reptiles also often lay their eggs in areas where a quite constant temperature obtains or, in the case of some pythons, actually incubate their eggs by raising their own temperature above ambient, an increase shown to be necessary for proper development (Vinegar, 1974). Since in squamate reptiles (lizards and snakes) oviparous, ovoviviparous, and truly viviparous forms can all occur (Bellairs, 1970), two of these even occurring within closely related species in the same genus (e.g., *Lacerta, Sceloporus*), opportunities are available for studies on the ecological and behavioral consequences, direct and indirect, of such variation.

A second type of study is observation and experiment on the embryos themselves. Given the wealth of findings and techniques from the many studies on bird embryos and the controversies that dominate discussions of ontogeny and the nature-nurture debate resulting from such work (see, for example, Chapter 2; Burghardt, 1977b), the ignoring of reptiles is deplorable. Controversy based on similarities and differences between chick, frog, and mammal development might be illuminated by looking at reptiles, but there are few studies of behavioral embryology using reptile material. Decker (1967) found considerable similarity between the chick and snapping turtle motility patterns. The main difference is that the turtle has a lower maximum activity rate which declines after the midpoint of incubation, whereas the chick embryo is more active and stays so until shortly before hatching. *Lacerta vivipara* embryos are more active than the snapping turtle, but none of the movements seem at all similar to those of the hatchlings (Hughes *et al.*, 1967). Obviously, more avian and reptilian species need to be studied before any class distinctions are drawn.

DIET AND FOOD PREFERENCES

Careful examination of food availability, food choice, and food utilization might be quite profitable and integrate behavior and physiology (Burghardt, 1977d). Populations of the same species may eat different foods; changes with age may also occur. We do know that in several reptilian species, as compared to birds and mammals, growth, size and age at sexual maturity vary greatly with food availability, and perhaps population density. Digestive limitations and nutrition should also be studied. The work of Hahn and Koldovsk (1965) with mammals could serve as a model.

PATTERN AND COLOR

Neonate crocodilians and turtles generally look similar to the adults, although both patterning and coloration are usually brighter and more distinct in the young. This is true of the American alligator, *Alligator mississippiensis,* and the South American mata mata turtle. More pronounced changes are seen in many lizards and especially in snakes. Guides to snakes of any area in the world will point out the often drastic changes in pattern and color with age that occur in some species, but not others. Familiarity with adults will allow identifying a newborn garter snake, *Thamnophis,* down to the species while in rat snakes, *Elaphe,* it is much more difficult. Young of some South American pit vipers, *Bothrops,* are brown while the adults are bright green. The female viper in Britain, *Vipera berus,* is brown at birth, like the male, but turns bright red at about a year old, turning to brown again when larger. This change is so remarkable these immature red females were considered a separate species even into the present century (Appleby, 1971). Such color changes must have some effect on survival in relationship to predators, if nothing else, so I think the full explanation of such ontogenetic changes will have to tie in with the entire life history of the species. The color and pattern polymorphism of many serpents may also have behavioral consequences. Temperament differences may occur in newly-hatched morphs of Japanese rat snakes, *Elaphe climacophora,* which also show ontogenetic change of one morph into the other (Hadley and Gans, 1972).

HABITAT

Evidence is accumulating that many young reptiles live in different areas from the adults and in many instances the whereabouts of juveniles are virtually unknown, as in the ubiquitous box turtles, *Terrepene,* of the United States. Even where the young live in the same general area as the adults, the use of space and access to prime areas may differ. Food resource partitioning by age and sex also occurs, but the actual mechanisms (food preferences, social competition) are rarely determined. Careful work on the determinants of habitat choice, change, and correlated behavioral, physiological, and predatory consequences is needed.

SOCIAL FACTORS

The processes of dispersion of hatchlings and juveniles need to be assessed. Simon and Middendorf (in press) have made a start with the lizard *Sceloporus jarrovi.* The related spiny lizard, *S. undulatus,* has been the subject of an experimental study on effects of density and crowding on juveniles (Tubbs and Ferguson, 1976). Courtship and copulation have not been described for most reptiles so it is not surprising that developmental studies are rare. Carpenter (1977) estimates courtship has been described in only 3% of all species of snakes. Reptiles reared in social isolation do cften appear to court and mate appropriately in captivity. But the situation in natural social groups may be different. While the sexual "drive" and behavior patterns may be endogenously organized and hormonaly induced via external stimuli such as temperature and day length, social factors, including experience, may play an important but subtle role (e.g., *Anolis carolinensis,* Crews, 1975). Evans (1961) reports that testosterone injections enhanced agonistic behavior and dominance in young turtles. Sexual dimorphism can be marked or absent, and more study, especially of its ontogenetic aspects, might illuminate many areas of reptilian social behavior.

As humans we are biased toward visual distinctions between animals. It is also important to consider chemical, auditory, and other modalities. There is some evidence that the sexes can distinguish each other by odor (e.g., snakes) and visually (e.g., *Sceloporus* lizards, Noble and Bradley, 1933). Other physical differences also appear (see for example Fitch and Henderson, 1977).

While some aspects of agonistic behavior and territoriality occur at or shortly after birth, this is not always the case (as in iguanas). I am aware of no developmental information on combat rituals in snakes (Carpenter, 1977), whose elaborate "dances" seem restricted to males.

Parental behavior or, more generally, solicitous behavior of adults directed toward young needs to be studied. Turtles are considered the oldest living reptile group and seem to be without such behavior either before or after the eggs hatch. In all crocodilians it probably occurs; Meyer (1977) has evidence for adult-young nurseries that last up to two years. In between, in some lizard and snakes the females brood and guard the eggs. Indeed, in some snakes the male also guards the eggs (e.g., cobras, M. A. Smith, 1943). These phenomena are often ignored in comparative treatments of parental care and investment (e.g., Maynard Smith, 1977).

One could list many other phenomena that we should know about before even trying to make generalizations and predictions about the evolutionary ontogenetic strategies and tactics employed and exploited by reptiles. Descriptive mathematical approaches (e.g., Singer and Spilerman, 1976) might be useful. The fact that young natricine snakes seem to be considerably less capable than adults, physiologically, of sustained activity (Plough, 1977, unpublished) may have far-reaching implications, pointing to the need for an area of reptilian developmental physiology (see also Gans, 1976, for crocodilian examples). Intra- and inter-individual variability may also be greater in newborns as compared to adult reptiles. Certainly this appears to be the case in responses to prey chemical stimuli in some snakes (Burghardt, 1975; Gove and Burghardt, 1975). Object play and curiosity, often considered nonexistent in reptiles and highly important in mammals, may need to be reassessed as naturalistic observations of reptiles (e.g., Lazell and Spitzer, 1977) and appropriate sensory considerations (e.g., Chiszar et al., 1976) accumulate. Thus investigations of ontogenetic processes in reptiles will gradually move to detailed long-term studies of field populations, comparative studies of closely related species (e.g., following King, 1961), and sophisticated experimental studies of mechanisms, all eventually leading to the formulation of specific and testable functional evolutionary ideas to explain the diverse ways reptiles go from zygote to senescence. In addition, studies of carefully selected reptiles may allow testing the limits of theories about development which differentially stress processes such as parental care, socialization, food habits, encephalization, habitat, or developmental rate.

Summary

Newly born or hatched reptiles engage in many behavioral patterns apparently typical of adults. A sampling of the diverse evidence on feeding, maintenance, and social activities was presented. Precociousness and apparently minimal or nonexistent parental care leads to the conclusion that, except for minor refinements in topography and stimulus control, the behavior of reptiles, in contrast to that of birds and mammals, does not really "develop" postnatally. Problems with this viewpoint were outlined. Changes associated with age, such as size, body pattern and coloring, physiological mechanisms, habitat, home range, availability and appropriateness of food, vulnerability to predators, and social behavior, may necessitate remarkable, but often subtle, physiological and behavioral alterations. Studies that focus directly on these aspects are needed.

Acknowledgments

Support for research and writing came from NSF and NIMH grants and a John Simon Guggenheim Foundation Fellowship. I thank my students and colleagues for making unpublished material available to me, and The Rockefeller University for hospitality and intellectual ambience. Hugh Drummond, Harry W. Greene and William S. Verplanck made helpful suggestions on the original manuscript.

Note

1. See Burghardt (1970) for references not cited in this paragraph.

References

Appleby, L. G. 1971. British snakes. London: John Baker.

Arnold, S. J. 1977. Polymorphism and geographic variation in the feeding behavior of the garter snake. Science 197:676–678.

Bellairs A. d'h. 1970. The life of reptiles. Vol. 2. New York: Universe.

Bogert, C. M. 1941. Sensory cues used by rattlesnakes in their recognition of ophidian enemies. Ann. N. Y. Acad. Sci. 41:329–343.

Burghardt, G. M. 1964. Effects of prey size and movement on the feeding behavior of the lizards *Anolis carolinensis* and *Eumeces fasciatus*. Copeia 1964:576–578.

Burghardt, G. M. 1967. The primacy effect of the first feeding experience in the snapping turtle. Psychon. Sci. 7:383–384.

Burghardt, G. M. 1970. Chemical perception in reptiles. Pages 241–308 In Communication by chemical signals, eds., J. W. Johnston, Jr., D. G. Moulton, and A. Türk. New York: Appleton-Century-Crofts.

Burghardt, G. M. 1973. Chemical release of prey attack: extension to naive newly hatched lizards, *Eumeces fasciatus*. Copeia 1973:178–181.

Burghardt, G. M. 1975. Chemical prey preference polymorphism in newborn garter snakes, *Thamnophis sirtalis.* Behaviour 52:202–225.

Burghardt, G. M. 1977a. Learning processes in reptiles. In The biology of the reptilia. Vol. 7. eds., C. Gans and D. W. Tinkle, pp. 555–681. New York: Academic Press.

Burghardt, G. M. 1977b. Of iguanas and dinosaurs: social behavior and communication in neonate reptiles. Am. Zool. 17:177–190.

Burghardt, G. M. 1977c. Ontogeny of communication. In How animals communicate, ed., T. A. Sebeok, pp. 67–93. Bloomington: Indiana Univ. Press.

Burghardt, G. M. 1977d. The ontogeny, evolution, and stimulus control of feeding in humans and reptiles. In The chemical senses and nutrition, eds., M. Kare and O. Maller, pp. 253–275. New York: Academic Press.

Burghardt, G. M., H. W. Greene, and A. S. Rand. 1977. Social behavior in hatchling green iguanas: life at a reptile rookery. Science 195:689–691.

Burghardt, G. M., and E. H. Hess. 1966. Food imprinting in the snapping turtle, *Chelydra serpentina.* Science 151:108–109.

Burghardt, G. M., and C. H. Pruitt. 1975. Role of the tongue and senses in feeding of naive and experienced garter snakes. Physiol. Behav. 14:185–194.

Bustard, H. R. 1967. Mechanism of nocturnal emergence from the nest in green turtle hatchlings. Nature 214:317.

Carpenter, C. C. 1977. Communication and displays of snakes. Am. Zool. 17:217–223.

Carr, A., and H. Hirth. 1961. Social facilitation in green turtle siblings. Anim. Behav. 9:68–70.

Chiszar, D., T. Carter, L. Knight, L. Simonson, and S. Taylor. 1976. Investigatory behavior in the plains garter snake (*Thamnophis radix*) and several additional species. Anim. Learn. Behav. 4:273–278.

Chiszar, D. and C. W. Radcliffe. 1977. Absense of prey–chemical preferences in newborn rattlesnakes (*Crotalus cerastes, C. enyo.* and *C. viridis*). Behav. Biol. 21:146–150.

Chiszar, D., K. Scudder, and L. Knight. 1976. Rate of tongue flicking by garter snakes (*Thamnophis radix haydeni*) and rattlesnakes (*Crotalus* var. *viridis, Sistrurus catenatus tergeninus* and *S. c. edwardsi* during prolonged exposure to food odors. Behav. Biol. 18:273–283.

Cooper, W. E., Jr. 1971. Display behavior of hatchling *Anolis carolinensis* Herpetologica 27:498–500.

Crews, D. P. 1975. Psychobiology of reptilian reproduction. Science 189:1059–1065.

Crook, J. H. 1965. The adaptive significance of avian social organizations. Symp. Zool. Soc. Lond. 14:181–218.

Czaplicki, J. A. 1975. Habituation of the chemically elicited prey-attack response in the diamond-backed water snake, *Natrix rhombifera rhombifera.* Herpetologica 31:402–409.

Decker, J. D. 1967. Motility of the turtle embryo, *Chelydra serpentina.* Science 157:952–954.

Desmond, A. J. 1976. The hot-blooded dinosaurs. New York: Dial.

Dullemeijer, P. 1961. Some remarks on the feeding behavior of rattlesnakes. K. Ned. Akad. Wet. Versl. Gewone Vergad. Afd. Natuurkd. Série C 64:383–396.

Dundee, H. A., and M. C. Miller, III. 1968. Aggregative behavior and habitat conditioning by the ringneck snake, *Diadophis punctatus arnyi*. Tulane Stud. Zool. Bot. 15:41–58.

Evans, L. T. 1959. A motion picture study of maternal behavior of the lizard, *Eumeces obsoletus* Baird and Girard. Copeia 1959:103–110.

Evans, L. T. 1961. Structure and behavior in reptile populations, in Vertebrate speciation, ed., W. F. Blair, pp. 148–178. Houston: Univ. Texas.

Fitch, H. S. 1975. A demographic study of the ringneck snake (*Diadophis punctatus*) in Kansas. Misc. Publ. Univ. Kans. Mus. Nat. Hist. 62:1–53.

Fitch, H. S., and R. W. Henderson. 1977. Age and sex differences in the Ctenosaur (*Ctenosaura similis*). Milw. Public Mus. Contr. Biol. Geol. No. 11. 11 pp.

Froese, A. D., and G. M. Burghardt. 1974. Food competition in captive juvenile snapping turtles, *Chelydra serpentina*. Anim. Behav. 22:735–740.

Gans, C. 1961. The feeding mechanism of snakes and its possible evolution. Am. Zool. 1:217–227.

Gans, C. 1976. Questions in crocodilian physiology. Zool. Afr. 11:241–248.

Garrick, L. D., and J. W. Lang. 1977. Social signals and behavior of adult alligators and crocodiles. Am. Zool. 17:225–239.

Gehlbach, F. R. 1970. Death-feigning and erratic behavior in leptotyphlopid, colubrid, and elapid snakes. Herpetologica 26:24–34.

Gove, D., and G. M. Burghardt. 1975. Responses of ecologically dissimilar populations of the water snake Natrix S. sipedon to chemical cues from prey. J. Chem. Ecol. 1:25–40.

Greenburg, N., and D. Crews. 1977. Introduction to the symposium: social behavior in reptiles. Am. Zool. 17:153–154.

Greene, H. W. 1973. Defensive tail displays in snakes and amphisbaenians. J. Herp. 7:143–161.

Greene, H. W. 1976. Scale overlap, a directional sign stimulus for prey ingestion by ophiophagous snakes. Z. Tierpsychol. 41:113–120.

Greene, H. W. 1977. Phylogeny, convergence, and snake behavior. Unpublished Ph.D. dissertation, University of Tennessee, Knoxville, Tennessee.

Greene, H. W., and G. M. Burghardt. 1978. Behavior and phylogeny: constriction in ancient and modern snakes. Science. 200:74–77.

Greene, H. W., G. M. Burghardt, B. A. Dugan, and A. S. Rand. Predation and the defensive behavior of green iguanas. J. Herpetal. 12:169–176.

Greene, H. W., and J. A. Campbell. 1972. Notes on the use of caudal lures by arboreal green pit vipers. Herpetologica 28:32–34.

Hadley, W. F., and C. Gans. 1972. Convergent ontogenetic change of color pattern in *Elaphe climacophora* (Colubridae: Reptilia). J. Herpetal. 6:75–78.

Hahn, P., and O. Koldovsky. 1966. Utilization of nutrients during postnatal development. Oxford: Pergamon Press.

Hall, C. R. 1965. Social organization of the old-world monkeys and apes. Symp. Zool. Soc. Lond. 14:265–289.

Heatwole, H., and E. Davison. 1976. A review of caudal luring in snakes with notes on its occurrence in the Saharan sand viper, *Cerastes vipera.* Herpetologica 32:332–336.

Heckrotte, C. 1962. The effect of the environmental factors in the locomotory activity of the plains garter snake (*Thamnophis radix radix*) Anim. Behav. 10:193–207.

Herzog, H. A., Jr., and G. M. Burghardt. 1977. Vocal communication signals in juvenile crocodilians. Z. Tierpsychol. 44:294–304.

Hughes, A., S. V. Bryant, and A. d'h. Bellairs. 1967. Embryonic behavior in the lizard, *Lacerta vivipara.* J. Zool. Lond. 153:139–152.

Jerison, H. J. 1973. Evolution of the brain and intelligence. New York: Academic Press.

Jones, S. D. 1974. Investigations on the circadian nature of activity rhythms in newborn and older snakes. Unpublished M.A. thesis, University of Tennessee, Knoxville.

Kalmus, H. 1965. Origins and general features. Symp. Zool. Soc. Lond. 14:1–12.

King, J. A. 1961. Development and behavioral evolution in *Peromyscus.* In Vertebrate speciation, ed., W. F. Blair, pp. 122–147. Austin: Univ. Texas.

Klein, J., and M. S. Loop. 1975. Headfirst prey ingestion by newborn *Elaphe* and *Lampropeltis.* Copeia 1975:366.

Lazell, J. D., Jr., and N. C. Spitzer. 1977. Apparent play behavior in an American alligator. Copeia 1977:188.

Loop, M. S., and L. G. Bailey. 1972. The effect of relative prey size on the ingestion behavior of rodent-eating snakes. Psychon. Sci. 21:189–190.

Loop, M. S., and S. A. Scoville. 1972. Response of newborn *Eumeces inexpectatus* to prey-object extracts. Herpetologica 28:254–256.

Lorenz, K. Z. 1965. The evolution and modification of behavior. Chicago: Univ. Chicago Press.

Mahmoud, I. Y., and N. Lavenda. 1969. Establishment and eradication of food preferences in red-eared turtles. Copeia 1969:298–300.

Maynard Smith, J. 1977. Parental investment: a prospective analysis. Anim. Behav. 25:1–9.

Meyer, G. R. M. 1977. Alligator behavior, ecology, and populations in Okefenokee marshes. Paper read at the Gainesville, Fla. meeting of the Amer. Soc. Ichth. Herp. June.

Morris, D. 1967. The naked ape. New York: McGraw-Hill.

Mrosovsky, N. 1968. Nocturnal emergence of hatchling sea turtles: control by thermal inhibition of activity. Nature 220:1338–1339.

Naulleau, G. 1966. La biologie et les comportements prédateurs de Vipera aspis au laboratoire et dans la nature. Thesis. Paris: P. Fanlac.

Noble, G. K., and H. J. Clausen. 1936. The aggregation behavior of *Storeria dekayi* and other snakes with special reference to the sense organs involved. Ecol. Monogr. 6:269–316.

Noble, G. K., and E. R. Mason. 1933. Experiments on the brooding habits of the lizards *Eumeces* and *Ophisaurus*. Amer. Mus. Nov. 619(1):1–29.

Oppenheim, R. W. 1973. Prehatching and hatching behavior: a comparative and physiological consideration. Behav. Embryology 1:163–244.

Pough, F. H. 1977. Ontogenetic change in blood oxygen capacity and maximum activity in garter snakes (Thamnophis sirtalis). J. Comp. Physiol. B. 116:337–345.

Pope, C. H. 1961. The giant snakes. New York: Knopf.

Prestrude, A. M., and F. T. Crawford. 1970. Tonic immobility in the lizard, *Iguana iguana*. Anim. Behav. 18:391–395.

Raun, G. G. 1962. Observations on behavior of newborn hog-nosed snakes, *Heterodon p. platyrhinos*. Tex. J. Sci. 14:3–6.

Seba, A. 1734. Locupletissimi Rerum Naturalium Thesauri, Vol. I. Amsterdam: Janssonio-Waesbergios, J. Wetstenium, and Gul. Smith.

Simon, C. A., and G. A. Middendorf. 1977. Territory establishment by juvenile lizards (*Sceloporus jarrovi*). Anim. Behav. (In press.)

Singer, B., and S. I. Spilerman. 1976. Mathematical representation of development theories. Inst. Res. Poverty discussion papers. Univer. Wisconsin, Madison.

Smith, A. K. 1974. Incidence of tail coiling in a population of ringneck snakes (Diadophis punctatus.) Trans. Kans. Acad. Sci. 77:237–238.

Smith, M. A. 1943. Fauna of British India. Reptilia and amphibia, Vol. 3. London: Serpentes, Taylor and Francis.

Stamps, J. A. A field study of the ontogeny of social behavior in the lizard *Anolis aeneus*. Behaviour. (In press.)

Thomson, J. A. 1926. The gospel of evolution. New York: Putnam.

Tinkle, D. W. 1972. The role of environment in the evolution of life history differences within and between lizard species. In A symposium on eco-systematics, eds., R. T. Allen and F. C. James, pp. 77–100. University of Arkansas Museum Occasional paper No. 4.

Tubbs, A. A., and G. W. Ferguson. 1976. Effects of crowding on behavior, growth and survival of juvenile spiny lizards. Copeia 1976:821–823.

Vinegar, A. 1974. Evolutionary implications of temperature induced abnomalies of development in snake embryos. Herpetologica 30:72–74.

FEATHERED REPTILES

Members of the class Aves have been well represented in classical and modern studies of behavioral ontogeny. Furthermore, C. O. Whitman and O. Heinroth, independently, first recognized the value of using behavioral phenotypes in studying taxonomic relationships based on their observation of birds. The chapter by Marc Bekoff considers the development of comfort behavior in Adélie penguins. He is mainly concerned with progressive changes in development of these birds, and demonstrates the applicability of two multivariate techniques to studies of behavioral ontogeny. Because of the difficult field conditions under which the penguins were observed, individual differences in behavior were not stressed. However, as pointed out by Colin Beer in comments presented at the symposium, the possibility exists that while a group of animals may not show consistent sequential patterning in grooming when viewed collectively, each individual may show some behavioral consistencies. Howard Hoffman discusses the results of laboratory studies of imprinting. He suggests that some features of imprinting are the product of well-known behavioral processes such as (1) the tendency to exhibit innate filial reactions to certain aspects of parental stimulation, (2) classical conditioning, and (3) the development of fear responses. Imprinting is a most controversial area, but the comments by Eckhard Hess illustrate the convergence beginning to be found in workers approaching the phenomenon with divergent views; note also, however, the exchange between H. Hoffman and K. Immelmann in *American Scientist* (1977, 65:398). The reader should also be made aware of recent advances in studies of imprinting and social attachment by P. P. G. Bateson, G. Gottlieb, K. Immelmann, J. L. Shapiro, B. Tschanz, and others. In the last chapter in this section, Donald Kroodsma stresses the importance of learning in song development by various songbirds. The questions posed in the second part of his title reflect his areas of concentration. Intra- and interspecific variability seem to be the "name of the game."

Chapter Nine

A Field Study of the Development of Behavior in Adélie Penguins: Univariate and Numerical Taxonomic Approaches

MARC BEKOFF

University of Colorado
Department of Environmental, Population
and Organismic Biology
Boulder, Colorado 80309

> *Of the problems confronting the student of behaviour*
> *that demand attention in the next decade or so, none is*
> *more pressing than the understanding of behavioural*
> *development* (Marler, 1975, p. 254).

Introduction

Despite considerable interest in avian comfort behavior and its relationship to such topics as ritualization, displacement behavior, social communication, and the evolution of behavior in general (e.g., Kortlandt, 1940; Andrew, 1956; von Iersel and Bol, 1958; McKinney, 1965; Delius, 1969; Ainley, 1970, 1974; Baerends, 1975; Borchelt, 1975), little still is known about comfort behavior in birds from a developmental perspective (e.g., Kortlandt, 1940; Fabricius, 1945; Nice, 1962; Kruijt, 1964; Dawson and Siegel, 1967; Horwich, 1969; Borchelt and Overmann, 1974; Spurr, 1975a; Bergmann, 1976; Brown, 1976; Mergler, 1976; Bekoff *et al.*, 1978; Borchelt, 1977). With respect to Adélie penguins, *Pygoscelis adéliae,* only some general information is available concerning the ontogeny of comfort (and other) behavior (Taylor, 1962; Penney, 1968; Thompson, 1974; Spurr, 1975a). This study was undertaken in order to answer questions dealing with processes of behavioral development in Adélie chicks and to compare chick and adult behavioral patterns. Specifically, we were concerned with (1) the ways in which individual motor acts became incorporated into behavioral sequences during ontogeny, (2) the ways in which the behavior of chicks changed during early ontogeny (before fledging), and (3) the relationship between behavioral variability and age. Comfort activities are a good group of behaviors for a developmental analysis because they appear early in life, are readily identified as individual acts that change little (if at all) in appearance during ontogeny (Thompson, 1974; Spurr, 1975a) and are repeated often.

In addition to the objectives stated above, this chapter then demonstrates the use of some quantitative procedures in ethological research. In order to perform rigorous quantitative analyses, two techniques from the numerical taxonomy literature (Jardine and Sibson, 1971; Sneath and Sokal, 1973; Clifford and Stephenson, 1975) were employed, namely, principal components and discriminant function analyses. Although these multivariate techniques are only beginning to be applied to ethological (Dudziński and Norris, 1970; Schnell, 1973; Michener, 1974; Bekoff et al., 1975; Aspey and Blankenship, 1976, 1977; Mihok, 1976; Bekoff, 1977a; Ringo and Hodosh, 1977; Sparling and Williams, 1977; Thelen and Farish, 1977) or ecological (James, 1971; Connor and Adkisson, 1976, 1977; Harner and Whitmore, 1977) problems, they provide powerful tools with which to attack a variety of questions (see also John et al., 1977, and Kertesz and Phipps, 1977) and their use is entirely appropriate. With respect to their application to problems of behavioral ontogeny, both principal components (see Miller et al., 1977) and discriminant function analyses can provide detailed information about progressive changes in behavior over time. In addition, these techniques are helpful in highlighting both individual and intergroup differences and then pinpointing the source of the differences (i.e., identifying the variables that are most important in discriminating between the groups that are being compared).

Methods

Identified penguin chicks of known age (dye–marked [picric acid] offspring of banded adults) and adults (over two years of age) were studied at the Cape Crozier Rookery, Ross Island, Antarctica, during the austral summer, 1974–1975. Chicks were observed from hatching until the last week of January, shortly before fledging and going to sea for the first time. The animals were observed through binoculars or with the naked eye. Sequences of behavior and a time indication every 15 s were read into a cassette tape recorder and later transcribed. All observations were made when winds were light, since wind conditions may affect the performance of some comfort movements (Ainley, 1974). Temperatures ranged from −7° to 0° C. Furthermore, the feathers of the adult birds were dry (for a comparison with oiling by penguins who had just exited from the water, see Ainley, 1970, 1974).

The following six groups of individuals were observed:

1. Chicks aged 7–13 d, non-oiling (n = 12)
2. Chicks aged 14–20 d, non-oiling (n = 15)
3. Chicks aged 21–28 d, non-oiling (n = 7)
4. Adults, non–oiling (n = 26)
5. Chicks, oiling (n = 6)
6. Adults, oiling (n = 15)

Non-oiling sequences (Groups 1–4) were those in which no bill contact was made with the uropygial (oil) gland. *Oiling* sequences (Groups 5 and 6) were those in which bill contact was made with the uropygial gland and oil was distributed over the dry feathers.

Due to a number of factors (e.g., mortality, inactivity, weather conditions, visibility), it was rarely possible to observe the same chicks throughout the course of the study. Therefore, for analyses, it was assumed that the samples were independent.

Actions Observed

The actions observed and their day of first occurrence in adult form (see also Spurr, 1975a) are listed in Table 1. The 16 variables listed in Table 2 were used in the multivariate analyses of oiling sequences. For analyses of non-oiling sequences, only the 11 starred (*) behaviors (Table 2) were used because the other five variables were invariant (equaled zero) for some of the groups of chicks. Some behaviors associated with oiling by adults, and wing–stretching by a 19 d chick, are shown in Figure 1.

Multivariate Techniques

The two techniques that were used in the analysis of our data were principal components analysis (PCA) and discriminant function analysis (DFA). PCA is a type of factor analysis that reduces the dimensionality of

TABLE 1. List of actions (and code) observed and day of first occurrence

ACTION*	DAY
Yawn (Y)	1
Head-shake (HSH)	2
Leg-stretch (LS) (L,R,B)	5
Rapid/Slow-wing-flap (RWF, SWF)**(L,R,B)	6
Sneeze (SN)	6
Wing-extend (WE) (L,R,B)	7
Nibble-preen of various body regions	7 (see below)
Wing-rub (WR; Fig. 1c) (L,R)	7
Phase II wing-stretch (PII)	9
Tail-wag (TW)	9
Ruffle-shake (RS)	9
Bite-wing (BWg) (L,R)	9
Foot-shake (FS)	10
Wing-stretch with yawn (WSY)***(L,R,B)	12
Wing-stretch (both, with body-shake) (WSBS)	13
Head-scratch (HSC) (L,R)	14
Rapid-wing-flap leading to Phase II wing-stretch	14
Wing-shake (WSh) (L,R,B)	15
Bill-to-wing-edge, (BW; Fig. 1b) (L,R)	16
Shoulder-rub (SR) (L,R)	33
Oil (Fig. 1a) (L,R)	35
Preening (L,R):	
Breast (BR)	7
Belly (BE)	7
Back (BA)	9
Side/Flank (SFL)	10
Wing	12
Shoulder (SH)	13
Leg	14
Cloaca (CL)	15
Tail-base (TB)	21

* The following actions were not observed in chicks: body-shake (BS), neck-stretch (NS), ruffle-feathers only (RF).

** The mean rate of rapid-wing-flapping for adults (n=27) was 5.5 cycles/s and that for chicks (n=18) was 3.7 cycles/s (determined by movie analysis; $t = 2.12$, $df = 43$, $p < 0.05$).

*** Can occur without yawn; Fig. 1d.

Source: For full descriptions see Ainley (1974) and text.

L = left; R = right; B = both left and right.

TABLE 2. Group means of 16 variables used in principal components and discriminant function analyses of comfort sequences in six groups of Adélie penguins

GROUP	D***	R*	SS*	SA*	SSA*	BW	WR*	SR	BR*	BE*	BA*	HSH*	SFL*	OIL	LEG	SH
7-13 d chicks***	0.6	3.3	65.5	48.3	25.9	—	8.2	—	30.0	11.7	12.2	5.4	6.1	—	—	—
14-20 d chicks***	6.3	4.5	75.3	27.6	18.1	—	4.9	—	21.6	7.2	9.8	7.3	9.4	—	—	—
21-28 d chicks***	8.4	4.7	77.6	21.3	12.7	—	3.7	—	24.8	7.4	5.4	11.7	8.6	—	—	—
Adults***	7.1	6.2	83.7	12.8	10.3	—	13.8	—	15.9	3.6	5.8	3.8	13.6	—	—	—
Chicks, oil	16.5	6.9	70.0	14.1	13.5	2.6	4.8	12.9	12.0	5.6	16.0	7.9	12.1	3.7	6.1	12.7
Adults, oil	9.4	8.5	83.3	8.8	7.7	5.0	14.3	9.8	13.6	4.8	5.2	2.8	14.7	9.1	2.2	6.3

*All 16 variables were used in the analyses of oiling sequences. Only the 11 asterisked (*) behaviors were used in the analyses of non-oiling sequences.

**D = duration; R = rate; SS = individual remained in the same area of the body during a transition from one act to the next; SA = individual remained on the same side of the body during a transition from one act to the next; SSA = individual remained on the same side of the body and in the same area during a transition from one act to the next; see Table 1 for code for other variables.

***Non-oiling sequences.

Note: The numbers represent the relative percentage of occurrence of the behaviors, except for D (duration) and R (rate), which are recorded in minutes and number of acts per minute, respectively.

FIG. 1. *Behaviors associated with the distribution of oil from the uropygial gland (a–c). (a) Oiling by an adult Adélie penguin. The bird reaches over a shoulder and pinches the base of the gland feather tuft with its mandibles. This causes the expulsion of oil. The penguin then draws the tuft through its mandibles and pushes its chin across it with a forward movement of the head. The oil gland becomes functional between 30–33 d of age. The first dry-oil sequence was observed on day 35. (b) After gathering oil, the bird grasps the top edge of a wing between its mandibles (almost always on the same side of the body to which it turned to make*

a multivariate data set to fewer (usually two or three) dimensions. It can be used to distinguish between individuals in a heterogeneous population (Aspey and Blankenship, 1976, 1977) based on age, size, weight, social behavior, etc. The first principal axis (x) or eigenvector accounts for the largest amount of sample variance; the second principal axis (y) accounts for the second largest amount of variance; etc. In the present study, the PCA and DFA were performed using NTSYS (Numerical Taxonomy System) programs and the three-dimensional plots were drawn using VUDATA (Kaltenhauser and Welner, 1975), a program available at the University of Colorado Computing Center (a program for 3–D graphics soon will be readily accessible; Aspey *et al.*, 1977).

DFA is a method by which various characters from previously identified populations are compared to one another. The resulting axis, or discriminant function, is calculated from pooled variances and covariances of the characters. *The discriminant function defines the axis that maximally separates the groups and identifies the variables most important in defining the discrimination.* This process relies not only on differences between group means, but also on the ways in which variables covary with one another. Therefore, eyeballing the means presented in Table 2 does not necessarily highlight those characters that are the best discriminators of the populations being compared. The larger the discriminant function coefficient, the better a given character is for discriminating between the groups. Simply, DFA allows an investigator to pinpoint those characters that best discriminate the populations being compared. In addition, when the results are plotted on a linear axis (Fig. 5 and 6), it is possible to fit other groups on the axis to determine where they fall with respect to the original populations.

---

contact with the uropygial gland; Ainley, 1974; Bekoff, Ainley, and Bekoff, 1978)
and draws the bill along the wing's edge towards its body. This is called bill–to-wing–edge (BW), *and is the first step in the transfer of oil to the head.* (c) *After performing* BW, *the penguin then typically performs a wing-rub* (WR) *to transfer oil from the wing-edge to the head. The wing is raised from the side to project straight out or slightly upwards and backwards. The top, side, or back of the head is then rubbed on the wing's leading edge.* (d) *Wing–stretching* (WS) *by a 19-day old chick. When performed by adults, the legs and body are stretched and the feathers are sleeked. The neck is turned upward, the tail is stretched downward, and yawning often occurs. The wings are moved behind the bird until the tips almost touch one another. Note that gray down is being replaced with white feathers.*

Results

The results section will be divided into two sections. First, I shall briefly review some general findings (details may be found in Bekoff *et al.*, 1978). Second, I shall discuss the results of the multivariate analyses.

GENERAL RESULTS

Age at First Performance of Comfort Behaviors

These data are presented in Table 1. By day 21, all actions except shoulder-rub (SR) and oil had been observed. No oiling was observed prior to 35 d. The oil gland became functional between 30–33 d and the feather tuft extending out of the gland did not reach full length until 34–38 d.

Comfort Sequence Analysis

Data for duration, rate, relative frequencies of occurrence of various actions, and continuity with respect to side and area of the body will be discussed below. With respect to the temporal distribution of comfort movements, only oiling and bill–to–wing–edge (BW) showed nonrandom distribution for adults (Bekoff *et al.*, 1978). For both adults and chicks, oiling showed nonrandom distribution in the first one–third of oiling sequences. BW showed a similar distribution for adults and chicks, although in chicks, the distribution was not quite significant. The distribution of all other comfort movements was random (Bekoff *et al.*, 1977).

Two-Act Transitions

Data for all 6 different groups were cast into six 15 × 15 transition matrices (see Ainley, 1974 and Bekoff *et al.*, 1978 for the way in which actions were combined). The degree of similarity between the six matrices were analyzed by calculating Pearson's product-moment correlation coefficients (*r*) for transition probabilities in corresponding cells (cells representing the same two-act transition; see also Pearson and Parker, 1973). These results are presented in Table 3. The most significant correlations were derived when comparing the two groups of non-oiling chicks and non–oiling by adults and dry–oiling by adults. Non–oiling by chicks and dry–oiling by chicks were not significantly correlated. It is of interest to note that dry–oiling by chicks and dry–oiling by adults were significantly correlated, however, the relationship was not as strong as that for non–oiling by adults and dry–oiling by adults.

TABLE 3. Correlations (r) between corresponding cells of comfort behavior transition matrices

	CHICKS (21–28 DAYS)	NON-OIL ADULTS	DRY-OIL CHICKS	DRY-OIL ADULTS
Chicks (14–20 d)	0.64***,[+]	0.43**,[+]	X	X
Chicks (21–28 d)	—	0.31*,[++]	X	X
Non–oil adults	—	—	—	0.73***,[++]
Dry–oil chicks	—	—	—	0.20**,[+++]

X = non-significant
* = $p < 0.05$
** = $p < 0.01$
*** = $p < 0.001$
[+]$n = 56$
[++]$n = 42$
[+++]$n = 176$

TABLE 4. The variability in the coupling of two-act transitions during comfort behavior by Adélie penguins of different ages

	GROUP	COEFFICIENT OF VARIATION
Non–oil chicks	7–13 d ($n = 9$)	90.10
	14–20 d ($n = 50$)	84.16
	21–28 ($n = 25$)	76.59
	Non–oil adults ($n = 88$)	96.66
Dry–oil	Chicks ($n = 89$)	109.55
	Adults ($n = 132$)	143.76
Wet oil*	Adults ($n = 130$)	135.03

* From Ainley (1970)

Variability of Transitions

The variability in the coupling (transition probability) of all two-act transitions that occurred greater than 15 times was calculated for each of the six groups of birds (Table 4). Since we could not assume that our data were stationary (Fentress, 1972; Slater, 1973, 1975; Dow et al., 1976; Bekoff, 1977a), information–theoretic measures of variability could not be employed. Instead, coefficients of variation were calculated. For three non–oil groups, there were no significant differences in the coefficients of variation (there were insufficient data for non–oiling by the 7–13-day chicks). However, dry–oiling by chicks was significantly more variable than either non–oiling by 14–20-day chicks ($c = 2.16$, $df = 137$, $p < 0.05$, see Dawkins and Dawkins, 1973 for the formula for c) or non–oiling by 21–28-day chicks ($c = 2.43$, $df = 112$, $p < .02$). Dry–oiling by adults was significantly more variable than dry–oiling by chicks ($c = 2.83$, $df = 219$, $p < .01$). Finally, wet–oiling by adults (i.e., the birds had just come out of the water, see Ainley, 1974) was not significantly more variable than dry–oiling by adults.

The Order of Appearance of Comfort Movements During Ontogeny and the Construction of Sequences Later in Life

We also analyzed the relationship between the order of appearance of individual comfort movements during ontogeny and the order in which they were incorporated into sequences later in life. Of the 58 two-act transitions that were analyzed, there were only 29 instances in which the first act of the two–act sequence appeared earlier in life.

PRINCIPAL COMPONENTS AND DISCRIMINANT FUNCTION ANALYSES

Principal Components Analyses

The summarized means for the six groups of animals are presented in Table 2. Four principal components analyses were run (see below) and interesting descriptive trends were detected, indicating progressive development from one age group to another. In all cases, Factor I was related to age (gross behavioral development), however, it was not possible to label the other factors (this is not uncommon; Sneath and Sokal, 1973, p. 247).

(1) *Non–oiling chicks* (*Groups 1–3, n = 34*): These results are presented in Figure 2, in which factor scores for Factors I and II (x- and y-axes) are plotted for each group. The three groups were not separated on the z-axis (Factor III). Factor I accounted for 28.27% of the variance (eigenvalue = 3.11), II for 16.16% (1.83), and III for 15.09% (1.66). The groups are nicely separated on the x-axis (Factor I). The 21–28-day chicks (oldest) are almost totally contained ($5/7 = 71\%$) within the plot of the 14–20-day chicks (middle–aged); however, there is good separation between these two groups, with the 21–28-day chicks clustering on the right of the plot. The 7–13-day chicks overlapped with the 14–20-day chicks in $2/12$ of the cases (17%) and with the 21–28-day chicks only $1/12$ of the times (8%). There was 36% ($8/22$) overlap between the 14–20- and 21–28-day chicks. The variables having their highest loadings on Factor I were duration (+0.61), same side (+0.82), same area (−0.75), breast (−0.73), and side/flank (+0.64). The variables loading high on Factor II

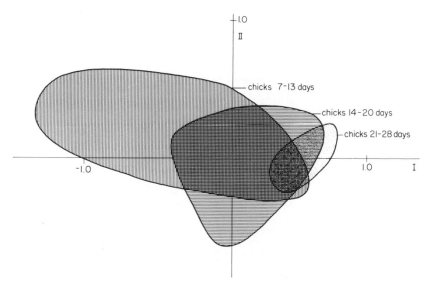

FIG. 2. *A two-dimensional principal components plot of factor scores for three groups of non-oiling chicks. Note the age progression as one moves across the axes (see text).*

were breast (+0.63), and head–shake (−0.94). Only wing–rub (−0.68) had a high loading (>±0.50) on Factor III.

(2) *Non–oiling chicks and non–oiling adults (Groups 1–4, n = 60)*: These data are presented in Figure 3, top. I have presented only the 2–D plot because the 3–D plot was rather cramped and not notably different. Factor I accounted for 28.07% of the variance (eigenvalue = 3.08), II for 16.16% (1.78) and III for 14.23% (1.56). Once again, the 7–13-day chicks were nicely separated from the other groups. There was 17% (2/12) overlap between the youngest chicks and the adults, 27% (6/22) overlap between the 14–20-day and 21–28-day chicks, and 36% (12/33) overlap between the 21–28-day chicks and the adults. Reading the 2–D plot from left to right, there is a progression from the youngest chicks to the adults, with good separation on the *x*–axis (Factor I) and slight separation on the *y*–axis (Factor II). Duration (+0.52), same side (+0.84), same area (−0.73), breast (−0.77), and side/flank (+0.62) loaded high on Factor I. Duration (−0.52), wing–rub (+0.57), and head-shake (−0.57) loaded high on Factor II. Only back (+0.58) had a high loading (>±0.50) on Factor III. It is interesting that breast and side/flank are separated from back.

(3) *Oldest non–oiling chicks (21–28 d) and non–oiling adults (Groups 3 and 4, n = 33)*: These data are presented in Figure 3, bottom. Factor I accounted for 25.10% of the variance (eigenvalue = 2.76), II for 20.77% (2.28), and III for 19.13% (2.10). The two groups were best separated on

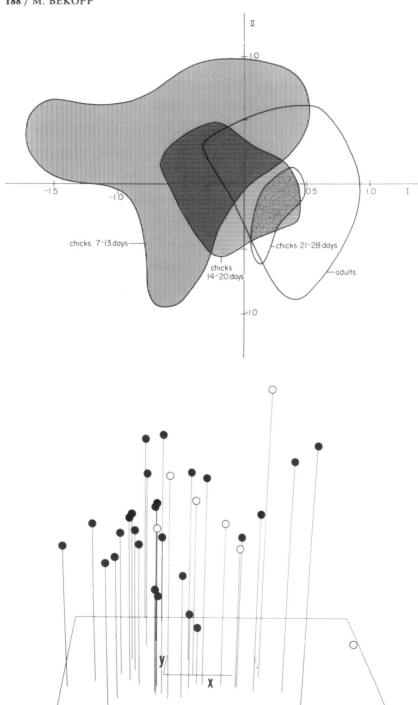

the x–axis (overlap = 39%). Those variables loading highest on Factor I were duration (+0.56), rate (−0.86), same side (−0.63), wing-rub (−0.68), and head–shake (+0.57). Same area (+0.77), breast (+0.63), and side/flank (−0.64) loaded highest on Factor II. Once again, back preening (+0.62) loaded highest on Factor III (no other variable had a loading of >±0.50) and was separated from breast and side/flank.

(4) *Oiling by chicks and adults (Groups 5 and 6, n = 21):* These data are presented in Figure 4. Factor I accounted for 30.76% of the variance (eigenvalue = 4.92), II for 18.78% (3.01), and III for 10.23% (1.65). There was no overlap between the groups on the x–axis with fair separation between them on the y–axis. Duration (+0.66), rate (−0.53), same side

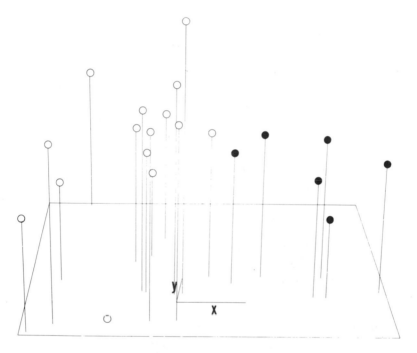

FIG. 4. *A 3–D plot of individual factor scores for oiling by chicks (black circles) and oiling by adults (open circles). Note the total separation of the 2 groups of the x–axis (see text).*

FIG. 3. Top: *A 2–D plot for three groups of non–oiling chicks and non–oiling adults. Note once again the age progression as one goes from left to right (see text). Bottom: A 3–D plot of individual factor scores for non–oiling by 21–28 d chicks (open circles) and non–oiling adults (black circles; see text).*

(−0.71), same side and area (+0.50), wing–rub (−0.65), shoulder–rub (−0.55), back (+0.86), leg (+0.60), and shoulder (+0.88) loaded high on Factor I. Shoulder–rub (+0.66), breast (+0.65), head–shake (+0.81), side/flank (−0.67), and oil (+0.61) loaded high on Factor II. Bill–to–wing–edge (+0.72) and belly (−0.63) loaded highest on Factor III. The following trends were interesting: (i) back and breast were separated, (ii) those behaviors associated with oiling and the distribution of oil (oil, shoulder–rub, bill–to–wing–edge) loaded highest on Factors II and III, and (iii) shoulder–rub and wing–rub (also associated with the distribution of oil) had negative loadings on Factor I.

Discriminant Function Analyses

Principal components analyses provided general pictures of how the age groups could be separated. Discriminant function analyses were next performed in order to determine those variables that contributed the most toward discriminating between the age groups in pair–wise comparisons. When taken alone, some characters are useful in differentiating between the groups. However, in pooled comparisons, such as DFA, it is not always clear which variables are the best discriminators between the two groups. Four DFA were run as follows.

(1) *Youngest chicks (7–13 d) versus oldest chicks (21–28 d), non-oiling.* These data are presented in Figure 5, top. There was no overlap between the groups. Rate (discriminant weight = 0.37) and duration (discriminant weight = 0.18) were the only variables that could be used to separate the two groups on this axis, and it should be noted that their discriminant weights are not very large (none of the weights for the other nine variables were greater than 0.04). The sequences performed by the 7–13-day chicks were of shorter duration and they also performed fewer acts per minute (Table 2). Although none of the other characters could be used to separate the two groups when pooled with one another, when taken alone, some of them are useful in comparing the groups (Bekoff *et al.*, 1978). When the 14–20-day chicks and adults were fit onto this linear axis, there was no overlap with the 7–13-day chicks. However, there was overlap between the 14–20-day chicks, the 21–28-day chicks, and the adults. Reading Figure 5 (top) from right to left, there is a notable age trend from the youngest to the oldest individuals.

(2) *Youngest chicks (7–13 d) versus adults, non-oiling.* There was no overlap between these two groups (Figure 5, bottom). Although the behavior of the youngest chicks is very different from the behavior of adults (see above and also Bekoff *et al.*, 1978), once again only rate (discriminant weight = 0.41) and duration (discriminant weight = 0.14)

Chicks 7-13 days / chicks 21-28 days:

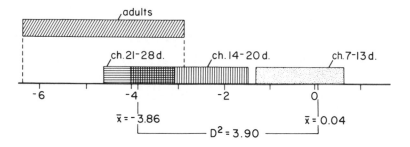

Chicks 7-13 days / adults:

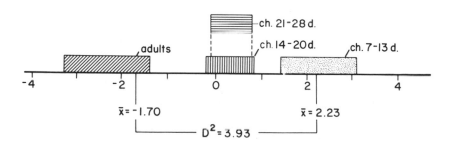

FIG. 5. Top: *Linear discriminant values for 7–13 d chicks and 21–28 d chicks cast on a 7–13 d chick/21–28 d chick discriminant axis. The distance between the groups in discriminant function units,* D^2, *= 3.90. Note the way in which the middle-aged chicks (14–20 d) overlap the oldest chicks and also the overlap of the adults with the two oldest groups of chicks (see text).* Bottom: *Linear discriminant values for 7–13 d chicks and adults cast on a 7–13 d chick/adult discriminant axis* (D^2 = 3.93). *Note the way in which the 14–20 d and 21–28 d chicks fit in between the youngest chicks and the adults (see text).*

Chicks 21-28 days / adults:

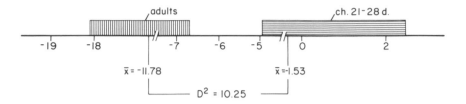

$\bar{x} = -11.78$ $\bar{x} = -1.53$

$D^2 = 10.25$

Chicks (oil) / adults (oil):

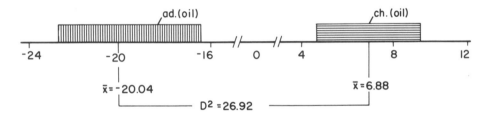

$\bar{x} = -20.04$ $\bar{x} = 6.88$

$D^2 = 26.92$

FIG. 6. Top: *Linear discriminant values of 21–28 d chicks and adults cast on a 21–28 d chick/adult discriminant axis* ($D^2 = 10.25$). *The other groups could not be discriminated on this axis (see text).* Bottom: *Linear discriminant values for oiling by chicks and oiling by adults cast on a chick/adult discriminant axis. Note the large separation of the groups* ($D^2 = 26.92$). *None of the other groups could be discriminated on this axis (see text).*

were the major characters separating the two populations (none of the other nine characters had weights greater than 0.05). The chicks performed fewer acts per minute and their sequences were of shorter duration than the adults' (Table 2). When the two other groups of non-oiling chicks were fit onto this axis, they fell nicely between the youngest chicks and the adults.

(3) *Oldest chicks (21–28 d) versus adults, non–oiling.* These data are presented in Figure 6, top). There was no overlap between the two groups. Rate, same area, wing–rub, and side/flank (discriminant weights = 0.78, 0.66, 0.32, and 0.42), respectively) were the characters that best discriminated between the two groups. The other groups could not be discriminated on this axis.

(4) *Oiling by chicks and adults.* There was no overlap between the two groups (Fig. 6, bottom) and the other groups of individuals could not be discriminated on this axis. The characters that best separated these two groups (and their discriminant weights) were: rate (1.65), same area (1.61), bill–to–wing–edge (1.35), shoulder–rub (1.40), head–shake (3.14), oil (1.25), and leg preening (1.25). The comparative means are presented in Table 2.

Discussion

Both univariate and multivariate numerical taxonomic procedures have helped to elucidate progressive age changes in the development of comfort behavior among the different age groups of Adélie penguins. However, the way in which comfort behaviors are incorporated into chick and adult displays (for descriptions see Ainley, 1975; Spurr, 1975b) was not apparent from this or other (Thompson, 1974; Spurr, 1975a) studies. By the time that Adélie chicks fledge (approximately eight weeks of age), they perform a large number of adult–like displays (D. Ainley, personal communication).

Adélie penguins are semi–altricial birds (Nice, 1962). In comparison with more precocial fowl in which some comfort activities occur within a few days of hatching, or even during hatching (Nice, 1962; McKinney, 1965; Dawson and Siegel, 1967; Brown, 1976), the comfort movements of Adélie chicks first occur later in life. The earliest performed comfort movements were those directed to the body areas (breast, belly) comprising major portions of the bird's surface area. In addition, it is these areas of the body that make the most contact with the ground when the chick lies down. As the chick begins to stand upright for a greater period of time and also begins to walk around, dirt collects on other body areas as well. These changes in posture and activity may play some role in the relative distribution of comfort behaviors during early ontogeny. Other factors may also play a role in the appearance and distribution of comfort activities. At approximately 20 d of age, contour feathers begin to emerge (Taylor, 1962; Spurr, 1975a). This then corresponded with the marked increase in the number of acts per sequence (Bekoff *et al.,* 1978). In addition, as down was lost from the legs at approximately 25 d of age, the proportion of leg preening increased. Similar trends have been observed during early life in Greylag geese, *Anser anser* (Mergler, 1976).

The last actions to emerge ontogenetically were shoulder rubbing (day 33) and oiling (day 35), just after the uropygial gland became functional and just prior to fledging (approximately eight weeks of age). The appearance in Adélie penguins of both shoulder–rubbing and oiling at approximately the same age is interesting in that shoulder–rubbing distributes oil from the head to the shoulder after the oil has been transferred to the head by wing–rubbing (Ainley, 1974). However, wing-rubbing, which is also used in other contexts, first appeared on day 7. Kruijt (1964) also noted a similar relationship in Burmese Junglefowl (*Gallus gallus spadiceus*). Head–rubbing, an action used to distribute oil from the oil gland, appeared on the same day of life (day 11) that the oil gland became functional.

THE DEVELOPMENT OF COMFORT SEQUENCES

During non–oiling by chicks and adults, the distribution of comfort movements was random (Bekoff *et al.,* 1978). However, during dry–oiling sequences, both oiling and bill–to–wing–edge, the major route by which oil is transferred from the bill to the wing and then to the head, were nonrandomly distributed in the first one–third of the sequences. This relationship was strongest in the adult group. This finding is not suprising, since once oil is gathered in the bill, there would be minimal loss (through dripping or swallowing) if the bird soon after distributed it to hard–to-reach areas.

The different age groups were also compared with respect to the degree of similarity of two–act transitions. That is, the six different transition matrices were compared with one another by calculating correlation coefficients for transition probabilities in cells representing the same two–act transition (see also Pearson and Parker, 1973). The 14–20-day and 21–28-day chicks showed the highest similarity when they were compared to one another (and also to all other groups). Both groups of chicks were also significantly correlated with non–oiling adults. The differences between chick non–oiling sequences and chick dry–oiling sequences are indicated by the lack of a significant correlation between the groups. On the other hand, dry–oiling by chicks and dry–oiling by adults were significantly correlated. The major differences between the groups in which there were non-significant correlations involved behaviors that are important in the gathering and distribution of oil, as well as changes in the distribution of breast preening.

Another difference among the groups of birds that was not highlighted by correlational information involved shoulder preening. During dry–oiling by chicks, the shoulder appeared to be a "pivot" point from which the birds would cross from one side of the body to the other. For example, for transitions involving the shoulder (excluding shoulder–shoulder movements), only 8/115 (7%) occurred on the same side of the body. In contrast, in the other groups of penguins, over 90% of transitions involving the shoulder were on the same side of the body. There is no obvious reason(s) for these differences. During dry–oiling by chicks, shoulder preening did not occur at a specific point in the on–going sequence, and the posture that the chicks assumed while preening their shoulders during dry–oiling did not appear to differ from the posture that the non–oiling chicks assumed when they preened their shoulders. It does remain possible that a more subtle factor was involved. Perhaps the dry–oiling chicks were trying to make the transition from shoulder preening to gathering oil and in doing so they began to lose their balance. The best way to regain balance would be to shift their weight to the other side of the midline. After this shift they might simply continue preening. The fact that the oil gland in the younger chicks was not functional makes this suggestion plausible and the large difference between the groups worthy of further study.

The Order of Appearance of Comfort Movements and the Construction of Sequences

A number of investigators have suggested that there is a direct relationship between the order in which behavioral units appear during ontogeny and the order in which they are incorporated into behavioral sequences later in life (Coghill, R. Oppenheim, personal communication; Tinbergen, 1951; Horwich, 1972). Borchelt (1977) found this to be the case in Bobwhite (*Colinus virginianus*) and Japanese quail (*Coturnix coturnix japonica*). However, in other published studies in which data are available for reanalysis (M. Bekoff, unpublished data) and in the present study, the relationship does not hold. Of the 58 two–act transitions that were analyzed in this study, the first act appeared earlier in life than the second act in only 29/58 cases. In addition, oiling is the last comfort activity to appear during ontogeny and occurs at its highest frequency in the beginning of oil sequences. Along this line, Kortlandt (1940) suggested that actions appear during ontogeny in the reverse order to that of their appearance in behavioral sequences later in life (Tinbergen, 1951, p. 141). With the exception of oiling and behaviors used to distribute oil, our data do not support this idea either. There does not appear to be any relationship between the order of emergence of comfort movements during ontogeny and the order in which they are performed sequentially later in life. Actually, this is not a suprising finding in that it would be to an individual's advantage to be able to vary its movements sequentially to attend to peripheral stimuli, for example, rather than to be "locked" into a "fixed" sequence from which there is little chance of escape. Indeed, with few exceptions, comfort movements were randomly distributed.

Behavioral Variability and Age

The question of behavioral variability involves analyses at two different levels (Barlow, 1977; Bekoff, 1977a,b), the motor acts themselves and sequential dependencies between individual acts. There have been very few studies that have addressed the question of the relationship between variability and age (Chapter 1; Schleidt and Shalter, 1973; Wiley, 1973; Bekoff, 1977b) and none have been concerned with age changes in sequential variability. Our results for non–oiling suggest that there are no significant differences in the variability of coupling of two–act transitions among the age groups while the results for dry–oiling indicate that the adult sequences are more variable. Since all of the coefficients of variation are very high, it is difficult to assign any significance to the differences. However, when the results for non–oiling are combined with the findings of Schleidt and Shalter (1973), Wiley (1973), and Bekoff (1977b), additional evidence is available to show that behavior does not necessarily become more stereotyped as a result of maturation, practice, or other forms of learning (Schleidt and Shalter, 1973, p. 36).

It should be mentioned that high coefficients of variation may indicate that there is a bimodal distribution of the data (Schleidt, 1974). Indeed, both Poole and Fish (1975) and Schoen, Banks, and Curtis (1976) reported that sequences of play that they observed in rodents and ponies, respectively, were nonrandomly ordered. The coefficients of variation that I calculated for the two–act transitions that they reported were also very high (108.5%, Poole and Fish; 89.6%, Schoen, Banks, and Curtis). In the penguins, especially among the dry–oiling adults, behaviors associated with the gathering and distribution of oil showed very highly significant transition probabilities (Bekoff et al., 1978). Although other behaviors were tightly linked in all groups of penguins, it was not possible to decipher a "typical" sequence.

Age Changes as Determined by Principal Components and Discriminant Function Analyses

The use of these multivariate procedures was very helpful in highlighting age progressions in comfort activities. For example, in the principal components analysis of the three groups of non–oiling chicks, one can see a progression along the x-axis (Factor I, gross behavioral development; see also Miller *et al.,* 1977) from the youngest to the oldest chicks. A similar relationship was found when comparing the three groups of chicks to the non–oiling adults.

A more detailed comparison was provided by use of discriminant function analyses. When the youngest chicks were compared to the oldest chicks, there was no overlap between the groups. Furthermore, when the middle-aged chicks (14–20 d) and the non-oiling adults were fit onto the youngest chick/oldest chick linear discriminant axis, the middle chicks fit between the youngest and oldest chicks (overlapping with the latter group) and the adults overlapped with both the middle and oldest chicks, but their distribution also extended beyond that of the oldest chicks. A similar age progession was found in the discriminant analysis of the youngest chicks versus the adults. In addition, when the oldest chicks were run against the adults and oiling chicks were matched with oiling adults, there was no overlap in their distribution on their respective linear discriminant axes.

Although there are many differences between the groups of penguins, the use of these multivariate procedures, especially discriminant function analyses, has provided a clearer picture of the developmental process. In particular, discriminant function analyses have allowed us to pinpoint the variables that contributed the most toward discriminating between the different age groups. As mentioned above, some of the differences between the groups can be explained by directly observable events such as the emergence of feathers, the differential distribution of dirt, a functioning oil gland, and possibly by postural changes. Yet, when one compares the results of the DFA analyses in which there was an age comparison, *rate* of performance of comfort behaviors was one of the variables contributing the most to the separation of the groups. The youngest chicks performed fewer acts per minute than did the oldest chicks, the oldest chicks performed fewer acts per minute than did the adults during non–oiling, and the dry–oiling chicks performed fewer acts per minute than did dry–oiling adults. In addition, duration of the sequences was important in discriminating between the youngest chicks and both the oldest chicks and adults during non–oiling.

The importance of rate (and duration) in discriminating the different age groups might be related to neural maturation and muscular (physical) development. The progressive change among the groups is consistent with this argument. In this case, level of maturation would be a unifying variable and could operate as an internal constraint (Bateson, 1976a,b) on chick behavior. Adélie penguins are semi-altricial (nidicolous) birds and maturation processes most probably continue throughout early postnatal life (see Sutter, 1951; Pearson, 1972, Chapter 16). When one observes the youngest chicks, it appears as if these birds fatigue more easily than do the older chicks or adults. In addition, the movements of the youngest chicks (in fact, up until the birds are about 10–12 d of age) are much slower and less coordinated and the birds more easily lose balance. Lastly, there was much less variability in duration and rate measures among the chicks. Detailed studies relating neural maturation (and growth) to postnatal behavior, similar to embryological studies (Chapter 2; Hamburger, 1963; Pearson, 1972, Chapters 4 and 16; Gottlieb, 1973; Oppenheim, 1974), will be necessary to test the above hypothesis (see also Bronson, 1965; Bekoff, 1977c; Gibson, 1977). Yet, it is very interesting that one of the most important variables in discriminating between the age groups is the one that may be closely tied to a relatively invariant (less so, perhaps, than the other variables that could affect comfort behavior) process.

Because of the life style of the Adélie penguin, it is not possible to observe chicks after they are about eight weeks old, the age at which they fledge and go to sea for the first time. Consequently, it is impossible to determine how wet feathers affect their comfort behavior and how wet-oiling by chicks compares with wet-oiling by adults. Nonetheless, I hope that the present study has shed some light on the development of behavior in these birds, and also, that the methods of analysis that were used will find their way into other studies of behavioral ontogeny, and behavioral processes, in general.

Acknowledgments

Supported in part by NSF grants OPP-74-08677 to D. Müller-Schwarze and OPP-74-15582 to D. G. Ainley. The personnel of Operation Deep Freeze and the United States Antarctic Research Program (USARP) provided necessary logistical support required to get to Cape Crozier. Anne Bekoff and David G. Ainley helped with data collection. Jeffry B. Mitton literally went beyond the call of duty in helping with the principal components and discriminant function analyses. I am certain that he often wished that he had never introduced me to the ways of NTSYS! I would like to thank Richard N. Conner, Frances C. James, and Robert Plomin for comments on an earlier draft of this paper. Ronald Oppenheim provided some pertinent references. The Department of EPO Biology, through the gracious hand of David W. Crumpacker, chairperson, provided financial support for preparation of the figures.

References

Ainley, D. G. 1970. Communication and reproductive cycles of the Adélie penguin. Ph.D. dissertation, John Hopkins Univ. Baltimore, Md.

Ainley, D. G. 1974. The comfort behaviour of Adélie and other penguins. *Behaviour* 50:16–51.

Ainley, D. G. 1975. Displays of Adélie penguins: a reinterpretation. In *The biology of penguins*, ed., B. Stonehouse, pp. 503–534. Baltimore: University Park Press.

Andrew, R. J. 1956. Normal and irrelevant toilet behaviour in *Emberiza* sp. *Anim. Behav.* 4:85–91.

Aspey, W. P., and J. E. Blankenship. 1976. *Aplysia* behavioral biology: I. A multivariate analysis of burrowing in *A. Brasiliana. Behav. Biol.* 17:279–299.

Aspey, W. P., and J. E. Blankenship. 1977. Spiders and snails and statistical tails: applications of multivariate analyses to diverse ethological data. In *Quantitative methods in the study of animal behavior*, ed., B. A. Hazlett, pp. 75–120. New York: Academic Press.

Aspey, W. P., R. W. Zears, and J. E. Blankenship. 1977. "3-D": an interactive computer-graphics program for plotting scores on three coordinate axes. *Behav. Res. Meth. Instr.* (In press.)

Baerends, G. P. 1975. An evaluation of the conflict hypothesis as an explanatory principle for the evolution of displays. In *Function and evolution in behaviour*, eds., G. Baerends, C. Beer, and A. Manning, pp. 187–227. Oxford and New York: Oxford University Press.

Barlow, G. W. 1977. Modal action patterns. In *How animals communicate*, ed., T. A. Sebeok, pp. 98–134. Bloomington: Indiana University Press.

Bateson, P. P. G. 1976a. Specificity and the origins of behavior. *Adv. Study Behav.* 6:1–20.

Bateson, P. P. G. 1976b. Rules and reciprocity in behavioural development. In *Growing points in ethology*, eds., P. P. G. Bateson and R. A. Hinde, pp. 401–421. New York: Cambridge University Press.

Bekoff, M. 1977a. Quantitative studies of three areas of classical ethology: Social dominance, behavioral taxonomy, and behavioral variability. *In Quantitative methods in the study of animal behavior*, ed., B. A. Hazlett, pp. 1–46. New York: Academic Press.

Bekoff, M. 1977b. Social communication in canids: Evidence for the evolution of a stereotyped mammalian display. *Science* 197:1097–1099.

Bekoff, M. 1977c. Socialization in mammals with an emphasis on nonprimates. *In Primate bio-social development*, eds., S. Chevalier-Skolnikoff and F. E. Poirier, pp. 603–636. New York: Garland Publishing.

Bekoff, M., D. G. Ainley, and A. Bekoff. 1978. The ontogeny and organization of comfort behavior in Adélie penguins, *Pygoscelis adéliae. Wilson Bull.* (In press.)

Bekoff, M., H. L. Hill, and J. B. Mitton. 1975. Behavioral taxonomy in canids by discriminant function analyses. *Science* 190:1223–1225.

Berg, C. J. 1976. Ontogeny of predatory behavior in marine snails (Prosobranchia: Natacidae). *Nautilus* 90:1–4.

Bergmann, H.-H. 1976. Ontogenese und Koordination der Streckbewegungen junger Grasmücken (*Sylvia* sp., Sylviidae, Passeriformes). *Zool. Jb. Physiol.* 80:346–359.

Borchelt, P. L. 1975. The organization of dustbathing components in bobwhite quail (*Colinus virginianus*). *Behaviour* 53:217–237.

Borchelt, P. L. 1977. Development of dustbathing components in bobwhite and Japanese quail. *Dev. Psychobiol.* 10:97–103.

Borchelt, P. L., and S. R. Overmann. 1974. Development of dustbathing in bobwhite quail. I. Effects of age, experience, texture of dust, strain, and social facilitation. *Dev. Psychobiol.* 7:305–313.

Bronson, G. 1965. The hierarchical organization of the central nervous system: Implications for learning processes and critical periods in early development. *Behav. Sci.* 10:7–25.

Brown, N. S. 1976. The development of grooming behavior in the domestic chicken (*Gallus gallus domesticus*). *Paper presented at the AIBS meetings, New Orleans.*

Clifford, H. T. and W. Stephenson. 1975. *An introduction to numerical classification.* New York: Academic Press.

Conner, R. N., and C. S. Adkisson. 1976. Discriminant function analysis: A possible aid in determining the impact of forest management on woodpecker nesting habitat. *Forest Sci.* 2:122–127.

Conner, R. N., and C. S. Adkisson. 1977. Principal components analysis of woodpecker nesting habitat. *Wilson Bull.* 89:122–129.

Dawkins, R., and M. Dawkins. 1973. Decisions and the uncertainty of behaviour. *Behaviour* 45:83–102.

Dawson, J. S., and P. B. Siegel. 1967. Behavior patterns of chickens to ten weeks of age. *Poult. Sci.* 46:615–622.

Delius, J. 1969. A stochastic analysis of the maintenance behaviour of skylarks. *Behaviour* 33:137–178.

Dow, M., A. W. Ewing, and I. Sutherland. 1976. Studies on the behaviour of cyprinodont fish. III. The temporal patterning of aggression in *Aphyosemion striatum* (Boulenger). *Behaviour* 59:252–268.

Dudziński, M. L., and J. M. Norris. 1970. Principal components analysis as an aid for studying animal behaviour. *Forma et functio* 2:101–109.

Fabricius, E. 1945. Zür Ethologie junger anatiden. *Acta Zool. Fennica* 68:1–178.

Fentress, J. C. 1972. Development and patterning of movement sequences. *In The biology of behavior,* ed., J. A. Kiger, pp. 83–131. Corvallis: Oregon State University Press.

Gibson, K. R. 1977. Brain structure and intelligence in macaques and human infants from a Piagetian perspective. *In Primate bio-social development,* eds., S. Chevalier-Skolnikoff and F. E. Poirier, pp. 113–157. New York: Garland Publishing.

Gottlieb, G., ed. 1973. *Behavioral embryology.* New York: Academic Press.

Hamburger, V. H. 1963. Some aspects of the embryology of behavior. *Q. Rev. Biol.* 38:342–365.

Harner, E. J., and R. C. Whitmore. 1977. Multivariate measures of niche overlap using discriminant analysis. *Theor. Pop. Biol.* 12:21–36.

Horwich, R. H. 1969. Behavioral ontogeny of the mockingbird. *Wilson Bull.* 81:87–93.

Horwich, R. H. 1972. The ontogeny of social behaviour in the gray squirrel (*Sciurus carolinensis*). *Z. Tierpsychol. Suppl.* 8.

van Iersel, J. J. A., and A. C. A. Bol. 1958. Preening by two tern species: a study on displacement behaviour. *Behaviour* 13:1–88.

James, F. C. 1971. Ordinations of habitat relationships among breeding birds. *Wilson Bull.* 83:215–236.

John, E. R., B. Z. Karmel, W. C. Corning, P. Easton, D. Brown, H. Ahn, M. John, T. Harmony, L. Prichep, A. Toro, I. Gerson, F. Bartlett, R. Thatcher, H. Kaye, P. Valdes, and E. Schartz. 1977. Neurometrics. *Science* 196:1393–1410.

Kaltenhauser, J., and S. Welner. 1975. Vudata: a program for graphic data display. (Unpublished paper, University of Colorado Computing Center.)

Jardine, N., and R. Sibson. 1971. *Mathematical taxonomy.* New York: Wiley.

Kertesz, A., and J. B. Phipps. 1977. Numerical taxonomy of aphasia. *Brain and Language* 4:1–10.

Kortlandt, A. 1940. Eine Übersicht der angeborenen verhaltensweisen des mitteleuropäischen kormorans (*Phalacrocorax carbo sinensis* Shaw and Nodd.), ihre funktion, ontogenetische entwicklung und phylogenetische herkunft. *Arch. Neerl. Zool.* 4:401–442.

Kruijt, J. P. 1964. Ontogeny of social behaviour in Burmese red junglefowl (*Gallus gallus spadiceus*). *Behaviour Suppl.* 12.

Marler, P. 1975. On strategies of behavioural development. *In Function and evolution in behaviour,* eds., G. Baerends, C. Beer, and A. Manning, pp. 254–275. Oxford and New York: Oxford University Press.

McKinney, F. 1965. The comfort movements of anatidae. *Behaviour* 25:121–217.

Mergler, M. A. L. 1976. The development of preening patterns in greylag geese (*Anser anser anser*). (Unpublished paper from Die Österreichische Akademie der Wissenschaften, Institut für vergleichende Verhaltensforschung, Grünau, Austria.)

Michener, C. D. 1974. *The social behavior of the bees.* Cambridge, Mass.: Harvard Univ. Press.

Mihok, S. 1976. Behaviour of subarctic red-backed voles (*Clethrionmys gapperi athabascae*). *Can. J. Zool.* 54:1932–1945.

Miller, I. L., J. M. Whitsett, J. G. Vandenbergh, and D. R. Colby. 1977. Physical and behavioral aspects of sexual maturation in male golden hamsters. *J. Comp. Physiol. Psychol.* 91:245–259.

Nice, M. M. 1962. Development of behavior in precocial birds. *Trans. Linn. Soc. N. Y.* 8:1–211.

Oppenheim, R. 1974. The ontogeny of behavior in the chick embryo. *Adv. Study Behav.* 5:133–172.

Pearson, R. 1972. *The avian brain.* New York: Academic Press.

Pearson, R. G., and G. A. Parker. 1973. Sequential activities in the feeding behaviour of some Charadriiformes. *J. Nat. Hist.* 7:573–589.

Penney, R. L. 1968. Territorial and social behavior in the Adélie penguin. *In Antarctic bird studies.* ed., O. L. Austin, pp. 83–131. Washington: American Geophysical Union.

Poole, T. B., and J. Fish. 1975. An investigation of playful behaviour in *Rattus norvegicus* and *Mus musculus*. *J. Zool.* 175:61–71.

Ringo, J. M., and R. J. Hodosh. 1977. A multivariate analysis of behavioral divergence among closely related species of endemic Hawaiian *Drosophila*. *Evolution*. (In press.)

Schleidt, W. M. 1974. How "fixed" is the fixed action pattern? *Z. Tierpsychol.* 36:184–211.

Schleidt, W. M., and M. D. Shalter. 1973. Stereotypy of a fixed action pattern during ontogeny in *Coturnix coturnix coturnix*. *Z. Tierpsychol.* 33:35–37.

Schnell, G. D. 1973. A reanalysis of nest structure in the weavers (Ploceinae) using numerical taxonomic techniques. *Ibis* 115:93–106.

Schoen, A. M. S., E. M. Banks, and E. Curtis. Behavior of young shetland and welsh ponies (*Equus caballus*). *Biol. Behav.* 1:192–216.

Slater, P. J. B. 1973. Describing sequences of behavior. *Perspect. Ethol.* 1:131–153.

Slater, P. J. B. 1975. Temporal patterning and the causation of bird behaviour. *In Neural and endocrine aspects of behaviour in birds,* eds., P. Wright, P. G. Caryl, and D. M. Vowles, pp. 11–33. Elsevier: Amsterdam.

Sneath, P. H. A., and R. R. Sokal. 1973. *Numerical taxonomy: the principles and practice of numerical classification.* San Francisco: W. H. Freeman.

Sparling, D. W., and J. D. Williams. 1977. Multivariate analysis of avian vocalizations. (Submitted for publication.)

Spurr, E. B. 1975a. Behaviour of the Adélie penguin chick. *Condor* 77:272–280.

Spurr, E. B. 1975b. Communication in the Adélie penguin. *In The biology of penguins,* ed., B. Stonehouse, pp. 449–501. Baltimore: University Park Press.

Sutter, E. 1951. Growth and differentiation of the brains in nidifugous and nidicolous birds. *Proc. 20th Int. Ornithol. Congr.* 636–644.

Taylor, R. H. 1962. The Adélie penguin *Pygoscelis adéliae* at Cape Royds. *Ibis* 104:176–204.

Thelen, E., and D. J. Farish. 1977. An analysis of the grooming behavior of wild and mutant strains of *Bracon hebetor* (Braconidae: Hymenoptera). *Behaviour* 62:70–102.

Thompson, D. H. 1974. Mechanisms limiting food delivery by Adélie penguin parents exclusively to their genetic offspring. Ph.D. dissertation, University of Wisconsin, Madison.

Tinbergen, N. 1951. *The study of instinct.* Oxford and New York: Oxford University Press.

Wiley, R. H. 1973. The strut display of male sage grouse: A "fixed" action pattern. *Behaviour* 47:129–152.

Chapter Ten

Laboratory Investigations of Imprinting

HOWARD S. HOFFMAN

Department of Psychology
Bryn Mawr College
Bryn Mawr, Pennsylvania 19010

The term *imprinting* is the English rendering of the German word *Prägung,* which Lorenz (1935) used to describe the process by which a newly hatched gosling forms a social attachment to the first moving object it encounters (Hess, 1972). Because *Prägung* means stamping or coinage, its use in this context seemed quite appropriate. If, for example, the first object encountered was Lorenz himself, the gosling would immediately follow him and would thereafter approach and stay near him in preference to its own mother. It was as if the image of the first moving object encountered was somehow instantly stamped or imprinted upon the nervous system. Noting this apparently unique effect Lorenz (1937) and Hess (1959, 1973), among others, have taken the position that imprinting is a special process that cannot be readily interpreted using the theoretical constructs commonly employed by students of learning. They have argued, for example, that imprinting occurs too rapidly and its effects are too permanent to be merely an example of association learning. They have also noted that imprinting is most effective when it occurs during a brief "critical" period early in the subject's life. And they point out that there are few, if any, examples of learning that are known to operate in such a temporally constrained fashion. Another reason, they argue, for treating imprinting as a special process is that aversive stimulation enhances its effects. If subjects receive electrical shock during their first encounter with an imprinting stimulus, their tendency to approach it is usually enhanced. Aversive stimulation has the opposite effect in most learning situations. Numerous experiments reveal that subjects usually withdraw from stimuli that are associated with electrical shock.

Since, on the face of it, these are all rather telling arguments, and since they have been widely recognized, it is easy to see why the view of imprinting as a unique phenomenon has enjoyed considerable credibility. This is not to say that there have been no efforts to interpret imprinting as an example of learning, but such efforts have invariably encountered roadblocks. The problem has largely been one of conceptualization. If imprinting is to be interpreted as a form of learning, it is necessary to identify exactly what is learned and how. But even the most careful analysis of the circumstances under which imprinting occurs fails to provide much in the way of clues.

In the typical imprinting experiment a given moving object is presented to a newly hatched precocial bird. If the subject is the right age (for ducklings, approximately 16 h old), it will immediately react filially and will emit distress calls if the object is subsequently withdrawn. Moreover, if the subject is tested when older (for example, at five days of age) it will approach and follow the object it originally encountered and will try to escape from objects that are sufficiently different. Clearly, these observations imply that during imprinting something is learned, but since the procedures include no obvious reinforcer such as food, water or warmth, the question of exactly what is learned and how is open to speculation.

Several investigators (Bateson, 1966, 1971; Dimond, 1970; Salzen, 1970) have suggested that subjects form a neuronal model (i.e., learn the features) of a given imprinting stimulus, simply as a consequence of being exposed to it. According to this view once such learning has occurred, filial reactions are restricted to objects that match the neuronal model because sources of stimulation that produce mismatches generate aversive reactions. While this kind of "perceptual" or "exposure" learning can account for why older subjects withdraw from unfamiliar objects, it leaves unanswered the question of why young subjects should react so affirmatively on their first exposure to an appropriate stimulus. Moreover, it fails to explain why the subsequent withdrawal of that stimulus should be so aversive.

An even more telling argument against the proposition that imprinting is merely "perceptual" or "exposure" learning comes from a recent experiment (Hoffman and Ratner, 1973) in which day-old ducklings, *Anas platyrhynchos,* that were already imprinted to a given stimulus were nonetheless found to react affirmatively when the original stimulus was withdrawn and a second, novel stimulus was presented. This, of course, was not a new finding. As anyone who has handled very young organisms knows, infants of many species react affirmatively to strangers, and ducklings are no exception. What was new in this study was that it included a test for whether or not the ducklings were able to recognize that the novel stimulus was different from the one to which they had just previously been imprinted. Since that test provided unambiguous evidence that the birds were making this discrimination, it is clear that very young (day-old) ducklings are able to recognize (and in this sense from a neuronal model of) a stimulus to which they are imprinted, and that this fact, of itself, does not prevent them from also reacting affirmatively to a novel, but otherwise adequate stimulus.

This conclusion is similar to one reached by Schaffer (1966), who examined the ability of human infants to discriminate between their mother and strangers. He found that this ability became evident several months before the infants began to display fear of strangers. Apparently an "age gap exists when the stranger is experienced as different, yet not as frightening" (p. 103). On this basis Schaffer concluded that while the ability to make distinctions between the mother and strangers may be a necessary condition for fear of strangers, it is not a sufficient condition.

All of this suggests that in the duckling, and in the human infant, some factor (or factors) in addition to perceptual learning, per se, must contribute to the focused attachments these organisms display when they are older. One possibility is that a young organism's filial reactions become fixated on a given stimulus through a process of classical conditioning. This would explain why these reactions are so strong and so permanent, but as noted earlier, it raises the question of exactly what is conditioned and how. A part of the answer to this question comes from Fabricius (1962), who pointed out that the stimulation provided by visual motion was innately reinforcing to young precocial birds. A second portion of the answer stems from the finding that the nervous systems of many organisms include highly refined detector mechanisms that are only responsive to specific forms of stimulation. Thus in the cat, as well as a number of other organisms such as the pigeon, the rabbit, and the frog, it has been possible to isolate neural sites that respond only when the subject is exposed to stimulation provided by the motion of an object across the visual field. Since these sites are silent when the same object is stationary, they have been identified as "motion detectors" (McIlwain, 1972). Their relevance to imprinting becomes apparent when one considers the finding (Hoffman, Eiserer, and Singer, 1972) that while an appropriate imprinting stimulus (e.g., a moving object) immediately elicited a strong filial reaction in newly hatched (17 h old) ducklings, the same object presented stationary failed to do so until the subjects had been exposed to the object in motion for about 20 min. In short, the static features of the imprinting object (e.g., its size, shape, and color, etc.) were unable to elicit innately filial reactions, but they acquired the capacity to do so, and in this sense serve as conditioned stimuli, by virtue of their temporal and spatial association with the stimulation provided by the object in motion. It was as if the sight of the object in motion was activating motion detectors and their activity was generating a filial reaction. The same object stationary on the other hand, may have activated color, shape and size detectors, but until the activity of these detectors was associated with activity of the motion detectors they were unable to generate a filial response.

This is admittedly a grossly oversimplified picture of the neurophysiological events that seem to mediate imprinting, but it has the virtue of providing an intuitively reasonable account of the otherwise paradoxical fact that when we deal with imprinting the same object can act like a neutral stimulus (when first presented stationary), an unconditioned stimulus (when first presented in motion), and a conditioned stimulus (when presented stationary after having been presented in motion).

These considerations enable us to see how the processes imprinting entails could operate to insure that a newly hatched duckling would become attached to its biological parent in a natural setting. Like the imprinting object used in most laboratories, a living mother moves and thus provides at least one kind of stimulation that innately evokes filial behavior. It is also clear that she has many other features that can function in this way. It is, for example, known that certain of her calls can innately evoke filial reactions (Gottlieb, 1965), and there is no reason why certain of her other features (her texture, her warmth, her color) might not also function in this way, as some reports have suggested (Ramsey and Hess, 1954; Hess, 1973, p. 199). Such redundancy would make it all the more certain that a newly hatched duckling would stay near and react filially to its mother.

One can also see how certain salient features of the mother that might initially be neutral would come to evoke a conditioned filial reaction as a result of their spatial and temporal association with those features that were innate (unconditioned) elicitors of this response. With sufficient exposure to the mother most, if not all, of her features would eventually be conditioned and hence be completely familiar to the duckling.

The importance of familiarity lies in the fact that for precocial birds (and many other organisms as well) the tendency to react fearfully to unfamiliar stimulation increases as ontogenetic development proceeds (Hess, 1959, 1972; Hess and Schaffer, 1959). In ducklings, for example, fear reactions to novel stimulation are quite weak during the first day or so posthatch especially if that stimulation incorporates features that innately evoke a filial response. By the time a duckling is five days old, however, its wariness of the unfamiliar is quite pronounced so that rather than behaving filially to a novel but otherwise appropriate imprinting stimulus, it runs away. This age-related growth in the tendency to react fearfully to novel stimulation explains why the classical conditioning process elaborated above is a key element in the imprinting phenomenon. By establishing a conditioned filial reaction to the distinctive but initially neutral features of an imprinting object the process insures that all features of the object will be familiar and hence incapable of generating novelty induced fear at later stages of ontogenic development.

Once imprinting is conceptualized as an example of classical conditioning it becomes clear that the apparent instantaneous fixation or "stamping in" of the filial response that has often been thought to be its hallmark is perhaps best described as an illusion. The only instantaneous part of imprinting is the unconditioned response to the features of an imprinting object that innately evoke filial behavior. Moreover, rather than being a limited interval in which imprinting can take place, the critical period seems best characterized as that interval before the subject's nervous system has matured to the point where novel objects produce strong fear reactions.

If these ideas are correct one should be able to obtain imprinting after the so–called critical period has passed provided that steps are taken to eliminate the fear reactions that the imprinting stimulus might produce. This has often proved to be the case (See Ratner and Hoffman, 1974). When a duckling older than the so-called critical period is first exposed to a novel imprinting stimulus it initially tries to escape. If, however, its attempts to escape are unsuccessful they gradually terminate and then rather than reacting with indifference, the duckling begins to display an increasingly strong filial reaction. It approaches and tries to stay near the stimulus and it emits distress calls if the stimulus is withdrawn. In short, it behaves as if it is now imprinted to the stimulus. Since this effect has been obtained in ducklings that had previously lived with companions, as well as in ducklings that had previously been imprinted to other objects (Stratton, 1971), it seems clear that despite its capacity to evoke fear in an older duckling, a novel, but otherwise appropriate imprinting stimulus maintains its potentiality to also evoke a filial response. This, of course, is what should happen if certain features of an appropriate imprinting object innately elicit filial reactions, and an older duckling's fear of it is basically a reaction to its novelty.

All of this implies that in post-critical period imprinting, as in many behavioral phenomena, one is dealing with a situation in which conflicting response tendencies are being evoked. Under such circumstances the relative strengths of these tendencies become important determinants of the behavior observed. It should, for example, be easier to imprint an older duckling to a stimulus that has many features that evoke filial behavior—for example, its natural mother—than to imprint it to a more or less arbitrarily selected object that may contain only one or at most a few such features, and this should be the case regardless of the prior imprinting experience of the subject.

These same considerations should apply to imprinting in younger subjects as well. As noted earlier, in a natural setting and with its biological parent a young subject would be exposed to a much larger amount of stimulation that innately evokes filial behavior than can be obtained in a laboratory. Under such circumstances the learning that occurs might be especially rapid and be especially effective. But rather than representing a qualitative difference between imprinting in the field and imprinting in the laboratory (as suggested by Hess, 1973), it would seem to represent a difference in degree.

Earlier it was noted that one of the factors that has appeared to set imprinting apart as a unique phenomenon is that aversive stimulation has been found to enhance its effects. If electrical shock is occasionally presented while a newly hatched (approximately 16 h old) duckling is being imprinted to a given stimulus, it tends to follow that stimulus more closely (Kovach and Hess, 1963; Barrett et al., 1971). In most situations other than imprinting, subjects tend to withdraw from stimuli that are present at the time shock is delivered. Moreover, they often continue to do so long after the use of shock has terminated—implying that they have formed an association between these stimuli and the occurrence of shock.

While such observations might seem to provide strong evidence that aversive stimulation has unique effects in the context of imprinting, there are several experiments which suggest that this conclusion is not only premature, it is wrong. For example, Moltz, Rosenblum, and Halikas (1959) found that the tendency to follow an imprinting stimulus is enhanced when shock is delivered prior to rather than during the imprinting session. Similarly, Ratner (1977) found it enhanced following even though shock only occurred when the imprinting stimulus had been briefly withdrawn. In other words, both of these studies found that shocks produced an increase in following despite the fact that shock was never paired with either the imprinting stimulus or the responses it evokes. These findings must mean that with imprinting, as with many other phenomena, the delivery of shock has energizing (e.g., motivational) effects that are independent of its association with either a given stimulus or a given response. In this regard, it is of interest that if one arranges to deliver shock only when a duckling is following the stimulus (e.g., one explicitly punishes following), the tendency to follow declines (Barrett et al., 1971; Barrett, 1972). Apparently, in the context of imprinting, as elsewhere, the delivery of shock can have complex effects. The occurrence of an aversive stimulus per se enhances the motivational substrate that supports filial behavior. But if the delivery of the aversive stimulus is restricted to occurrences of a given response, that response is often suppressed—even though (like following) it might be an aspect of the filial behavior that would otherwise be enhanced.

Aversive stimulation can influence the behavior that characterizes imprinting in another way. Barrett (1972) and more recently Ratner (1977) have both found that if newly hatched ducklings received response-independent shock in the presence of a given imprinting stimulus, they follow it more closely than subjects that are not shocked. This, of course, is consistent with the previously described finding that shock per se enhances the motivational background for the expression of filial behavior. Both Barrett and Ratner also found, however, that if the use of shock was discontinued and the subjects were offered a choice between their original imprinting stimulus and a simultaneously presented identical object that had been rendered perceptually distinctive (by painting black stripes on it) they approached and followed the new stimulus and they avoided their original stimulus. Since these studies included controls for any differences in preference that the two stimuli may have afforded, and since Ratner's study revealed that the effect can be eliminated by restricting shock delivery to periods of stimulus withdrawal, it is apparent that ducklings, like most other organisms, can form associations between aversive events and the stimulation that prevails at the time. If that stimulation includes an object that elicits filial behavior and the duckling has no alternative, its tendency to approach and stay near the object is often enhanced; but if offered an alternative object that is not too novel, the alternative will be chosen. Clearly, rather than setting imprinting apart as a unique phenomenon, the effects engendered by aversive stimulation have, on close examination, turned out to be essentially the same as are obtained in the context of traditional learning.

This conclusion, in conjunction with the conclusions set forth earlier, have a number of broad implications that help to place imprinting into perspective. As noted by Francois Jacob (1977), "Evolution does not produce novelties from scratch. It works on what already exists, either transforming a system to give it new functions or combining several systems to produce a more elaborate one" (p. 1164). Our conclusions suggest that imprinting is no exception to this principal. They imply that the several seemingly distinctive features of imprinting are in fact the product of a limited number of well known behavioral processes. This in turn implies that the essential features of imprinting should be seen in any species whose behavior is based on those processes.

One of these processes involves the tendency to exhibit an innate filial reaction to certain aspects of the stimulation provided by a parent. If this stimulation is highly specialized, so that it only occurs in the mother, the expression of filial behavior would necessarily be restricted to encounters with her alone. But what if a young organism is so constructed as to react filially to features of its mother that are also exhibited by other organisms? What if, as with ducklings, the stimulation provided by movement itself is sufficient to elicit a filial response? What if, as with primates (Harlow, 1958), certain kinds of tactile stimulation evoke a filial response? One can see how an innate reaction to such widely occurring stimulation would make for great plasticity in the attachment phenomenon. One can also see how such plasticity might have a distinct adaptive advantage if individual members of a given species vary appreciably. Of course, given sufficient plasticity, predators would also provide stimulation that evokes filial behavior and there is little to be gained (except a quick demise) if a young organism follows and attempts to stay near a predator.

The experiments discussed here suggest that there are two behavioral processes that prevent this counterproductive event from happening. (1) During exposure to its mother a young organism undergoes a form of classical conditioning whereby it comes to react filially to an increasing number of her distinctive, but otherwise neutral, features. This makes it all but certain that she will be completely familiar on later occasions. (2) As a young organism matures it comes to be increasingly fearful of any novel stimulus. It is apparent that these two processes would eventually prevent a developing organism from reacting filially to a stranger, while still preserving the filial reaction to its mother (and other companions). In this regard, it seems relevant to note that members of different species mature at quite different rates, and the age at which they first begin to exhibit strong fear of novel stimulation also differs. In ducklings, fear of novelty begins to get strong at about two days posthatch. In human infants, on the other hand, fear of strangers begins to appear at about eight months. As Hess (1959b) has suggested, it may be no accident that in both species the capacity to locomote well and hence bring oneself into contact with novel and possibly dangerous stimulation also occurs at these times.

Acknowledgment

The preparation of this manuscript and much of the research upon which it is based was supported by National Institute of Mental Health Grant MH 19715.

References

Barrett, H. E. 1972. Schedules of electrical shock presentations in the behavioral control of imprinted ducklings. *J. Exp. Anal. Behav.* 18:305–321.

Barrett, J. E., H. S. Hoffman, J. W. Stratton, and V. Newby. 1971. Aversive control of following in imprinted ducklings. *Learn. Motiv.* 2:202–212.

Bateson, P. P. G. 1966. The characteristics and context of imprinting. *Biol. Rev. Cambridge Philos. Soc.* 41:177–220.

Bateson, P. P. G. 1971. Imprinting. *In The ontogeny of vertebrate behavior,* ed., H. Moltz, pp. 369–387. New York: Academic Press.

Dimond, S. J. 1970. Visual experience and early social behavior in chicks. *In Social behavior in birds and mammals: essays on the social Ethology of animals and man,* ed., J. H. Crook, pp. 441–466. New York: Academic Press.

Fabricius, E. 1962. Some aspects of imprinting in birds. *Symp. Zool. Soc. London* 8:139–148.

Gottlieb, G. 1965. Imprinting in relation to parental and species identification by avian neonates. *J Comp. Physiol. Psych.* 59:254–356.

Harlow, H. F. 1958. The nature of love. *Am. Psych.* 13:673–685.

Hess, E. H. 1959a. Imprinting. *Science* 130:133–141.

Hess, E. H. 1959b. Two conditions limiting the critical age for imprinting. *J. Comp. Physiol. Psych.* 52:515–518.

Hess, E. H. 1972. "Imprinting" in a natural laboratory. *Sci. Am.* 227:24–31.

Hess, E. H. 1973. *Imprinting: early experience and the developmental psychology of attachment.* New York: Van Nostrand Reinhold.

Hess, E. H., and H. H. Schaffer. 1959. Innate behavior patterns as indications of the "critical period." *Z. Tierpsychol.* 16:155–160.

Hoffman, H. S., L. A. Eiserer, and D. Singer. Acquisition of behavioral control by a stationary imprinting stimulus. *Psychon. Sci.* 26:146–148.

Hoffman, H. S., and A. L. Ratner. 1973. Effects of stimulus and environmental familiarity on visual imprinting in newly hatched ducklings. *J. Comp. Physiol. Psych.* 85:11–19.

Jacob, F. 1977. Evolution and tinkering. *Science* 196:1161–1166.

Kovach, J. K., and E. H. Hess. 1963. Imprinting: Effects of painful stimulation upon the following response. *J. Comp. Physiol. Psych.* 56:461–464.

Lorenz, K. 1935. Der kumpan in der umwelt des vogels. *J. Ornith.* 83:137–213.

Lorenz, K. 1937. The companion in the bird's work. *Auk* 54:245–273.

McIlwain, T. 1972. Central vision: visual cortex and superior colliculus. *Annu. Rev. Physiol.* 34:291–314.

Moltz, H., L. Rosenblum, and N. Halikas. 1959. Imprinting and level of anxiety. *J. Comp. Physiol. Psych.* 52:240–244.

Ramsey, A. O., and E. H. Hess. 1954. A laboratory approach to the study of imprinting. *Wilson Bull.* 66:196–206.

Ratner, A. M. 1977. Modification of ducklings' filial behavior by aversive stimulation. *J. Exp. Psych. Anim. Behav. Proc.* 2:266–284.

Ratner, A. M., and H. S. Hoffman. 1974. Evidence for a critical period for imprinting in Khaki Campbell ducklings. (*Anas platyrhynchos domesticus*). *Anim. Behav.* 22:249–255.

Salzen, E. A. 1970. Imprinting and environmental learning. *In Development and evolution of behavior,* eds., L. R. Aronson, E. Tobach, D. S. Lehrman, and J. S. Rosenblat, pp. 158–178. San Francisco: W. H. Freeman.

Schaeffer, H. R. 1966. The onset of fear of strangers and the incongruity hypothesis. *J. Child Psych. Psych.* 7:95–106.

Stratton, V. N. 1971. The control of social behavior in young ducklings. (Unpublished doctoral dissertation. Pennsylvania State University, State College.)

Commentary

ECKHARD H. HESS

Department of Psychology
University of Chicago
Chicago, Illinois 60637

My first reaction is that until now each author has been an advocate for using a particular species or group through which to look at behavioral problems, as well as for the use of a certain methodology. While this approach has obvious usefulness, and the chapters are certainly most interesting, they do not seem to address themselves to the specific topic of evolution.

However, I shall restrict my comments to the specific points raised by Howard Hoffman. First, as Konrad Lorenz has so often said, "I couldn't disagree less." I, too, began my investigations of imprinting with a learning approach. In fact, my original research grant was titled "Experimental Analysis of Imprinting: *A Form of Learning"* (italics added). However, I became cognizant of the discrepancy between results in our laboratory when using human imprinters as compared to those obtained using a mallard decoy, where the behavior of that decoy was reasonably similar to maternal actions of a real mallard (Hess and Hess, 1969). Only with such an attempted reproduction of the natural state of affairs did imprinting produce the strength which is characteristic of the filial response when it occurs in nature. I feel certain that the results obtained by Hoffman when he used a sponge would have been different had he used a more natural maternal object.

Hoffman talks about "innate feature detectors" that trigger the imprinting process and agrees exactly with my position that the greater number of stimulus aspects found in the natural mother trigger innate physiological reactions and behavioral responses. It is therefore *not a question of disagreement but rather one of emphasis.* Hoffman deemphasizes these innate factors and devotes considerable time explaining his laboratory findings and how they relate to imprinting. A silent sponge and a *real* maternal object can easily give different results. To sum up, I am now studying imprinting in a way which allows comparison with the natural events representing that process, as well as an examination of imprinting in the natural situation. Hoffman is studying how far one can distort the natural aspects of the imprinting process and still generate its main characteristics in the laboratory. In that he is not alone.

Reference

Hess, E. H., and Hess, D. B. 1969. Innate factors in imprinting, *Psychonomic Sci.* 14:129–130.

Chapter Eleven

Aspects of Learning in the Ontogeny of Bird Song: Where, From Whom, When, How Many, Which, and How Accurately?

DONALD E. KROODSMA

Rockefeller University, Field Research Center
Millbrook, New York 12545

A leisurely stroll on a spring day or a casual glance at the literature reveals an overwhelming and bewildering diversity of singing behavior among the songbirds. There are tremendous disparities in song repertoire sizes among birds—males of some species develop a single song (review in Bertram, 1970), some develop 100 songs (Verner, 1975; Kroodsma, 1975), but a brown thrasher (*Toxostoma rufum*) may sing thousands (Kroodsma and Parker, 1977). Continuity or rates of singing vary as does the sequencing of song themes during a song performance. There are musical and nonmusical songsters, superior and inferior singers, and birds that mock and those that don't (Hartshorne, 1973). Mates of some species duet (Thorpe, 1972) while in most species the females do not sing. Males of some species learn exact copies of model songs, but learning plays a minor, and improvisation plays a major, role in song development in other species. Song dialects, a consequence of song learning and dispersal, may be present or absent in congeners (e.g. *Zonotrichia* spp., Marler and Tamura, 1964; Lemon and Harris, 1974). Practically all possible combinations of times for sensitive periods exist: individuals of some species may learn throughout life (Laskey, 1944); some may be capable of adding new syllables to their repertoire at 18 mo (Rice and Thompson, 1968); chaffinches (*Fringilla coelebs*) are sensitive to song learning during their first fall and refine the details in the spring (Thorpe, 1958); cardinals

(*Richmondena cardinalis*) and meadowlarks (*Sturnella magna*) stop learning at a year (Lemon and Scott, 1966; Lanyon, 1957); and a number of species terminate their song learning after a couple months of age (White-crowned sparrow, *Zonotrichia leucophrys,* Marler, 1970; swamp sparrow, *Melospiza georgiana,* Kroodsma, unpublished data).

Selection seems to have run rampant in molding such a diversity of singing behaviors, and it is quite understandable that Marler (1967), in the first serious attempt at a synthesis on understanding the diverse developmental strategies involved, was quite "puzzled." But some progress has been made in the past ten years, and here I review some published and preview several unpublished recent advances in our understanding of the ontogeny of bird song. Since careful analyses of all experimental studies of song development among songbirds have revealed that juveniles must be exposed to normal adult songs in order for song development to proceed in truly normal pathways (Kroodsma, 1977a), the focus of this discussion will be on the various aspects of this song learning process. Specific topics will include where, from whom, when, how many, which, and how accurately songs are learned during song ontogeny.

Where and from Whom Songs Are Learned

Field data on this topic are very meager, with good data on only two species, the Bewick's wren (*Thryomanes bewickii*) and the indigo bird (*Vidua chalybeata*). In the Bewick's wren (Willamette Valley, Oregon), young males disperse from their home territory roughly at the age of five weeks; the birds are resident throughout the year and by 60 d of age juvenile males can establish a territory which will be maintained for the life of the individual. After recording juvenile males, their fathers, and neighboring males where the juvenile established his territory, I concluded that juveniles are probably capable of learning from their father (see also the details of the sensitive period of the long-billed marsh wren, *Cistothorus palustris,* below), but a premium is placed on matching the song types of neighboring males with whom interactions will occur throughout life; thus, song types of the father may be modified or new ones learned in order to match more closely the songs of neighboring males. So, the songs retained in the repertoire of the juvenile male are those which are acquired from and reinforced by neighboring territorial males where the juvenile is on his own territory (Kroodsma, 1974).

In the indigo bird flexibility to change song types in the repertoire of the individual is apparently maintained throughout life; a premium on song type matching by interacting conspecifics remains, for if a male moves from one area to another, his songs also change to match the songs of the other males at the new display center (Payne, 1975).

Data for several species are available from laboratory experiments, but one must be cautious in relating such data directly to what may be occurring in nature. In the zebra finch (*Poephila guttata*), the social bond with the father appears a strong social constraint on song learning, for songs of natural fathers are learned preferentially to those of conspecifics in the next cage, and songs of foster fathers of other species are learned preferentially to the songs of conspecifics which may be singing nearby. There is a hint, however, of a predisposition to learn songs of conspecifics, since young zebra finches, raised by females under conditions where they could see and hear both conspecific males and males of other grassfinch species in nearby cages, developed songs consisting only of species-specific elements (Immelmann, 1969). Laboratory data for the bullfinch (*Pyrrhula pyrrhula*) also implicate similar social constraints on song learning (Nicolai, 1959). Vocal learning in a social context also occurs among cardueline finches, where mates tend to converge on similar call notes in their repertoires, and where birds in a cohesive winter flock may also tend to share similarities in their call repertoire (Mundinger, 1970). Finally, it is the social interaction with and song learning from mates which produces the finely synchronized performance among species where mates duet with one another (Thorpe, 1972).

Timing of Song Learning

We have a rough idea of the timing of learning in a number of species. For example, Marler (1970) found that a white-crowned sparrow produced a good copy of a training song if he heard it between 8 and 28 d or between 35 and 56 d of age; exposure before eight days and after 50 d had little effect. Song sparrows (*Melospiza melodia*) have a peak of sensitivity between five and ten weeks of age, when two birds each learned seven song components (Mulligan, 1966). In the zebra finch the juvenile male gradually learns the fine structure of the song elements (through roughly day 66) and develops the proper timing and sequencing of those elements by day 80, when song learning is complete (Immelmann, 1969). The timing of song learning in some species may be hormonally related, for Nottebohm (1969) found that castration of a juvenile chaffinch delayed song learning until exogenous testosterone was administered in the bird's second year; on the other hand, Arnold (1975) demonstrated convincingly that gonadal androgens are not essential for normal song development in zebra finches.

Thus, we know roughly when birds learn their songs, but the very details of the sensitive period are lacking. What is the fine anatomy of this period of learning? Is there a peak, or are there peaks, of sensitivity? Is there a sharp or gradual onset and termination of sensitivity? What are the hormonal and neurophysiological correlates of this sensitive period among different species? Why do species differ? And lastly, how do the fine points of the sensitive period correlate with dependency on parents, dispersal, establishment of the juvenile male's own territory, or other unique features of a given species' exploitation system? Do these fine points vary geographically as these parameters of the population biology vary? In an effort to tackle some of these questions I have begun a study of song learning in the long-billed marsh wren; these insectivorous birds can be hand-reared in the laboratory (though with some difficulties compared to sparrows) and develop large song repertoires, allowing one to glean a great deal of information from a single male.

In a pilot experiment, two males were exposed to nine different song types between the ages of 15 and 65 d, another nine between 65 and 115 d, and still another nine the following spring. Both males learned all nine song types to which they were exposed before 65 d of age, but none of the other 18 songs. In refining this pilot experiment, nine juvenile males were used, but the presentation of tutor songs was refined to such an extent as to allow a very detailed determination of exactly when songs were learned. Over a 72 d period, beginning at ages 6 to 15 d, the males were exposed to a total of 44 different song types. On any given day males heard 1000 repetitions of each of three different song types. However, some song types were presented over 9-d periods (9000 exposures/9 d; $n = 8$), some were presented over six-day periods (6000 exposures/ 6 d; $n = 12$), while other songs were presented over three-day periods (3000 exposures/ 3 d; $n = 24$ song types).

The nine long-bill males learned an average of 16 different songs (see Fig. 1), and were adept at learning song types presented over nine-day as well as three-day periods. Song learning occurs roughly between 15 and 60 d of age, and when data for all nine males are graphed together, two peaks of sensitivity appear, one near day 35 and the other near day 53 (Fig. 2). All the birds were housed together and some improvised song types did occur in the repertoires of several of the males; however, several internal controls in the experiment indicate that the influence of males upon each other played a relatively minor role in shaping the sensitive period as depicted in Figure 2.

In spite of the detail in the picture of the sensitive period, caution must be exercised in attempting to relate this sensitive period to events occurring in nature. Relating this period of sensitivity to (1) gaining of independence by the juvenile male, (2) dispersal, (3) establishment of

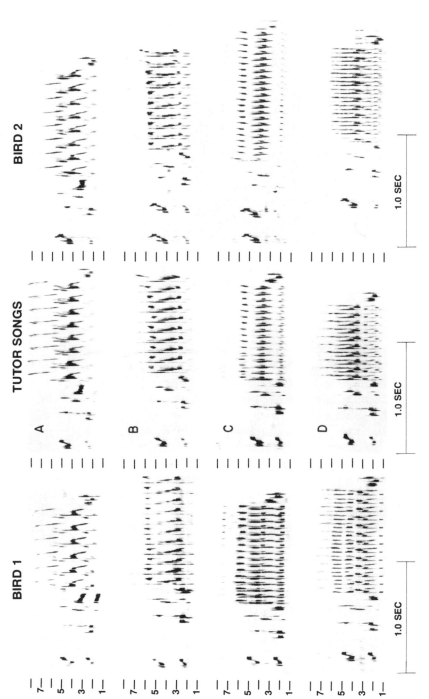

FIG. 1. Examples of songs that two male long-billed marsh wrens learned from the training tapes. The vertical scale on each sonogram is kHz, the horizontal is seconds.

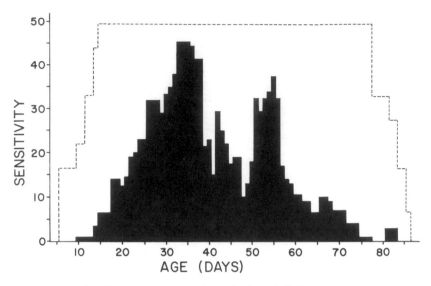

FIG. 2. *The sensitive period for the long-billed marsh wren, a composite of individual graphs from each of nine males.*

territory, (4) the role of hormonal levels, or possibly the (5) postjuvenile molt would be an exciting advance. The clear-cut sensitive period for learning from tutor tapes in the laboratory does provide a nice model for establishing various hormonal and neurophysiological correlates of song learning, but further data on song learning in the long-bill suggest that events in nature may be considerably more complicated than as presented in Figure 2: (1) in Welter's (1935) study at Ithaca, N.Y., the last peak of hatching occurred after the adult males had stopped singing for the year! Can these juveniles postpone the sensitive period until the following year, possibly by a mechanism similar to that which Nottebohm (1969) simulated through castration of a male chaffinch? (2) Five New York juvenile males learned no songs from nine other New York males to which they were exposed after 120 d of age; however, one Michigan male, under the same conditions, learned five of a possible 15 songs. Population differences in degree or timing of song learning present ideal opportunities for a comparative approach to further our understanding of the selective forces behind different strategies. (3) Finally, two males exposed to tape recorded model songs until day 45, and then exposed to a live tutor, rejected those songs which were heard from the tutor tape and learned the songs from the live bird. Social interactions may play a crucial role in determining exactly where, when, and from whom songs are learned in nature.

How Many Songs Are Developed

The number of song types developed varies from species to species (see introduction), but may also vary intraspecifically, both within and between populations. Within a population, for example, three rock wrens (*Salpinctes obsoletus*) had 119, 85, and 69 song types ($n = 1234$, 1276, and 2262, respectively; Kroodsma, 1975). In Bewick's wrens, the number of song types or song components that a male develops is dependent upon his date of hatching; song learning is completed during the first fall, and birds hatched early in the breeding season develop more songs than birds hatched later in the season (Kroodsma, 1972b). In other species where birds can add or modify their vocal repertoire throughout life, repertoire size may continue to increase and be age-correlated. Experiments with canaries (*Serinus canarius*) indicate that females during mate selection are probably attentive to this intraspecific variability in male song (Kroodsma, 1976).

Variation in song repertoire sizes among different populations of a given species may be striking. Long-billed marsh wrens near Seattle, Washington, may have 150 song types, while an Illinois male may have as few as 30 (Verner and Kroodsma, in preparation). Whether such a variation is a consequence of intensified or relaxed selection, or a result of climatic conditions which dictate a resident or migrant status which could in turn control time available for song development, must await further study.

Those wren species which develop large song type repertoires deserve further mention, for several population parameters seem well correlated with the repertoire size and singing behaviors of these species. Individual long- and short-billed marsh wrens (*Cistothorus platensis*) and rock wrens have been studied with over 100 song types in their repertoires, but repertoire sizes for house (*Troglodytes aedon*), winter (*T. troglodytes*), Bewick's, Carolina (*Thryothorus ludovicianus*), canyon (*Catherpes mexicanus*), and probably cactus (*Campylorhynchus brunneicapillus*) wrens do not approach these values. Furthermore, these three species with the larger song repertoires use frequently or exclusively song organizations involving more immediate variety (i.e., ABCABC . . . MNOMNO . . ., or ABCDEFGHIJKL . . .) rather than the more monotonous organization where a given song type is repeated many times in succession before switching to another type (i.e., AAAAAA . . . BBBBBB . . .), as occurs in the other 6 wren species. These three wren species occur in higher densities in communities of lower avifaunal diversity, where intraspecific interactions are likely to be more frequent and intense; two of the three are highly polygynous, implicating epigamic selection as possibly playing a role (Kroodsma, 1977b).

In the long-billed marsh wrens of eastern Washington, where males synchronize themselves in long sequences and countersing with like songs, the repertoire sizes of males are surprisingly similar: for 5 males studied the number of song types was 114, 114, 114, 109, and 107 ($n = 1257$, 730, 771, 446, and 461, respectively; Verner 1975)! With a total of only 127 different song types among the 5 males, the intricate communication system of the long-bill seems to place a premium on neighboring males having identical song type repertoires.

With two captive long-billed marsh wrens, the ontogeny of song matching (or countersinging) was quite evident. Two males, each with identical repertoires of 9 song types, were housed in adjacent cages and allowed to interact freely. Bird 2 lagged behind bird 1 in song development, and remained in plastic song while bird 1 was in full song. Over all, bird 2 matched the song type of bird 1 nearly 1/3 of the time (1/9 expected; see Table 1), while bird 1 matched the song type of bird 2 no more than expected by chance alone. However, when the songs of bird 2 were amplified over a loudspeaker system, bird 1 matched the songs of bird 2 to a much greater extent (18.5%). Furthermore, during the 40 d that these two birds were studied, there was a gradual reduction in the percent of the time that bird 2 matched the songs of bird 1. It is conceivable that such an ontogeny of song matching and countersinging, coupled with loudness of delivery, could betray the overall vigor or age of a bird, as it seemed to under these experimental conditions, and might even be used by conspecifics in assessing the quality of potential mates or rivals.

TABLE 1. A summary of the singing behaviors and interactions of two long-billed marsh wren males which were hand-reared and studied in the laboratory

		FOLLOWING EVENT			
		Bird 1		Bird 2	Expected
		FREE SINGING		COUNTERSINGING	
	Bird 1	* Matching Self	15.8%	* Match Bird 1 31.1%	11.1%
		* Tutor Sequence	31.8%	* Tutor Sequence 24.6%	12.3%
PRECEDING EVENT		COUNTERSINGING		FREE SINGING	
	Bird 2	* Tutor Sequence	22.6%	* Tutor Sequence 25.9%	12.3%
		Match Bird 2	11.4%	* Matching Self 14.6%	11.1%

* Different from expected by random singing; chi-square, $p < 0.001$

INFLUENCE ON THE NATURE OF SONG DIALECTS

Various constraints on vocal learning often insure that interacting conspecifics will possess like vocalizations. Thus, a male Bewick's wren is probably highly capable of learning from its father before 35 d of age; however, he may reject or modify those paternal song patterns after he disperses (35–100 d; *when*) in order to learn from and match the songs of conspecific adult males (*which* songs and *from whom*) who hold territories adjacent to the territory which the juvenile male declares (*where*). Close examination of the microgeographical distribution of song patterns or song elements reveals that in the majority of songbirds neighboring males have songs more similar to one another than to more distant males; males usually respond the strongest to songs most like their own (i.e., songs of the same "dialect"), and if they have several different songs in their repertoire, they often countersing with like songs as well.

Yet the nature of the "dialect" in different species may be largely dependent on *how many* different song patterns are in the repertoire of the individual. In the white-crowned sparrow, the great majority of males have a single song type, and in California chapparel, boundaries between different dialects may be very abrupt, even in the absence of habitat barriers, enabling one to map "populations" where all males possess like songs (Baptista, 1975). On the other hand, a Bewick's wren male may have 16 different song types; *each* song type may have a unique microgeographical distribution, often with very sharp boundaries and hybrid songs at contact zones of different song patterns. Emphasis appears to be on neighboring males having like songs, as in the white-crowned sparrow, but in the wren one is not able to map "populations" of birds having identical song repertoires, for the entire system appears as 16–20 *z. 1. nuttalli* dialects superimposed on one another.

As more studies emerge, it is becoming increasingly apparent that most songbirds have more than a single song type in their song repertoires, suggesting that something quite unique may be occurring in a species where males have a single, precisely copied song. Instead of the usual question as to "Why do some species develop such large song repertoires?", the opposite question may be appropriate here: "What is unique to the reproductive biology of a species like the white-crowned sparrow that selection has so limited their song repertoires?" Sharp dialect boundaries where all songs in the birds' repertoires change is far more feasible when only one song is involved than when larger numbers of songs are involved (given similar habitat conditions and degree of continuity). Have advantages of sharp dialect boundaries selected for essentially one learned song type in the white-crown?

Baker's (1975) demonstration of genetic differences of birds in neighboring dialect areas is an exciting step; he postulates that dialect areas may have a historical origin, with colonizing individuals establishing new dialect areas. Efforts to determine whether dialect boundaries can actually repel dispersing juveniles is under way (Baker, in preparation).

Selectivity of Song Models

Among mimics such as the mockingbird, much song learning from heterospecifics occurs, but even among other species, the nonmockers, occasional reports do occur of some learning from other species (Kroodsma, 1972a, 1973; Baptista, 1972). Yet, for the most part, species identity in the wild is totally unambiguous, partly because of predispositions to learn only conspecific song.

Two experiments at the Rockefeller University Field Research Center have tested the selectivity in song learning between conspecific pairs, two *Cistothorus* wrens and two *Melospiza* sparrows. For the wrens, a tutor tape with 67 different songs was prepared. On this tutor tape were (1) songs of normal short-billed marsh wrens ($n = 9$), which consist of an introduction of several notes and a trill, (2) songs of normal long-billed marsh wrens ($n = 9$), which consist of an introduction, trill, and several concluding notes, (3) normal songs of Bewick's wrens ($n = 4$), (4) normal swamp sparrow songs ($n = 5$), and (5) all manner of song combinations with (a) either long- or short-bill introductions, (b) trills of long-bills, short-bills, swamp sparrows, or Bewick's wrens, and (c) presence or absence of a long-bill conclusion ($n = 40$). The short-bills learned nothing from the tapes, whereas the long-bills learned preferentially those songs which contained long-bill trills; this species difference will be explored further below. Interesting here is the fact that the probability that a long-bill would learn a given trill was directly related, exponentially, to the number of long-bill components (0 to 3) that were in the song.

Song and swamp sparrows are closely related members of the genus *Melospiza*. The normal songs of the two species are of similar length but very different temporal organization. The swamp sparrow song is relatively simple, consisting of a phrase of repeated syllables, but the song of the song sparrow is more complex, usually with at least four phrases. As in the *Cistothorus* wrens, the preferred micro-habitats differ, but territories of the two species may abut or overlap. Data from the laboratory indicate that learning of song models plays an important role in song development in the males of each species. In order to test the selectivity of the song learning process, artificial songs were made from the elements of natural song of the two species. Swamp sparrow features in the training songs included single phrase songs consisting of sequences of identical syllables delivered at a steady rate, while song sparrow features involved variable rates of syllable delivery (accelerating or decelerating) and a multi-partite song structure (two parts). Natural syllables from each species song were edited from field recordings and spliced together to make the training songs, 22 of which were presented to each male of the two species.

Eight male swamp sparrows learned 12 syllables from the tutor tapes, and all twelve were of swamp sparrow origin. Four of the learned syllables were from one-phrase songs, while eight were from two-parted songs; five came from phrases of a steady rate, while seven were learned from either accelerated or decelerated syllable series. The swamp sparrow clearly can recognize conspecific syllables regardless of their context in the training songs. The five song sparrows, on the other hand, learned 22 syllables from the training tapes; 11 were of swamp and 11 of song sparrow origin, with no particular preferences exhibited for learning syllables of a given rate of delivery or from one- or two-parted songs (Marler and Peters, 1977).

The evolution of such different ontogenies and song selectivities in the two *Melospiza* sparrows is puzzling. Why is there such a difference, or, to what features of each species' life history are these ontogenies adapted? Evolutionary strategies may become clearer as song development and life history parameters of other closely related species are studied. Such hope is encouraged from studies of the *Cistothorus* wrens, where the differing song ontogenies of the two wren species may be well suited to each species' exploitation system.

Role of Improvisation vs. Learning Among Cistothorus Wrens

The two North American marsh wrens have recently been reclassified as congeneric (*Cistothorus* spp., Thirty-third Supplement to the A.O.U. Checklist, 1976), and are an ideal species pair for a comparative approach to studying vocal ontogenies. The two species are alike in so many ways: (1) males of each species are polygynous (Verner 1965, Welter 1935, Kale 1965; Crawford 1977); (2) the density of breeding pairs is very high compared to all other North American wrens; (3) the preferred habitats are marshes or wet meadows (essentially monolayers without an appreciable vertical dimension) in which the overall avifaunal diversity of the community is very low, usually making the wren the most abundant species present (Kroodsma, 1977b). To an extent, their singing behaviors are also very similar: males of both species sing a relatively high percentage of the time (30 to 40% singing, 60 to 70% in silence during an actual singing performance), use large song-type repertoires that may number well over 100 in some populations, and continue to sing (presumably for additional mates), often day and night, throughout much of the breeding season.

However, the details of the singing behavior differ markedly. All long-bill males in a population possess nearly identical song-type repertoires (Verner, 1975); in the laboratory, males between the ages of 20 and 80 d readily learn the precise details of songs presented to them over loudspeakers. On the other hand, short-bill males in a natural population have very dissimilar song types (except for a stereotyped introduction; see below) and unequal repertoire sizes; they cannot countersing with like songs (though they definitely do influence one another while singing); and males in the laboratory between 15 and 90 d of age demonstrate little to no ability to learn tutor songs presented over loudspeakers. Yet these experimental short-bills are capable of producing sizeable song repertoires which contain, for the most part, songs which are very similar to normal wild-type songs.

In the long-billed marsh wren, juvenile males learn the songs of adults and then, in order to maintain the observed homogeneity of song patterns within a population, must settle to breed nearby and then return yearly to that locality. Juveniles in eastern Washington do disperse a considerable distance from their father's territory ($\bar{x} = 1951$ m), but fidelity to the site of breeding (and perhaps the site of song learning?) is quite strong: seven of 13 adults returned to the same or adjacent territories in successive years (Verner, 1971). On the other hand, the occurrence of the short-bill from year to year can be very unpredictable; birds may appear on territory or desert their territory in the middle of the normal breeding season. Of ten males banded in Minnesota, none returned to the same locality in the following year (J. Burns, personal communication).

The key to the nomadic-like movements of the short-billed marsh wren, as opposed to the long-bill's, may be found in the habitat stability of preferred breeding sites. The short-bills breed in wet meadows, while the long-bills occur in marshes with more standing water; during periods of insufficient rainfall, the wet meadows are the first to dry up, and even if the two species are breeding in the same vicinity, the short-bills are the first to lose their breeding sites (R. E. Stewart, personal communication). It is probably a direct result of this relative habitat instability and unpredictability that the short-bill demonstrates far less site tenacity than does the long-bill; the fact that no subspecific differentiation has occurred in the short-bill is further suggestive evidence of a relatively high degree of movement within the breeding range of the species.

The song ontogeny in the two species seems nicely correlated with their respective ways of life. The short-bill does not learn precise details of the song, but develops a generalized song (actually a large number of them) which is recognizable as that of the short-billed marsh wren throughout the range of the species in North America. Furthermore, the brief introduction of the song, involving at most four to five different notes, is practically identical among all birds studied, including birds from Michigan, Minnesota, Illinois, North Dakota, and the wintering grounds in Florida; this same introduction did not develop normally in the hand-reared birds, but they did develop introductions in common with one another, suggesting that this very stereotyped introduction is learned. So the short-bill male develops a very stereotyped introduction and a very generalized conclusion to his song, both of which label the species; however, the details offer little clue as to the geographic origin of the bird.

While the song of the long-bill is unmistakable throughout the range of the species, neighboring males tend to have identical songs. Also, considerable geographical variation does exist, and experiments would probably reveal that long-bill males discriminate local versus more distant songs (as has been demonstrated in so many other species where songs are learned and local "dialects" exist). Subspecific differentiation is extensive among the long-bills, though this occurs predominantly in salt marshes which are outside the geographical range of the short-bill.

Summary

Learning of adult motor patterns plays a role in the song ontogeny of all oscines that have been studied in detail, but there are many different facets to this learning process. (1) Field data reveal that in some species a male learns the songs of adults (*from whom*) at the location where that male will breed (*where*); neighboring males thus can interact with nearly identical songs. (2) In most species, song learning is limited to the first couple months of life (*when*), but within a given species this may vary among populations and is undoubtedly very dependent upon social interactions. (3) Males of some species develop only 1 song pattern per individual, while males of other species may develop more songs (*how many*) and interact in complex ways; the number of songs learned by each male in a population may affect the nature of song dialects. (4) Among *Melospiza* sparrows, the swamp sparrow rejects song sparrow syllables and learns to sing only conspecific song syllables, but the song sparrow will learn song elements of either species with equal facility; selectivity (*which* songs are learned) of song models differs markedly even among closely related species. (5) The *accuracy* of copying models during song learning among *Cistothorus* wrens may have coevolved with the relative stability of the habitats occupied by the two species; short-billed marsh wrens occupy relatively unstable wet meadows, have a generalized song, and populations appear highly mobile, whereas the long-billed marsh wren lives in more stable marshes, has population specific song patterns and/or sequences, and demonstrates more site tenacity than does the short-bill.

Acknowledgments

Financial support has been provided by the Frank M. Chapman Memorial Fund and by NSF Grant No. BNS76-07704A01. I thank Jared Verner and colleagues at the Rockefeller University Field Research Center for valuable discussion and comment, Roberta Pickert for her unwavering assistance in the research, Cathy Quimby for helping with field work, and Melissa Kroodsma for her unlimited patience, encouragement, and assistance in all phases of these labors.

References

American Ornithologists' Union. 1976. Thirty-third supplement to the American Ornithologists' Union Check-list of North American birds. *Auk* 93:875–879.

Arnold, A. P. 1975. The effects of castration on song development in zebra finches (*Poephila guttata*). *J. Exp. Zool.* 191:261–277.

Baker, M. C. 1975. Song dialects and genetic differences in white-crowned sparrows (*Zonotrichia leucophrys*). *Evolution* 29:226–241.

Baptista, L. F. 1972. Wild house finch sings white-crowned sparrow song. *Z. Tierpsychol.* 30:266–270.

Baptista, L. F. 1975. Song dialects and demes in sedentary populations of the white-crowned sparrow (*Zonotrichia leucophrys nuttalli*). *Univ. Calif. Publ. Zool.* 105:1–52.

Bertram, B. 1970. The vocal behavior of the Indian hill mynah, *Gracula religiosa*. *Anim. Behav. Monogr.* 3:79–192.

Crawford, R. D. 1977. Polygynous breeding of short-billed marsh wrens. *Auk* 94:359–362.

Hartshorne, C. 1973. *Born to sing. An interpretation and world survey of bird song.* Bloomington: Indiana Univ. Press.

Kale, H. W., II. 1965. Ecology and bioenergetics of the long-billed marsh wren (*Telmatodytes palustris griseus* [Brewster]) in Georgia salt marshes. *Publ. Nuttall Ornithological Club* No. 5.

Kroodsma, D. E. 1972a. Variation in songs of vesper sparrows in Oregon. *Wilson Bull.* 84:173–178.

Kroodsma, D. E. 1972b. Singing behavior of the Bewick's wren: development, dialects, population structure, and geographical variation. Ph.D. dissertation, Oregon State Univ., Corvallis, Oregon.

Kroodsma, D. E. 1973. Coexistence of Bewick's wrens and house wrens in Oregon. *Auk* 90:341–352.

Kroodsma, D. E. 1974. Song learning, dialects, and dispersal in the Bewick's wren. *Z. Tierpsychol.* 35:352–380.

Kroodsma, D. E. 1975. Song patterning in the rock wren. *Condor* 77:294–303.

Kroodsma, D. E. 1976. Reproductive development in a female songbird: differential stimulation by quality of male song. *Science* 192:574–575.

Kroodsma, D. E. 1977a. Correlates of song organization among North American wrens. *Am. Naturalist.* 111:995–1008.

Kroodsma, D. E. 1977b. A reevaluation of song development in the song sparrow. *Anim. Behav.* 25:390–399.

Kroodsma, D. E., and L. D. Parker. 1977. Vocal virtuosity in the brown thrasher. *Auk* 94:783–785.

Immelmann, K. 1969. Song development in the zebra finch and other Estrildid finches. *In Bird vocalizations,* ed., R. A. Hinde, London and New York: Cambridge Univ. Press.

Lanyon, W. E. 1957. The comparative biology of the meadowlarks (*Sturnella*) in Wisconsin. *Publ. Nuttall Ornithological Club* No. 1.

Laskey, A. R. 1944. A mockingbird acquires his song repertory. *Auk* 61:211–219.

Lemon, R. E., and M. Harris. 1974. The question of dialects in the songs of white-throated sparrows. *Can. J. Zool.* 52:83–98.

Lemon, R. E., and D. M. Scott. 1966. On the development of song in young cardinals. *Can. J. Zool.* 44:191–197.

Marler, P. 1967. Comparative study of song development in sparrows. *Proc. 14th Int. Ornithol. Cong.* 1966:231–244.

Marler, P. 1970. A comparative approach to vocal development: song learning in the white-crowned sparrow. *J. Comp. Phys. Psychol.* 71(No. 2, Part 2):1–25.

Marler, P., and S. Peters. 1977. Selective vocal learning in a sparrow. *Science* 198:519–521.

Marler, P., and M. Tamura. 1964. Culturally transmitted patterns of vocal behavior in sparrows. *Science* 146:1483–1486.

Mulligan, J. A. 1966. Singing behavior and its development in the song sparrow *Melospiza melodia. U. C. Publ. Zool.* 81:1–76.

Mundinger, P. C. 1970. Vocal imitation and individual recognition of finch calls. *Science* 168:480–482.

Nicolai, J. 1959. Familientradition in der gesangsentwicklung des gimpels (*Pyrrhula pyrrhula* L.). *J. Ornithol.* 100:39–46.

Nottebohm, F. 1969. The "critical period" for song learning in birds. *Ibis* 111:386–387.

Payne, R. B. 1975. Song dialects and population structure in the indigo birds of

Africa. *Proc. Symp. on Dialects in Bird Song*. Media services, St. Louis University.

Thorpe, W. H. 1958. The learning of song patterns by birds, with especial reference to the song of the chaffinch, *Fringilla coelebs*. *Ibis* 100:535–570.

Thorpe, W. H. (1972). *Duetting and antiphonal song in birds. Its extent and significance*. Leiden. E. J. Brill.

Rice, J. O., and W. L. Thompson. 1968. Song development in the indigo bunting. *Anim. Behav.* 16:462–469.

Verner, J. 1965. Breeding biology of the long-billed marsh wren. *Condor* 67:6–30.

Verner, J. 1971. Survival and dispersal of male long-billed marsh wrens. *Bird-banding* 42:92–98.

Verner, J. 1975. Complex song repertoire of male long-billed marsh wrens in eastern Washington. *Living Bird* 14:263–300.

Welter, W. A. 1935. The natural history of the long-billed marsh wren. *Wilson Bull.* 47:3–34.

Section Five

PUSHING ON TO THE PRIMATE PINNACLE

"Primates and other mammals"—how frequently this phrase can be found in literature dealing with behavior. "Primate" and "human" are also frequently contrasted. This dichotomy is simply a bad habit, as erroneous as talking about "fish and wildlife." In this section both nonhuman and human primates are considered. William Mason uses the results of studies in which rhesus monkeys were raised with different types of substitute mothers to answer questions concerned with cognitive development. In particular, he deals with the development of two kinds of "knowing"—"knowing how" (a specific competence) and "knowing that" (information about certain aspects of the world). Jeanne Altmann also considers mother-infant interaction in her chapter. In her field study of yellow baboons, she found that maternal style, either restrictive or laissez-faire, emerged as a major variable affecting the development of infant dependence by the first six months. Maternal dominance rank, not infant gender, was a very important predictor of maternal style. In their chapter, Robert Plomin and David Rowe stress the utility of the behavior-genetic approach to human social behavior, particularly social responsiveness in infants. They point out the importance of studying within-species variability and not discarding it as "noise in the system." Their data suggest that genetic influences may be more important in the development of social responsiveness to strangers than to the mother. Lawrence Harper and Karen Huie consider the development of sex differences in humans and suggest that such dimorphism can be understood as biologically based predispositions that evolved as a result of the hunter-gatherer lifestyle. Some dimorphic predispositions are evident in the free play of young children. Since human behavior has an evolutionary history, the use of phylogenetic arguments is appropriate to analyses of human social behavior. Also, the comparative approach, when used carefully and cautiously, is an invaluable tool in analyses of human behavioral evolution and ontogeny.

Chapter Twelve

Social Experience and Primate Cognitive Development

WILLIAM A. MASON

Psychology Department
and
California Primate Research Center
University of California
Davis, California 95616

I know how to ride a bicycle, eat with chopsticks, operate a typewriter without looking at the keys, and ask for a bottle of beer in Spanish. I also know that a bicycle is harder to pedal against the wind, chopsticks will not turn to limp pieces of spaghetti in my hand, a typewriter will not produce prose on its own initiative, and the beer will taste bitter.

Obviously, I am using *know* in these two sets of examples in two different senses. The first type of knowing refers to something I can do or claim to be able to do; at issue is a specific competence that can be inferred by assessing my actions against explicit criteria. The second type of knowing is broader and more generic. The examples do not refer to particular skills or achievements, but to information I claim to have about certain aspects of the world in which I live. Gilbert Ryle expresses a similar idea in his distinction between knowing *how* and knowing *that* (Ryle, 1949).

The difference between the two kinds of knowing is familiar to animal behaviorists. It is reflected, for example, in the use of such terms as *habit* to refer to a learned tendency for a specific stimulus to elicit a certain response and *expectancy* to refer to the anticipation that certain consequences are likely to follow the appearance of a particular stimulus or the performance of a particular response. Nevertheless, we are not entirely comfortable with the distinction and we often ignore it, particularly in research on the cognitive consequences of early social experience.

I believe that this is a mistake—at least when we are dealing with some species of nonhuman primates—and I hope to demonstrate why. The theme of this essay is that if we are to reach an understanding of the role of social experience in the cognitive development of certain species it is essential to distinguish between the organism's competences, which are reflected in its specific achievements, and its generalized expectancies or coping strategies, which are reflected in its basic stance toward the environment (Lewis and Goldberg, 1969). It is a distinction, if you will, between the individual's successes in dealing with problem situations and its style in doing so. Although both kinds of knowledge are dependent on experience, I believe they have different developmental antecedents and carry quite different implications for behavioral adaptability. A primary concern in what follows is to illustrate what I mean by generalized expectancies or coping strategies, and to show that the kind of strategy that a developing individual acquires can be heavily influenced by its relationship with its mother.

The evidence will be drawn chiefly from our findings on rhesus monkeys raised with different types of substitute mothers. I recognize that the attachment figure is only one of the many sources in the early environment that are likely to contribute to cognitive development. Nevertheless, the peculiar features of the filial bond give the attachment figure a special status, and they suggest with unusual clarity the kinds of experiences that are involved in the acquisition of coping strategies and the developmental processes through which they operate.

Consider the nature of the filial bond. Anyone who has observed an infant rhesus monkey with its mother probably shares my impression that she is the most interesting, important, and compelling object in its small world. Much of its early behavior is organized around her. She is its point of departure for forays into the surrounding environment and a haven to return to when things get out of hand. She is the focus of attention, a source of rewards and punishments. Her comings and goings, her actions on the environment, have a salience for the infant that no other object commands.

If this characterization is essentially correct, it appears that the attachment figure occupies a privileged and influential position, one particularly well suited to shape the early development of behavior. Yet, in spite of the amount of attention that has been devoted to mother–infant relations in nonhuman primates, we have little systematic information on the cognitive implications of the maternal role.

I believe that one reason our understanding of the mother's influence on cognitive development has not advanced further is our reluctance to give credence to the distinction between competence and coping strategies. The source of this reluctance is easy to appreciate. We are comfortable with the concept of competence because it is easily translated into operational terms. It has a concreteness that lends itself readily to empirical demonstration. In contrast, coping strategies are more patently and persistently hypothetical. One can never hope to find a specific instance—a point-at-able example—of a coping strategy. In order to draw an inference regarding its presence or nature in an individual, it is necessary to sample his behavior across many situations, to classify those situations in terms of a common pattern or theme, and to show that his reactions across the situations conform to a general plan that possesses potential biological utility for him, in the sense that it is likely to lead to an adaptive or "life–preserving" outcome. With so many problematic features, can there be any wonder why we have been unwilling to make a place for such a concept in our thinking?

Wild-Born and Laboratory-Raised Monkeys

Speaking for myself, I did not set out with any particular convictions regarding the distinction between the two kinds of knowing, and only gradually came to realize its importance to problems of primate socialization and cognitive development. My first inkling that the distinction might be useful occurred about twenty years ago. I was engaged in a series of experiments comparing the behavior of wild-born rhesus monkeys with the behavior of monkeys raised from birth in individual cages. The results showed clearly that laboratory-raised monkeys were not socially competent. This was most evident in sexual performance, particularly by males. In interpreting these findings I emphasized that laboratory-raised monkeys lacked many elementary social skills (Mason, 1960). I still believe that this interpretation is correct.

Nevertheless, certain observations and results coming out of this research were not readily explained on the basis of a simple lack of competence. For example, one test designed to measure the monkey's attraction to conspecifics required that the animals first learn to pull a simple latchstring. I used food to pretrain them for this task (Mason, 1961). The laboratory-raised monkeys required much more time than the wild–born animals to learn to pull the string, and it was evident that their whole approach to the problem was qualitatively different. Wild–born monkeys gave every indication that they perceived the situation as a "problem" in which some solution was possible. Some monkeys hit upon the latchstring straight off, but others in this group tried a variety of different responses until they discovered the correct one. In contrast, the laboratory-raised monkeys behaved as though they lacked any consistent strategy or plan. In order to train some of them to pull, I had to resort to various tricks, such as smearing banana on the string or giving personal "demonstrations." Although all monkeys eventually acquired the appropriate response and continued to perform efficiently once they had done so, the differences between groups in their initial approach to the problem made a lasting impression on me.

Another test in this series suggested even more clearly that something other than competence was involved in the differences between groups. The monkeys were observed while alone in a 12' × 14' room. As compared to laboratory-raised monkeys, those born in the wild more frequently engaged in gross motor activities (jumping, backward somersaults), had higher locomotion scores, defecated or urinated in more sessions, and more frequently touched the objects that were placed in the room. All these differences were statistically significant. They also vocalized more, although not reliably so (Mason and Green, 1962). Clearly, "skill" was not an important factor in this situation. There was no "problem"—at least from my point of view—inasmuch as I had set no task for the monkeys, had provided no visible goal that they could reach. Nothing they did changed the situation; it was simply a matter of the monkeys remaining in the room for the prescribed period, whereupon they were removed and returned to their living cages. Yet, even though it made no sense in this context to describe one group as more or less competent than the other, it was clear that the groups differed markedly in their styles of coping with confinement in a strange place.

Mobile and Stationary Artificial Mothers

The problem of coping strategies was raised again a few years later, but this time in an experiment that drew attention to the potential importance of the attachment figure. The research involved two groups of laboratory-raised rhesus monkeys, both maternally separated at birth and placed with cloth artificial mothers that were identical in construction except that for one group the surrogates were stationary and for the other they moved up and down and around the cage on an irregular schedule throughout the day (Mason and Berkson, 1975). The original purpose of the experiment was to test the hypothesis that the stereotyped body-rocking shown by most macaques raised alone or with inanimate surrogates was a response to the absence of maternal movement. This expectation was fully confirmed (Mason and Berkson, 1975).

It soon became apparent, however, that the mobile artificial mother was doing a great deal more than just carrying the infant about the cage. We had unwittingly created a social substitute that was capable of simulating some of the generic attributes of social interaction. The movements of the surrogate were not completely predictable: It could withdraw from the infant without warning, or sneak up behind it and deliver a gentle rap on the head; its comings and goings demanded adjustments that were not required of the monkey raised with a stationary device. The mobile mother also stimulated and sustained interaction: It was withdrawn from, pursued, pounced on, and wrestled with. Rough–and–tumble play, for example, was about three times more frequent in monkeys raised with moving surrogates than in those raised with stationary devices.

Naturally, we wanted to explore the nature and range of developmental effects that these unanticipated contrasts in the two types of artificial mothers might have produced. We observed the monkeys in many different situations, and where intergroup differences were found they suggested that animals raised with mobile mothers were more like wild–born monkeys than those raised with stationary devices. For example, when the monkeys were about nine months old they were tested in a novel room in a replication of Mason and Green's study comparing wild–born and maternally separated monkeys. Monkeys raised with mobile mothers more often entered the room without prompting than did those raised with stationary mothers; they also had higher scores for gross motor activities, contact with objects in the room, urination, and defecation. The measures that differentiated monkeys raised with mobile and stationary surrogates and the direction of the differences were the same as those that differentiated wild–born and laboratory–raised monkeys in the original experiment (Mason and Berkson, 1975).

When these monkeys were about two years old, more than a year after they were permanently separated from their artificial mothers, we measured their tendency to look at other monkeys. Various stimulus conditions were used, such as a mother with her infant, a juvenile male, and monkeys of other species. The data clearly demonstrated that the level of looking behavior was higher in monkeys raised with mobile surrogates than in those raised with stationary dummies (Eastman and Mason, 1975). Figure 1 presents the results by sessions, summed across viewing conditions, for the two laboratory–raised groups and a wild–born comparison group.

Tests of problem-solving and social behavior produced few dramatic intergroup differences, although the general pattern was consistent with

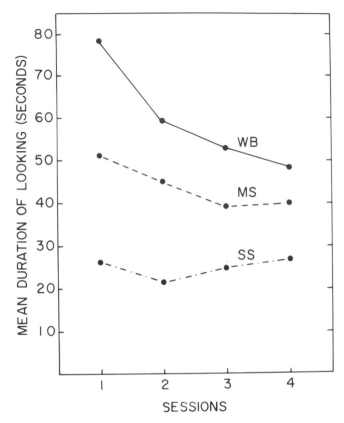

FIG. 1. *Duration of looking by wild-born monkeys (WB) and monkeys raised with mobile (MS) or stationary (SS) artificial mothers.*

that suggested by other results. For example, in the initial series of problem–solving tests, monkeys raised with mobile mothers contacted significantly more problems, even though they were no more successful in solving them (Anastasiou, 1970). The first social pairings (starting when the monkeys were about 14 months old) indicated that animals raised with mobile surrogates approached other animals more and withdrew from them less, and made fewer threats and attacks (Mason and Berkson, 1975). When the monkeys were 4–5 years of age they were tested for a second time in a social setting; once more we found that relations were less tempestuous in the mobile–surrogate group. Moreover, the only females to brace their legs and support the males during mounting attempts, and the only male to show the complete mounting pattern, were raised with mobile mothers (Anderson, Kenney, and Mason, in press).

I was convinced by the time this research was completed that we needed to look much more carefully at the attachment figure as a determinant of early cognitive development. It seemed obvious that a cloth–covered dummy, even a mobile one, was not the optimum vehicle for such a task. Another approach was required. We needed an attachment figure that was less mechanical, more truly social, and at the same time less "specialized" than the natural mother.

Dogs as Mother Substitutes

With this end in view we did some preliminary research with dogs as mother substitutes (Mason and Kenney, 1974). The monkeys formed strong attachments to the dogs and the dogs seemed to like the monkeys. They played together, slept together, groomed each other. The dogs conformed completely to the popular stereotype of tolerant, accepting, and highly social creatures. At the same time, however, there was no indication that the dogs (all female) responded maternally to the monkeys. Our hope was that such a "generalized" companion would as surrogate mother throw more light on the broader cognitive consequences of early social relationships than a natural mother, whose behavior has been shaped by evolution to complement, support and direct the development of her offspring along species–typical paths. It was also convenient to have a "mother" who was not undergoing developmental changes concurrently with those in the infant, and who would not object violently when we removed the infant for brief periods in order to test it.

Our current project includes monkeys separated from the mother at birth and raised with either inanimate surrogates or with dogs. The inanimate surrogates are plastic hobby horses mounted on wheels and they are covered with an acrylic fur saddle (Fig. 2); the dogs are female mongrels (Fig. 3). All monkeys are housed outdoors with their surrogates in kennels that afford them frequent visual contact with people, dogs, other monkeys, and a variety of other ongoing and occasional events. Moreover, from the third to the 15th month of life every monkey was routinely allowed to roam in several different complex outdoor enclosures containing a variety of playthings, puzzles, barriers, and climbing devices. Our aim was to provide each monkey with a varied and stimulating environment, and I am confident that their general experience was more diversified and "enriched" than that of the typical mother–raised laboratory macaque, to say nothing of the maternally separated infant raised in a nursery. We hoped in this way to obviate some of the confounding produced by the general environmental restriction present in

FIG. 2. *Rhesus monkey with inanimate mother substitute.*

most primate rearing studies and to gain a clearer view of the way in which interaction with an attachment figure contributed to cognitive growth.

Six monkeys were assigned to inanimate surrogates and six monkeys to each of two dog groups. The dog–raised monkeys differed in that in one group, the Free–dog group, dog and monkey could roam together in the complex environments, whereas in the other, the Restricted–dog group, the dog was confined to a small region in the exposure environment that was visually isolated from the whole. Monkeys raised with hobby horses were exposed to the complex environments precisely as the monkeys in the Restricted–dog group. The project is now in its fourth year and is continuing. Sufficient data have been collected, however, to establish differences between all groups. The most pervasive and abiding contrasts are between the dog–raised monkeys and those raised with inanimate surrogates and I will limit myself here to such comparisons.

These data provide the strongest suggestion that we have obtained thus far that the monkey's basic stance toward the environment, its "generalized expectancies" or characteristic "coping strategies," are

FIG 3. *Rhesus monkey with canine mother substitute.*

heavily influenced by the kind of attachment figure with which it has been raised. Our results indicate that monkeys raised with dogs are more attentive to the environment, more responsive, less likely to be indifferent when confronted with change, and more likely to achieve an adaptive outcome by acting on the environment than are monkeys raised with hobby horses. To document this interpretation I will draw upon the results of many different tests, and present the findings within broad functional categories. Although results are shown in longitudinal format, this is done in order to convey the consistency of group differences over time; no significance can be attached to suggested "developmental trends" inasmuch as the details of setting and procedures (e.g., number of exposures, duration of exposure periods, etc.) varied across tests.

A number of tests have been completed in which the monkeys were observed while alone in novel surroundings. The first observations were made when the animals were less than two months old and the most recent were completed when they were in their fourth year of life. Measures were routinely made of heart rate and distress vocalizations. Results for heart rate, presented in Figure 4, indicate a higher level for dog–

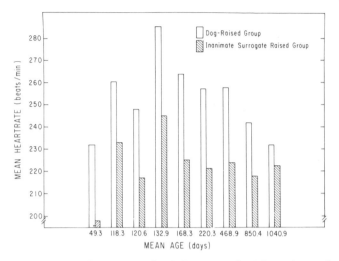

FIG. 4. *Mean heart rate of monkeys raised with canine and inanimate mother substitutes.*

raised monkeys than for the monkeys raised with inanimate surrogates on every occasion. Overall differences between groups are significant beyond the .01 level. The picture is similar for distress vocalizations, except for the two most recent tests, and overall differences are also significant ($p < .01$; Fig. 5). On a few occasions we have also assayed plasma cortisol to provide an additional measure of responsiveness. The results, presented in Figure 6, show higher levels in the dog–raised group on three of the four tests. For the combined tests the difference between groups is significant at the .05 level.

Do these results indicate that monkeys raised with hobby horses are simply calmer or less susceptible to stress than those raised with dogs? If this were the whole story, one might expect that in problem–solving situations, the monkeys raised with inanimate surrogates, being less agitated, would enjoy some advantage over monkeys raised with dogs. In fact, just the reverse seems to be the case.

Our first indication that dog-raised monkeys were likely to be better problem-solvers than monkeys raised with hobby horses occurred when the animals were about four months of age. They were tested in a delayed-response situation in which the reward for a correct response was contact with the substitute mother. As the monkey watched, a handler led the surrogate through one of four differently colored doorways whereupon it disappeared from view. The disappearance of the surrogate coincided with the beginning of a delay period ranging up to 45 s. At the end of the

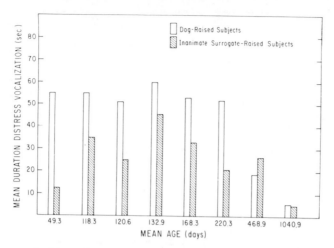

FIG. 5. *Mean distress vocalizations (*coo, scream*) of monkeys raised with canine and inanimate mother substitutes.*

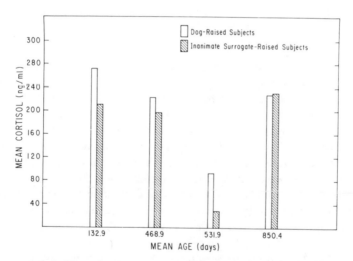

FIG. 6. *Mean levels of plasma cortisol in monkeys raised with canine and inanimate mother substitutes.*

delay the monkeys were allowed 30 s in which to select and enter one of the goal boxes; if they failed to enter within this period a "balk" was scored for the trial. The most striking difference between groups in the first phase of testing was the much higher level of balking in the inanimate surrogate group. They refused to respond on 46% of the trials, as compared to less than 2% for the dog-raised monkeys ($p < .001$; Fig. 7).

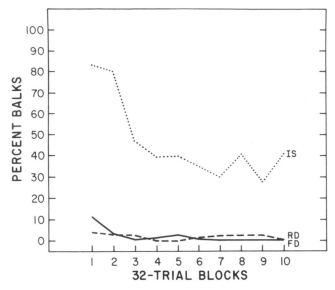

FIG. 7. *Percentage of trials in which monkeys raised with canine and inanimate mother substitutes failed to respond.*

These results might be interpreted in various ways: For example, the high level of balking by monkeys raised with hobby horses could reflect the fact that these animals had had no opportunity to learn to follow their substitute mothers before they were tested. In contrast, even the monkeys raised with restricted dogs could have learned to follow their companions in the kennels. Another possibility is that the strength of attachment was less in the inanimate surrogate group, that their motivation to regain contact with the surrogate was low. Actually, we have a great deal of information to the contrary. Although I cannot claim that the strength of attachment to the surrogate is equal in monkeys raised with hobby horses and with dogs, there is no question that the hobby horse was a powerful incentive and an effective security object.

Fortunately, there is no need to resolve these interpretive difficulties here, for we have other data showing striking contrasts between groups in problem–solving situations in which the attachment figure played no direct role. Although it was present during testing (to eliminate the potentially disruptive effects of separation), the situations were designed so that it could not contribute to successful performance.

The problems were for the most part relatively unstructured. We were not interested in controlling performance, or training the animals on a particular task, or measuring their specific problem–solving skills. Our aim was to place them in settings in which they could achieve some appropriate reward—typically a bit of preferred food—by interacting with the situation, attending to its relevant features, and performing some simple instrumental response, such as pulling, climbing, or pushing, which was within their normal repertoire. Furthermore, we made no attempt to maximize motivation. Although the monkeys were tested before the daily feed, scraps of food were generally lying about from the previous day's ration, even though they were not as highly preferred as the fruits, nuts and candies that were presented in the tests. Such a relaxed approach to the testing of problem-solving behavior is bound to encourage individual variability, and we have had our share of it. Nevertheless, the general pattern of results is remarkably consistent over more than a year of measuring performance on such tasks. The percentage of trials in which problems were contacted is presented for 11 different situations in Figure 8. This measure is consistently higher in the dog-raised group ($p < .05$); Figure 9 shows that successful performance is also substantially higher in this group ($p < .02$).

We have examined another aspect of behavioral adaptability and the results are completely congruent with the findings on problem–solving behavior. For the nonhuman primates, as for man, keeping in touch with the world is very much a visual affair. It will be recalled that the Eastman and Mason experiment indicated that the tendency to look at novel social stimuli was substantially stronger in monkeys raised with mobile mothers than in those raised with stationary surrogates. We were interested in whether similar contrasts would be found between monkeys raised with

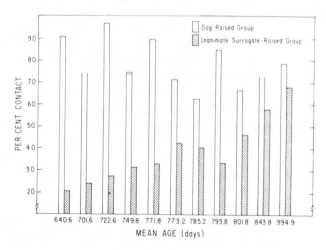

FIG. 8. *Percentage of trials in which problems were contacted by monkeys raised with canine and inanimate mother substitutes.*

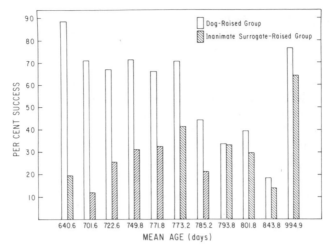

FIG. 9. *Percentage of problems solved by monkeys raised with canine and inanimate mother substitutes.*

dogs and hobby horses in an environment that provided a much richer visual world to both groups than was available to the monkeys in the original study. The apparatus was similar to the one used by Eastman and Mason. Essentially, the animals were placed in an enclosed chamber containing peepholes, through which they could look at projected color-transparencies. In the first test series we presented a single slide (e.g., landscape, interior of a room, etc.) for nine trials, and on the tenth trial introduced a new picture. The results are presented in Figure 10. It is evident that monkeys raised with dogs demonstrated a much higher level of looking behavior in this situation than did monkeys raised with hobby horses. Moreover, should there be any question that the dog–raised monkeys were actually looking at the projected pictures, their performance on the tenth trial presenting the novel picture removes all doubt. They not only showed a sharp increase in duration of looking, as compared to the immediately preceding trial, but evidenced a strong positive contrast effect: The duration of looking at the test slide was nearly twice that elicited by the repeated stimulus even on its first presentation, even though it was also completely novel on that trial.

Partly to confirm this incidental finding, and partly to determine whether or not additional experience with the situation would lead to increased looking behavior in the inanimate surrogate group, a second experiment was completed in which on half the sessions a different picture was presented on each of the ten trials (variable series) and on half the sessions the same stimulus was repeated for nine trials, with a novel

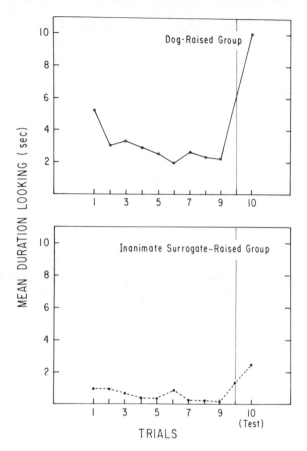

FIG. 10. *Duration of looking at projected color slides by monkeys raised with canine and inanimate mother substitutes. The same slide was presented on Trials 1-9 and on Trial 10 a novel slide was introduced.*

stimulus on Trial 10, as in the first experiment (constant series). The results, presented in Figure 11, confirm the contrast effect, show that a variable series produced essentially no intrasession decrement in looking behavior, and provide no evidence that the additional experience in the test situation had any strong positive effect on the behavior of the monkeys raised with hobby horses. The final test of looking behavior was completed about one year after the start of the first test. Transparencies were again used, but pictures were scaled to represent three levels of complexity. Although duration of looking increased somewhat in the monkeys raised with hobby horses, the level was less than half that of the

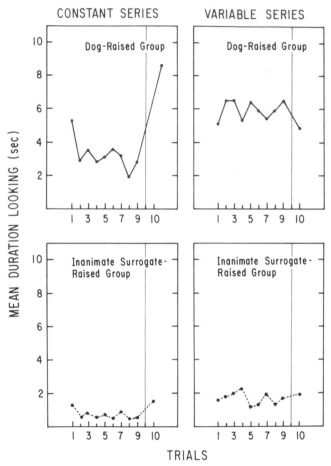

FIG. 11. *Duration of looking at projected color slides by monkeys raised with canine and inanimate mother substitutes. Constant series: The same slide was presented on Trials 1-9 and on Trial 10 a novel slide was introduced. Variable series: A different slide was presented on each trial.*

dog–raised groups. Furthermore, in contrast to dog–raised monkeys, they failed to differentiate reliably between the three levels of stimulus complexity. As a check against the possibility that looking was suppressed in monkeys raised with hobby horses because they were excessively stressed by the situation, measures were obtained of distress vocalizations, plasma cortisol, and heart rate. The only reliable difference between groups was in heart rate, and it was significantly lower in the inanimate surrogate group.

Coping Strategies and Early Social Influences

These results provide convincing evidence that the rhesus monkey's characteristic ways of coping with novel situations are profoundly affected by the nature of its relations with the attachment figure. This figure apparently has a kind of paradigmatic quality; it serves the developing individual as its first exemplar of what the larger world is like and exerts a powerful influence on how it is prepared to deal with it. Merely "enriching" the general environment was not sufficient to override this paradigmatic influence.

The fact that the effective attachment figure was a mechanical device in one experiment and a dog in the other is significant, for it suggests that the relevant dimensions of the early social environment for the development of coping strategies are not closely tied to the species–specific structure of the mother–infant relationship. This is not to say that the particularities of the natural relationship are inconsequential, of course, but only that they appear to be superimposed on more generalized developmental processes.

I believe that the critical distinction between attachment figures in these experiments is the presence or absence of response–contingent stimulation. Stationary surrogates and hobby horses surely provide few opportunities for the developing individual to experience the fact that his behavior has effects on the environment and to learn that the events going on around him are amenable to his control. Inert mother substitutes make no demands, occasion no surprises, do not encourage the development of attentional processes and the acquisition of the simple instrumental behaviors that are the fabric of social interaction.

The importance of response–contingent stimulation in human perceptual–cognitive development is illustrated by the research of J. S. Watson (Watson, 1971; Watson and Ramey, 1972). Watson demonstrated that infants will learn to control a mobile suspended above their cribs, and respond to its movements with smiling and cooing. Of greater interest in the present context is his finding that infants whose mobiles moved on a fixed schedule independent of their behavior learned that they could not control the device; moreover, its movements did not produce smiling and cooing. Later, when both groups were tested in the laboratory with a different mobile that all infants could activate, only those having previously experienced response–contingent stimulation with mobiles learned to control it. Watson concludes: ". . . the ability to learn that something is *uncontrollable* is probably nearly as adaptive a capacity as the ability to learn to control those things which can be controlled" (Watson, 1971, p. 149, italics mine). The fundamental importance of response-contingent stimulation is also emphasized by Seligman (1975) in his discussion of learned helplessness.

That mothers can be an important source of response–contingent stimulation hardly requires comment. However, the implications of this fact for cognitive-perceptual development are not intuitively obvious. The results presented here offer strong support for Lewis and Goldberg's suggestion that ". . . contingency is important, not only because it shapes acquisition of specific behaviors, but because it enables the child to develop a motive which is the basis for all future learning. The main characteristic of this motive is the infant's belief that his actions affect his environment . . . the mother is important because it is the contingency between the infant's behavior and her responses that enables the infant to learn that his behavior does have consequences" (Lewis and Goldberg, 1969, p. 81).

Acknowledgments

Support for this research was provided by National Institutes of Health Grants HD06367 and RR00169. Unpublished data reported here were obtained in collaboration with Mr. D. DeSalles, Ms. M. D. Kenney, Dr. E. Sassenrath, Mr. S. A. Williams, and Dr. B. S. Wood.

References

Anastasiou, P. J. 1970. Problem solving ability of differentially reared rhesus monkeys. (Unpublished doctoral dissertation, Tulane Univ., La.)

Anderson, C. O., A. McM. Kenney, and W. A. Mason. 1977. Effects of maternal mobility, partner, and endocrine state on social responsiveness of adolescent rhesus monkeys. *Develop. Psychobiol.* 10 (In press).

Eastman, R. F., and W. A. Mason. 1975. Looking behavior in monkeys raised with mobile and stationary artificial mothers. *Develop. Psychobiol.* 8:213–222.

Lewis, M., and S. Goldberg. 1969. Perceptual-cognitive development in infancy: a generalized expectancy model as a function of the mother-infant interaction. *Merrill-Palmer Q. Behav. Dev.* 15:81–100.

Mason, W. A. 1960. The effects of social restriction on the behavior of rhesus monkeys: I. Free social behavior. *J. Comp. Physiol. Psychol.* 53:582–589.

Mason, W. A. 1961. The effect of social restriction on the behavior of rhesus monkeys: II. Tests of gregariousness. *J. Comp. Physiol. Psychol.* 54:287–290.

Mason, W. A., and G. Berkson. 1975. Effects of maternal mobility on the development of rocking and other behaviors in rhesus monkeys: a study with artificial mothers. *Develop. Psychobiol.* 8:197–211.

Mason, W. A., and P. H. Green. 1962. The effects of social restriction on the behavior of rhesus monkeys: IV. Responses to a novel environment and to an alien species. *J. Comp. Physiol. Psychol.* 55:363–368.

Mason, W. A., and M. D. Kenney. 1974. Redirection of filial attachments in rhesus monkeys: dogs as mother surrogates. *Science* 183:1209–1211.

Ryle, G. 1949. *The concept of mind.* London: Hutchinson. (Reprinted by Barnes & Noble, New York, 1971.)

Seligman, M. E. P. 1975. *Helplessness: on depression, development, and death.* San Francisco: W. H. Freeman.

Watson, J. S. 1971. Cognitive-perceptual development in infancy: setting for the seventies. *Merrill-Palmer Q. Behav. Dev.* 17:139–152.

Watson, J. S., and C. T. Ramey. 1972. Reactions to response-contingent stimulation in early infancy. *Merrill-Palmer Q. Behav. Dev.* 18:219–227.

Chapter Thirteen

Infant Independence in Yellow Baboons

JEANNE ALTMANN

Allee Laboratory of Animal Behavior
5712 S. Ingleside Ave.
Chicago, Illinois 60637

Introduction

Baboons live in a complex and changing world. They cope by being adaptable. Like humans, they specialize in being unspecialized, a mode of life that depends on an ability to shift to alternatives when the primary mode of adaptation is no longer viable. This flexibility is clearly revealed in the development of infant baboons, for whom success often depends on their own abilities and those of their mothers to adopt alternative ontogenetic strategies.

In the area of motherhood and infancy, field studies have lagged behind laboratory research. They rarely have made use of suggestive laboratory results or provided essential quantitative information on the range and importance of naturally existing variability. The present field study was designed to test hypotheses arising from laboratory studies and from theoretical considerations. In this first report I summarize results of preliminary analyses of the extent of variability in some maternal and infant experiences and the sources and consequences of that variability. Emphasis is placed on data from those infants that were born after the first month of the study.

During 14 months (from July 1975 through July 1976 and October 1976), I carried out a field study of 18 mother–infant pairs. These baboons were members of Alto's Group, a group of yellow baboons (*Papio cynocephalus*) in Amboseli National Park, Kenya, where my co–workers and I have obtained longitudinal data on individuals since 1971 (cf. Altmann *et al.*, 1977, Altmann and Altmann, 1970, Hausfater, 1975, Post, 1976, Slatkin, 1975, Slatkin and Hausfater, 1976). Birth and death rates for Alto's Group (Altmann *et al.*, 1977, and in preparation) currently are those for a stationary population (Pielou, 1969) of approximately the current stable age distribution. Group size has fluctuated around 45 individuals for about four years.

Five females in the study group had infants that were under a year of age when the mother-infant study began. Thirteen infants were born during the study period. These 18 infants and their mothers were included in my sample. Three mothers were included twice, with two successive infants. Of the adult females in Alto's Group, only two were not included in the present study: Jane (whose youngest offspring was two years old), who died shortly after the study began, and Lulu, who has cycled without becoming pregnant since at least 1971. Consequently; my sample included the full range of age, parity, and dominance ranks among the adult females. Basic demographic information for the mother–infant pairs is presented in Table 1. Some relevant sociological and behavioral information is found in Appendix I.

Background

Let me begin by describing a few aspects of the milieu of these mothers and infants. The habitat of Amboseli baboons is short–grass savannah. The baboons live in semiclosed groups which descend from sleeping trees at about 0800 hour and travel approximately three miles each day before again ascending trees after 1800 hour. In this respect days of birth are no different than other days: the group makes no accommodation of routine for new mothers. The baboons spend about 50% of their day feeding, another 20% walking, both of these varying with the season and, for females, with reproductive status and perhaps with dominance rank. The presence of Maasai, lions, and especially leopards requires considerable vigilance. Adult female dominance ranks are linear and stable over time spans of the order of six years (Hausfater, 1975, and in preparation). It is within these constraints on time, energy, and attention that baboon females give birth and infants develop.

During the first half of the present study, four subadult males (Stu, Stiff, Russ, and Red) matured to adulthood. Fighting between them and other males increased. Stiff migrated from the group; each of the other three experienced a rapid rise in dominance rank. There were frequent, short-term male migrations out of and into the group and finally two

TABLE 1. Study infants, ordered by birthdate.

| | INFANT | | MOTHER | | MATERNAL FOCAL SAMPLE OBSERVATION MINUTES FOR EACH MONTH | | | | | | | | | | | |
NAME	BIRTHDATE	HEALTH AT BIRTH	NAME	PARITY	1	2	3	4	5	6	7	8	9	10	11	12
♀ Pooh	31 Oct 74	W?	Plum	P										154	140	120
♂ Ozzie	24 Dec 74	H?	Oval	M(≥4)								180	115			
♂ Fred	1 Jan 75	H?	Fem	P							289	180		180		
♀ Alice	1 Jan 75	H?	Alto	M(≥4)							283	180	120	105	120	
♀ Eno	22 Apr 75	H?	Este	M(≥4)				250	304	255	270	150	135			
♀ Summer	7 July 75	H?	Scar	M(≥3)	269	294	409	240	270	115	135	150	165			360
♂ Pedro	9 July 75	H?	Preg	M(≥4)	170	180	329	245	255	140	150	135				
♀ Misty	15 Aug 75	W	Mom	M(≥4)	620	540	135									
♂ Bristle	17 Aug 75	H	Brush	P*	621	435	375	465	480	336	150	270	390	135	240	135
♂ Hans	15 Oct 75	H	Handle	P	775	660	540	471	375	345	390	465	405	120		240
♂ Grendel	27 Jan 76	H	Gin	P	907	540	345	658	525	535			405			
♀ Sesame	2 Feb 76	W	Slinky	P*	675	675	345	525	660	390			405			
♂ Juma	28 Feb 76	H	Judy	M(≥4)	595	435	255									
♀ Safi	2 Mar 76	H	Spot	P	780	405	750	450	493			375				
♂ Moshi	1 June 76	H	Mom	M(≥5)	735	525			505							
♀ Oreo	28 July 76	H?	Oval	M(≥5)			525									
♀ Vicki	4 Oct 76	H?	Vee	P	650											
♀ Peach	8 Oct 76	H	Plum	M(2)	615											
Total focal sample in-sight time					7412	4689	4008	3304	3867	2116	1667	2085	2140	694	500	855
Total individuals sampled					12	10	10	8	9	7	7	9	8	5	3	4

Maternal parity is marked "P" for primiparous "*" if all previous pregnancies resulted in stillbirths or neo-natal death (see Appendix I), "M" for multiparous with actual parity indicated in parentheses. Infant health (H = healthy, W = weak) was judged on the basis of clinging ability, vigor, skin color (bright red as opposed to pale pink or grey toned), coat (thick and shiny considered healthy), eyes (clear considered healthy). Health assessments took several days of close observation. For those infants for whom this was not possible, assessment was based on the available information from more casual or brief observations and marked with a "?".

other "permanent" emigrations (Chip and Russ), and one death or emigration (adult male BJ), resulting in a return to a 2:1 adult female to adult male ratio in the group. Red, as a third-ranking male, continued to be involved in frequent fights and made repeated temporary emigrations from the group.

In the fall of 1972, a merger was completed between Alto's original group and a one–male group, High Tail's Group. During 1975–76, two clear subgroups could be detected within the present group. All female members of the former High Tail's Group (Jane, Este, Handle, Plum, Slinky, Brush) were in one subgroup along with Lulu, Fem, and Gin of Alto's original group. Lulu was the only high-ranking member of this subgroup. She, Este, and Jane were the only members of this subgroup who were fully mature at the time of the merger. Associated with this subgroup were males Even, Max, Ben, Red, and BJ. Chip and High Tail moved between the two subgroups as did juveniles.

These subgroups were identifiable in the use of two separate trees in the favorite sleeping grove, in grooming partnerships, and in neighbor relations. Likewise, mating partners and males associated with mother–infant dyads were almost entirely from within the subgroup. Some associative bonds were much stronger than others within each subgroup, but the basic division into these two subgroups seemed to account for much of the observed variation in tolerance between individuals and to identify many of the individuals that would constitute a new infant's social world.

Methods

Mothers with neonates appear to be the class of individuals in our population that are most sensitive to being observed. For this reason, and because most of our previous research has been done from atop a vehicle (Altmann and Altmann, 1970), the first month of the present study, and to some extent the second, was partially devoted to accommodating the females to observation on foot within 5–10 m. Much closer distances were possible for most mothers but an attempt was made always to stay at a distance that would not discourage the mothers' interactions with the most sensitive individuals in the group. Beyond that, some types of detailed data were considered sufficiently reliable to use only after several weeks or more of systematic sampling. Consequently, for some analyses I used data obtained from July 1975 onward, for others, not until later.

After various sampling schemes were tried, the following was established by September 1975. Each female was sampled during the last month of pregnancy for two days at least a week apart, on days one and five of infant life, on two days during the infant's second week, one day per week thereafter until the infant was six months old, then two or three days per month during the next six months. Whenever impassable mud, illness, or other factors reduced available observation days, sampling on mothers with older infants was sacrificed.

On each sample day two females were sampled alternately, each for 15 min in–sight time out of every hour from 0800 through the 1700 hour but excluding the 1200 hour. The females that were sampled on any one day were paired for proximity in age of their infants, thereby providing controlled pairs for examining effects of overall group or ecological variables (e.g., length of day journey).

At the beginning of each 15–min sample period and at the end of the three five-minute intervals thereafter, I took point (instantaneous) samples (Altmann, 1974), recording (1) the mother's behavior state (feeding, walking, grooming, other social interactions, resting), as in Slatkin (1975), (2) the identities of all her neighbors within two meters and five meters, (3) the distance between her and her infant (categorized as: in contact, within mother's arm's reach but not actually touching, within two meters but greater than arm's reach, two to five meters, five to ten meters, ten to twenty meters, or greater than twenty meters), and (4) whether the infant was playing with or in contact with any other individual.

Between these point samples, a focal sample (Altmann, 1974) was taken on the mother, during which I recorded social behaviors and changes in mother-infant spatial relations. Onsets and terminations of grooming and spatial states were timed. For other interactions, only the behavior and partner identity were recorded, retaining sequential order in the record. Details of specific behaviors will be discussed in the relevant sections of this chapter and in full report of the study (Altmann, in preparation). Between the scheduled maternal focal samples and on additional days, I did *ad libitum* sampling, some focal infant sampling, photo documentation, censusing and mapping of group movements.

The total number of maternal focal sample in-sight minutes for each infant during each month of life is indicated in Table 1. In the present chapter I shall be concerned with quantitative data primarily from the infants' first six months.

Results

MATERNAL CARE

During focal samples on a mother the behavior record included all instances of maternal restriction on infant attempts to leave the mother, all changes in spatial state (see below) and who effected the change, all instances of mother or infant following the other, and all instances of mothers directing punitive behaviors, such as biting, pushing, and hitting, toward their infants.

Parity

The first two months of life—the first few weeks, in particular—constitute a period of intense dependency for the infant and of stress for the mother. Although baboon infants cling to their mothers from the day of birth, most infants need some clinging assistance in the first few days, some for a week or more. The latter infants are more likely to die in the first year. Individual differences in maternal response in the first weeks of life seem to affect the likelihood of survival of these and other high–risk infants.

All baboon mothers seemed to clutch their infants automatically from birth and to respond to their neonates' distress cries with such "embraces." During these first few days of life the most conspicuous difference that I observed between primiparous and experienced mothers was in their subsequent responses. If an infant was already in contact and being held or cradled but the infant cried or gave rooting responses, experienced mothers usually lifted the infant higher on their ventrum by means of clutching movements, as a result of which the infant usually reached the nipple, clamped on with its mouth, and ceased its distress cries. The infant not only obtained its essential nutrition this way but was provided with a fifth anchor point when riding, an advantage that seemed appreciable in the first few days of life. In contrast, primiparous mothers only rarely gave repositioning or sucking aid even when, in the most extreme situations, the infant rode upside down and backward, and required constant maternal clutching. Infant contact seemed to be one undifferentiated state to these mothers. One often sensed that they were unresponsive to, or puzzled by, their infants' continued distress, despite the fact that nearby group members responded to the cries by watching the pair, moving closer, and increasing the repeated soft grunts that are given to mothers and infants. Most mothers learned quickly, though: increased responsiveness sometimes was observed by the end of the first day and certainly in the next few days.

Only in the case of the one totally incompetent mother, Vee (see Appendix I), did the incompetence seem to have obvious long–term consequences. Vee's care of Vicki was so poor on the day of birth, resulting in no nipple contact that day, that even Vee's appreciable improvement in care by day two seemed inadequate compensation. My speculation is that the early maltreatment had permanent consequences, perhaps leading to Vicki's death at three weeks of age.

Maternal Style

By the end of the second week of life, healthy infants climbed clumsily about in their mother's ventrum and made their first attempts to break contact. This was a period of close maternal attentiveness whenever mother and infant were not in contact and of rapid return to contact at the slightest disturbance of any sort. These first stages of separation were initiated primarily by the infant and were sometimes limited by the mother. Although most mothers were protective of their infants when intruders approached during the first months, mothers differed considerably in reaction to infant exploration during this period. The most restrictive mothers allowed virtually no break in contact for almost two months. The more "laissez–faire" mothers tolerated separation although they themselves seldom moved away from their infants during the first month. Finally, the most rejecting mothers frequently moved away from their infants during the first month and seldom watched or followed them, even when the infants were quite young.

By the end of the second month most mothers initiated some separation, and the infants oriented to and followed their mothers. Overt maternal attention to the infant and restrictiveness rapidly disappeared except during emergencies, whereas infant attention to and following of the mother became the norm.

Basically, mothers could be dichotomously characterized as being either in the range laissez-faire to rejecting or protective to restrictive in their behavior toward their infants, as summarized in Table 2 (detailed analysis in preparation). In comparing the first group (hereafter called "laissez–faire" mothers) to the second (hereafter called "restrictive" ones), we find that the laissez-faire mothers not only restricted their infants less and completely stopped doing so when their infants were younger (Table 2, col. 3), but they rarely if ever followed their infants. They seldom made contact with their infants (Table 2, col. 6) and at a younger infant age they increased the distance between themselves and their infants more often than they decreased it (Table 2, col. 5). At an earlier age they ignored their infants (Table 2, col. 4) or directed punitive behaviors toward them (Table 2, col. 7) when the infants attempted contact or suckling.

Seven of the 12 mothers observed during month one were classified as laissez-faire, five as restrictive. The average rank at the time of parturition was 6.1 for the laissez–faire mothers, 11.2 for restrictive ones. Six of the seven higher-ranking mothers were laissez–faire, four of the five lower-ranking ones were restrictive. Three of six female infants and four out of six male infants had restrictive mothers. Mom, the only mother who was observed during month one with each of two successive

TABLE 2. Age of infants at transitions in mother-infant relations.

INFANT	RATING	STOPPED RESTRAINING	REJECTED, IGNORED, AND STOPPED FOLLOWING INFANT	DISTANCE INCREASED MORE OFTEN THAN DECREASED	INFANT MADE ≥ 90% OF CONTACTS	BIT, HIT, OR PUSHED
♀ Safi	Laissez-faire	½	½	1½	2	2
♂ Moshi	Laissez-faire	½	1½	2	2½–3½	2–4½
♀ Misty	Laissez-faire	½	1½	2	a	a
♀ Vicki	Laissez-faire	½	0	—	—	—
♂ Pedro	Laissez-faire	1	2	1	b	1
♀ Summer	Restrictive	2½	4	2	3	5
♂ Grendel	Laissez-faire	1	1	1	2	1
♀ Sesame	Restrictive	2	2	3	4	3½
♂ Hans	Restrictive	3	5	4	5	5
♀ Peach	Restrictive	>1	>1	—	—	—
♂ Bristle	Restrictive	1½	2	2½	7	5
♂ Juma	Laissez-faire	½	½	1	2	a

[a]Not before infant death at 2½ months.

[b]Not before infant death at 8 months.

Note: Data for all mother-infant pairs observed at least during month 1. Infants ordered by maternal dominance rank at parturition. See text for further details and Table 1 for ages at which each infant was observed.

infants, was laissez–faire with both female Misty and male Moshi. Thus, maternal rank was a good predictor of maternal style, sex of infant was not.

Ill health and infant death occurred disproportionately among infants of laissez–faire mothers. It is not clear whether this association is a real one and, if so, what the nature and direction of causality is. As mentioned above, Mom was generally as laissez–faire with her weak infant Misty as she later was with her healthy infant Moshi, even tolerating juveniles carrying Misty around while Misty repeatedly bumped her head and often screamed. Likewise, no general difference could be detected in the behavior of the two most restrictive mothers, Handle and Slinky, one with a healthy infant, the other with a weak one. However, mothers did respond to worsening of infant health by increased protective behavior. This was observed with female Oval when her yearling Ozzie had a brief foot injury, probably caused by a thorn, with Spot when her infant Safi suffered severely from a presumed virus (see Appendix I), with Mom and Preg when Misty's and Pedro's health deteriorated.

SOCIAL INTERACTIONS

What are the social interactions that affect the infant, both indirectly through effects on the mother and directly, and that may account for some of these and other observed differences in maternal care? I shall first consider postpartum changes affecting all mothers similarly and then two factors, maternal dominance rank and adult male associations, that affect mothers differentially. The conclusions drawn in the following two sections must be considered tentative until all analyses are completed.

Postpartum Attraction

At parturition the life of a baboon female changes dramatically. She must not only nurse, carry, and protect the neonate while providing her own food, transport, and protection, as before, but in addition, she and her infant become a major focus of interest within the group (e.g., DeVore, 1963).

As measures of an individual's "interest" in an infant, I have utilized the amount of increase in the individual's approaches, interactions, grooming, and time spent near the mother–infant dyad, over the comparable level with the mother before the infant was born. By these criteria, virtually all group members exhibited some degree of interest in the infants, adult females and juveniles of both sexes more than adult males, older juveniles more than younger ones. However, the degree of interest varied considerably between individuals within age–sex classes, and these differences were consistent over time. High-interest adult females were scattered throughout the range of age and dominance ranks. Most of the variability between females in this interest was accounted for by subgroup membership (i.e., members of the mother's subgroup were, in general, more interested than those of the other) and by consistent individual differences—what might be considered personality differences. That is, these high–interest females exhibited interest in most new infants and maintained their interest in infants while cycling, pregnant, and after the birth of their own infants. No increase in attention to other infants occurred after the death of a female's own infant.

After parturition, mothers' involvement in social interactions soared, primarily at the initiation of others. Mothers were the recipients of much social grooming. Although during the last month of pregnancy they received less than one minute of grooming per hundred, during the first three months after parturition they received five to ten minutes per hundred.

Maternal Dominance Rank

Females and juveniles of both sexes approached higher-ranking mothers with caution, with approach-avoidance behavior, anxious glancing, and hesitant attempts to touch the infant. Such subordinates usually stayed one to two meters away from the higher-ranking mothers, watching the infant. In contrast, lower–ranking mothers were rapidly and directly approached, usually to the point of contact, and the infant was handled or muzzled. For a low–ranking mother her attractive infant resulted in her being placed in a position of frequent interaction with individuals from whom she ordinarily remained at a distance. For these low–ranking mothers, frequent avoidance, tensing, and submissive behaviors replaced occasional glancing and use of spacing.

Some individuals, whom I call "infant snatchers," most of whom were adult females, strongly and persistently pulled the clinging infant of mothers to whom they were dominant. During such interactions the mothers cowered, turned their heads away, and clutched their screeching infants. These stressful interactions, both the milder and the more extreme ones, occurred repeatedly to low–ranking mothers and probably were a major cause of some mothers' restrictiveness (e.g., Rowell et al., 1968). In the first two months, lower-ranking mothers often restrained or retrieved their infants, particularly at the approach of infant snatchers. Soon the infants themselves specifically avoided these individuals. Thus, we find in these first interactions the origins of the infant's assumption of its mother's relative dominance rank, a behavioral outcome of considerable long–term consequences.

"Godfathers"

Except for two circumstances, new mothers themselves rarely approached others during these early weeks. One exception was the approach made by mothers to other females with very young infants, an event that was more likely to occur if these other females were lower in rank than the mother. Second, some mothers, when they were repeatedly approached and their infant repeatedly yanked, would approach and sit next to a particular adult male associate.

Thus, relationships with adult males provided a second major source of variability in infant experience (Ransom and Ransom, 1971). Adult males, like females, showed varied degrees of interest in infants. For males, however, interest depended more strongly on the identity of the mother. Similarly, each mother's tolerance of male approaches depended on the identity of the male. That is, there were specific adult male-mother associations, and these seemed to reflect specific relationships between male and female that were established before the infant's birth. (See Appendix I.)

In most cases, at least two males attempted to associate with the mother and infant soon after parturition. The higher-ranking male sometimes exhibited herding behavior, followed the mother, threatened other males away, groomed the mother and, in fact, exhibited all elements of a sexual consortship except for actual mounting. As with sexual consortships, the role of the female in these interactions was not that of a passive observer. She did not follow a lower-ranking male while being herded by a higher-ranking one, nor did she ever threaten males who followed her. However, she selectively avoided certain males when they approached, reciprocated the following or not, selectively followed and groomed certain males, and so on.

Thus, some relationships of mothers and adult males were more reciprocal and enduring than were others. For convenience, I refer to such persistent and reciprocally associated males as "godfathers." Some mother–infant dyads had such a godfather, others did not. Usually, I could predict the existence and identity of such affiliations before the birth of an infant, commonly from its mother's mating, grooming, and neighbor associations (Appendix I). An affiliated male was likely to be the infant's father, i.e., the only male, or one of only two or three males, that copulated with the mother during the days that she conceived the infant.

The godfather, either through overt behavior or through his mere presence, often provided a buffer between the mother–infant dyad and other group members. When a godfather was within two meters of the mother–infant dyad, others approached more hesitantly; they gave repeated "anxious" glances toward the adult male, and veered in their approach to the mother so that they approached her on the side opposite the male. They sat farther from the dyad than usual, watching, and then suddenly trotted toward the mother and infant as soon as the male moved away. The godfather sometimes overtly threatened those who approached the dyad, particularly if the interloper pulled at the infant. One result of such male influence was that the rate of infant handling and snatching was lower in the presence of the godfather. Moreover, low–ranking mothers allowed more infant exploration if the godfather was nearby: the time these infants spent out of maternal contact was greater when their godfathers were within two meters. Low-ranking mothers and their infants had such godfather relationships more than did high-ranking ones.

These godfather relationships persisted beyond the neonatal period. Older infants rested against their godfathers, ran to them in time of distress, took greater liberties by feeding near them or by getting meat scraps from them. One infant, Grendel, who was the infant of the most rejecting mother, Gin, sought out and received virtually all of his nonnutritive nurturance from male High Tail, until High Tail was killed by a leopard. Another, older infant, Pooh, was repeatedly carried and protected by her godfather, Chip, such as at times when the group fled from Maasai tribesmen. Strikingly, a second male, Max, immediately took over this role when Chip migrated from the group. (See also Appendix I.)

There is another side to the godfather relationship. Godfathers sometimes derived obvious benefits from their relationships with infants. Males under attack sometimes take infants to their ventrum, an act thought to serve as an "agonistic buffer" (e.g., Deag and Crook, 1971; Ransom and Ransom, 1971), that is, to mitigate attack. However, only in the cases in which the male and the infant had an existing positive relationship did the infant remain riding and possibly serve this function. Otherwise the infant refused to cling. It then either dropped off or the male embraced it continuously, running three-legged. His locomotion was thus hindered and the infant usually screamed in distress, bringing other baboons to the infant's aid. Thus, the same male was likely to be an infant's possible father, its protector, and its exploiter (detailed analysis in preparation).

SPATIAL RELATIONSHIPS: DISTANCES

Contact Time

In the two parts of Figure 1, I have graphed the data for mother–infant contact time, plotted separately for infants of restrictive mothers and for

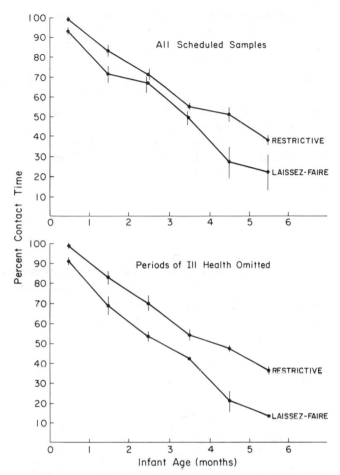

FIG. 1. *Percent of day time spent in contact, graphed separately for infants of restrictive mothers and for infants of laissez-faire mothers. For each point that is based on data from more than one infant, the point represents the mean for that class of infants and the line through the point indicates the standard error of that mean.*

infants of laissez–faire mothers. The upper graph includes all data, even those from periods in which infant ill health was appreciable. Even so, it is clear that infants of restrictive mothers spent more time in contact throughout the first half–year of life than did infants of laissez–faire mothers. For each month, their contact time was about the same as that for the other group one month earlier. The difference is more striking when data from periods of ill health are excluded, as shown in the lower graph, i.e., illness tended to increase contact time, particularly during the first few months of life.

A comparable analysis of contact time was done comparing male and female infants within each type of mothering. Most sickness occurred among female infants of laissez–faire mothers, resulting in slightly greater contact time (~ 0.06 per month) for these infants as compared to male infants of laissez-faire mothers. For infants of restricting mothers, however, differences were small and reversed in direction from month to month. When data from months of illness were removed, differences were small and reversed in direction from month to month both for infants of laissez–faire mothers and for infants of restrictive mothers. Thus in the present study there were no differences in contact time attributable to sex of infant.

Infants' Use of Space

The young infant's world can be viewed as a circle with its mother at the center. We might ask, what percent of its time does an infant of each age spend within circles of different radii? In Figure 2 monthly percentages

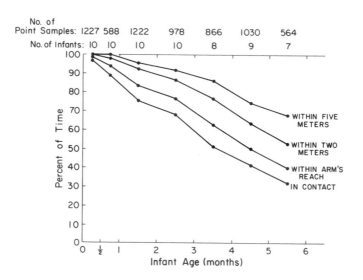

FIG. 2. *Percent of day time spent within increasing distances from the mother. Mean over the number of infants indicated. Data from all scheduled samples. (See text for details.)*

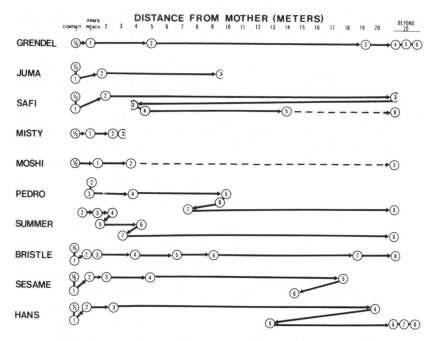

FIG. 3. *Distance from its mother required to account for 90% of each infant's day time. Infants ordered approximately by increasing maternal restrictiveness (see Table 2). Linear extrapolation was used between measured distances. The number in each circle indicates the month of life. The broken circles for month 3 for Juma and Misty indicate those infants' death during that month. The two broken circles for Safi in month 3 indicate data for Safi before and after the onset of illness. (See text for further details.)*

were calculated for each infant and were then averaged over all infants. In Figure 3 related data are presented in such a way as to highlight individual differences. The expansion of the infant's physical world is abundantly clear from these data.

Over the first six months, contact time dropped steadily from 100% in the first two weeks to about 32% in the sixth month. After the first two weeks, time spent out of contact but within arm's reach remained at about 8–10% of total time (but a decreasing proportion of the noncontact time). Time spent at each of the greater distances showed successive increases. By the sixth month, approximately a third (32%) of the time was spent more than five meters away, another third (36%) out of contact but within five meters, whereas during month two, only 5% of the time was spent more than five meters away, another 20% out of contact but within five meters.

Although infants made increasing use of space more distant from their mothers, their use of this space was not at all independent of the position of their mothers. Based only on available area, one would expect infants to spend approximately three times as much time between arm's reach and two meters as between contact and arm's reach, seven times as much between two and five meters as between arm's reach and two. Yet, at no time during the first year did the actual distribution even approach the random one. Always, infants used space nearer their mothers appreciably more than expected.

However, examination of Figure 3 reveals the considerable variability among the infants. In this figure I took as a criterion the distance from the mother that accounts for 90% of an infant's day time. Thus, during month two, a radius of two meters from the mother on the average accounted for 90% of an infant's day time, but for Grendel the radius required was five meters, for Moshi and Misty (Mom's two successive infants) it was between two and five meters, two meters for both Juma and Safi, arm's reach for Pedro. These are all the infants of laissez–faire mothers. By contrast, all infants of protective mothers were within arm's reach 90% of the time.

By the end of month eight all infants spent at least 40% of their time more than five meters from their mothers. By month six for most infants, and by month eight for all, the distance from their mothers required to encompass 90% of their time was no longer a useful measure—it was over 20 m and I often could not locate the two at the same time.

During the second half of the first year of life, contact time declined from 30% at month six to 10% by about month ten. Younger members of this class rode during part of long day journeys. Otherwise, contact occurred primarily during external alarms and for a long, dozing, nursing bout, commonly in the early evening, at times when mothers were either resting or involved in grooming interactions with others. Yet even at the age of 12 mo, infants spent more of their time (30%) within five meters of their mothers (including contact), than did any of their mothers' other associates.

SPATIAL RELATIONS: DYNAMICS

During the focal samples, I recorded the time and identity of the individual who effected a change in the mother-infant spatial relations (contact, arm's reach, greater than arm's reach) whenever a change occurred from one of these states to another and that same actor did not immediately (0.05 min) again change state; in the latter case the 0.05 min and actor identity criteria were again applied. For these spatial state records four categories of actors were distinguished: focal mother, her infant, mother and/or infant, but I could not tell which, and "other" (with the individual identified). Over 90% (usually well over 95%) of all transitions were effected by exactly one member of the mother–infant dyad. Transitions for which I could not distinguish between the mother and infant as actor accounted for less than 1% of the transitions for most months, with higher percentages occasionally occurring at times of group alarm or during months of highest rates of state changes. During the first two months "others" effected 2–5% of the transitions for infants of laissez–faire mothers, by carrying or transporting the infant. For restrictive mothers rates of transition were too low during these months to permit reliable estimation.

On average, 70–80% of all changes in spatial state that were made by one member of the mother-infant dyad were made by the infant. Infants made over 85% of the breaks in contact during the first five months, 60–80% thereafter. They made 50% of the contacts during month one, over 90% after month three. Again, the pattern of individual differences (Table 2, col. 6) reveals that Grendel, Juma, and Safi were making 90% of the contacts by month two, Summer in month three, Bristle, Hans, and Sesame not until month four or later. Data for Misty are quite variable, probably due to her poor health. Moshi made 61% of the contacts in month two, 99% in month five; no data are available for him for months three and four. Pedro made 89% of the contacts in month two but a steadily *decreasing* percentage thereafter, surely due to deteriorating health, especially impairment of clinging ability.

Data on changes among the three spatial states were reanalyzed in terms of increases or decreases in distance between mother and infant. As noted earlier, the first stages of independence were characterized by infants increasing the distance between the two, mothers decreasing it, that is, the earliest stages of spatial independence were infant initiated (Hinde and Spencer–Booth, 1967). Only later did mothers reverse the dynamics of the spatial relationship, increasing the distance between themselves and their infants more than decreasing it. Table 2, in col. 5, reveals that these stages were reached later for the restrictive mothers than for the laissez–faire ones. By this stage mothers had stopped following their infants, rarely if ever paused, looked at, or waited for them, usually ignored their infants' distress cries as the mother moved away (Table 2, col. 4).

Summary and Discussion

At this stage in my data analysis, and in this limited space, it would be inappropriate to attempt an integration of my results, either with the theoretical or with the voluminous empirical literature.

Maternal style, as dichotomized in the present report as either restrictive or laissez–faire, emerged as the major variable affecting the development of infant independence in the first six months. These effects were sufficiently strong as to swamp effects due to infant health and to render undetectable any effects due to infant gender. Most, but not all of the differences in maternal style could be predicted from maternal dominance rank but not from infant gender. Adult male associations represented a continuity with previous relations between the mother and male, served to provide partial protection for low–ranking mothers, and developed into independent relationships between their infants and these males. Both maternal restrictiveness and male protectiveness apparently are protection from other group members, not from members of other groups or from predators, and serve to buffer but not eliminate effects of maternal rank. The bonds with adult males formed in the early months only much later served more external functions.

Unrestricted infants were more in control of their early independence in that most of them had freedom but not rejection at this stage. They then were rejected and punished at an age when their restricted peers were finally being allowed to determine their own amount of independence. That is, with the exception of an early period of restriction, these two groups of infants followed broadly similar stages of development one to two months apart.

We do not yet know the consequences of absence of a restrictive phase, the effects of differences in timing of developmental processes, the differences in bonding that result from male protection, or of other possible sequelae of early experiential differences. One reason that these consequences are difficult yet particularly fascinating to trace is that both mothers and infants themselves seem to buffer potentially detrimental effects of extremes in any one variable. Low–ranking mothers are usually more restrictive. They make more use of godfathers than do high–ranking mothers. Rejected infants seek care from others, and so on. In this way, active restructuring of their social environment enables mothers and infants to avoid to some extent the disruptive effects of environmental extremes, "of being on the low end of a distribution."

Such complexities raise many questions about the long–term social and life–history consequences of the observed patterns of interaction and relationship. To what extent are females consistent in the mothering of successive infants? How enduring are the observed bonds between the adult males and females? Between the males and infants? Demographic patterns of maturation, mortality, and migration as well as changes in male dominance rank surely result in long–term change being imposed on short–term continuities. There is still much to untangle.

Appendix I

Selective case history descriptions of all mother-infant dyads with emphasis on adult male and kin associations. Descriptions of the first five infants depend heavily on non-focal sample data. Infants are ordered by age. See also Tables 1 and 2.

1. ♀ *Pooh.* Plum was very protective of her first infant, Pooh, who was small and exhibited locomotor disability when first observed to leave her mother (J. Scott, personal communication). There is no information on any early associations. When I first observed Pooh at the age of eight months, she was thin, smaller than the six-month-olds, retarded in her skin color maturation (Altmann *et al.,* 1977), and had a short, scraggy coat and impaired locomotion. She was undergoing severe weaning with frequent distress vocalizations and occasional biting by her mother, who otherwise usually ignored Pooh. At this age, Pooh was associated with male Chip, who sometimes carried Pooh, stayed behind the group with her, slept near or with her in the trees and fed near her.

The day after Chip migrated to another group, when Pooh was 15 mo old, Pooh appeared with what seemed to be a dislocated hip and could walk only very slowly and clumsily. Male Max immediately assumed Chip's role as godfather, staying considerable distances behind the group with Pooh, especially during the first few days. Pooh's condition eventually improved to its earlier state. Her relationship with Max continued. One day when Pooh was 24 mo old, male Even ran off with Pooh in his mouth, inflicting fatal wounds to her as he did so. Max chased Even until Even dropped Pooh. Max then stayed with Pooh, grooming and sitting nearby.

2. ♂ *Ozzie.* When Oval's healthy male infant, Ozzie, was born, her daughter, Fanny, was 1½ yr old. High Tail and Slim were Oval's consorts the week Ozzie was conceived (G. Hausfater, personal communication). No information is available on early adult male attachments, but at six

months of age Ozzie was often associated with male High Tail. Seven months later, High Tail and Slim were Oval's consorts during the period in which she became pregnant with her next infant, Oreo.

3. ♀ *Alice.* Alice, a healthy female infant, was born to the oldest and highest-ranking female, Alto. Although BJ, Crest, and Peter each mated with Alto during the week that Alice was conceived, Slim was Alto's consort during the days of most likely conception and was, therefore, probably Alice's father (G. Hausfater, personal communication). Alto had two other known living offspring in the group, five-year-old Spot and 1½-year-old Dotty, both of whom often groomed their mother. At six months of age Alice was associated with adult male Peter. An association between Alto and Peter has existed at least since 1971; during the present study an association existed between Dotty and Peter but not between Spot and Peter. In fact, Peter had grooming relations with the other family members, was often near them, including in the sleeping trees, and he supported the younger sisters and Alto in agonistic encounters with Spot. Alto died when Alice was 17 mo old, at which age she was still sleeping with and occasionally suckling from her mother. During the next few months, Alice slept in contact with Peter or her sister Dotty.

4. ♂ *Fred.* An early association was observed between male Max and Fem with her first infant, Fred (D. Post and J. Scott, personal communication). This strong, persistent relationship existed well into Fred's second year of life with Fred often taking advantage of it, feeding immediately next to Max. In this way Fred frequently obtained scraps of mammalian prey when others of his age did not. Max was most likely Fred's father, but BJ was also a consort during the last week of estrus (G. Hausfater, personal communication).

5. ♀ *Eno.* Eno's next-surviving sibling was five-year-old male Toto, who was not often associated with Eno or with his mother, Este. When Eno was first observed by me at 2½ mo only a slight association could be detected between Eno and male BJ, on whom she occasionally rode. No stronger relationship developed before BJ disappeared from the group nine months later. Before BJ's disappearance he and High Tail were Este's most common consorts when she resumed cycling.

6. ♀ *Summer.* Both Stubby and Slim repeatedly approached Scar when her female infant Summer was born, 17 mo after Scar's previous infant, Cete. Scar usually avoided Stubby and Slim, and neither she nor Summer developed an association with any other male. Perhaps because Summer and her mother remained in general rather isolated from other adults, Summer stayed in unusually close proximity to her mother as well as spending high amounts of time in actual contact. Cete associated with her younger sister only after Summer became independent six to eight months later.

7. ♂ *Pedro.* Peter, Slim, and briefly Stubby were mildly associated with Preg when her male infant Pedro was born. As with Slim's other such relationships, he did not carry the infant and soon stopped interacting with the dyad but remained a frequent neighbor of Preg, as did Peter. Two-year-old daughter Nazu was not associated with the dyad; four-year-old son Dogo was with them only occasionally. By three months of age Pedro's locomotion was decidedly abnormal and his coat and skin color development retarded. He deteriorated rapidly in the sixth month. At that age, his two-year-old sister, Nazu, began to associate with him and groom him. She continued to do so until his death at eight months. Slim was Preg's sole consort when she resumed cycling.

8. ♀ *Misty.* Slim and Peter were frequent neighbors in the period immediately after female infant Misty was born to Mom, an elderly, high-ranking female. No relationship developed with either male before Misty's death at 2½ mo. Partly due to Slim's threatening Striper away, Misty's two-year-old sister Striper was only slightly associated with her and her mother during Misty's brief life. Striper was much more closely associated with Mom when Mom's next infant, Moshi, was born. Slim was Mom's sole consort when she resumed cycling after Misty's death.

9. ♂ *Bristle.* Brush's first infant was born with webbed fingers and died in the first two weeks of life (D. Post and J. Scott, personal communication). Thereafter, Brush became pregnant in one cycle. Ben and Max both attempted to intensify their existing relationship with Brush when male infant Bristle was born six months later. Ben displaced Max and thereafter established one of the strongest and most persistent relationships observed, interrupted only for several months during which Ben had a severe shoulder wound.

10. ♂ *Hans.* Handle was, along with Slinky, one of the two most protective and restrictive mothers when her first infant, Hans, was born. Hans was kidnapped by adult female Gin when he was two days old and kept until dehydrated and unable to cling; he was retrieved by his mother the next morning. Like other primiparous females, Handle took over a month to accept and reciprocate or make use of male Even's attempts to remain nearby. Once established, however, this relationship was one of the most reciprocal, often including Handle, Hans, Even, and Even's consort female Lulu, as a common close subgroup unit. Red, who also attempted association with Hans and Handle in the first month, was displaced by Even and was repeatedly avoided by Handle.

As infants such as Hans, Bristle, Alice, and Fred became more independent, they were commonly associated with their godfathers at times that their mothers were not.

11. ♂ *Grendel.* Gin's association with young male Red persisted into male infant Grendel's first month. Gin was extremely rejecting and puni-

tive with this first infant and a relationship between Grendel and Red began to develop. At this time, Red was undergoing a period of rapid rise in rank and was frequently involved in agonistic interactions with other males, during which he probably inflicted Ben's serious wound (section 9, above). These interactions culminated in Red leaving the group, at first for two months, then sporadically thereafter. No relationship with Grendel was reestablished. Rather, during month two Grendel established a relationship with male High Tail, who had no obvious association with Gin at the time. High Tail was the only potential father recorded for Grendel but we have incomplete consort data for that cycle. Grendel rode on High Tail often and for long periods, fed next to High Tail and sat in his shade on hot afternoons, ran to him in times of alarm or other distress, until High Tail was killed by a leopard when Grendel was four months old. Thereafter, Grendel spent several weeks attempting to ride on other individuals, including a renewed attempt to spend more time in contact with his mother, but his mother remained rejecting and punitive. Most group members will readily carry only black infants; they only allowed Grendel to ride on them briefly. Grendel survived this difficult period, soon adjusting to his greater independence.

12. ♀ *Sesame.* Slinky's first pregnancy resulted in stillbirth, her second in a severely defective infant, who neither clung nor suckled, and who died two days after birth. Her third infant, Sesame, seemed weak from birth, but survived. After Sesame's birth, Max and Ben both sought association with highly protective Slinky, Max rapidly displacing the recently wounded Ben whom Slinky often avoided. In the first month, Even stayed nearby but noninteractive. Slinky neither approached nor avoided him and he lost interest in the dyad thereafter. Max remained moderately associated with the dyad but shifted much of his attention to Pooh when Chip emigrated. Max and Even were Sesame's most likely fathers.

13. ♂ *Juma.* A good example of the dynamics of the male associations is the case of Judy and her infant, Juma. We have only partial consort data for the cycle in which Judy became pregnant but her frequent associate in general was male Stubby, who died early in the present study, shortly after Judy became pregnant. During late pregnancy Judy was mildly associated with male Peter. Male Slim persistently followed Judy when Juma was born and even more persistently followed infant-snatching Spot who, with her own new infant, followed Judy. Spot sometimes groomed and stayed near Slim. Judy repeatedly moved away from both of them but was often in their presence due to their persistent following. Peter stayed somewhat farther away at these times. Peter and Judy did approach each other. Unlike her behavior with Slim, Judy appeared calm near Peter, did not restrain her infant or avoid Peter and

did groom him. Three-year-old Janet was often nearby and in grooming relations with her mother and younger brother.

When Juma was 2½ mo old, Judy and Juma disappeared overnight the first day of an apparent virus epidemic in the group. Peter was sick that day. He was one of the last two baboons to leave the sleeping grove, and as he slowly joined the group he repeatedly stopped and looked back at the grove. He was the only baboon to do so. We could detect no reaction to Judy and Juma's disappearance either by Slim or by Janet.

14. ♀ *Safi.* During early 1974–75, Spot had been associated with male Stubby (D. Post, personal communication), who died soon after Spot became pregnant. On the cycle in which she became pregnant her only consort recorded on days of likely conception was Stubby, but data are not available for all such days. When Spot gave birth to her first infant, female Safi, Spot ranked second only to her mother (Alto) in the female dominance hierarchy. Only her yearling sister Alice and male Slim intruded upon Spot and her infant. Slim was virtually the only animal who pulled Safi. Alice frequently pushed into Spot's ventrum, as she sometimes did with the other mothers, sitting on or pushing the infant Safi, to which Spot remained quite tolerant until she finally pushed Alice away if Alice persisted. Alice sometimes then enlisted support from her mother, Alto, from adult male Peter, or from her three-year-old sister, Dotty. At an early age Safi developed considerable independence.

During the apparently viral epidemic in May 1976, Spot and her infant, Safi, then seven weeks old, were two of the sickest baboons in the group. During the first few days of their illness, characterized by minimal, swaying movement, long rests and no feeding, juveniles and low-ranking females who had not ventured to do so before, repeatedly approached and handled the infant. There seemed to be a clear recognition of Spot's incapacity and what this implied. Spot recovered more quickly than Safi, who retained an appreciable limp for a month and never resumed dorsal riding.

15. ♂ *Moshi.* Moshi, a healthy male infant, took full advantage of Mom's nonrestrictive care and the relative lack of interference from others. He rapidly became one of the most independent infants. Moshi's sister, three-year-old Striper, spent considerable time with her mother and new brother. Slim was almost surely Moshi's father. Through repeated threats Slim displaced Peter as an associate of the dyad. The subsequent association was mild, did not last long, and included Slim's usual infant pulling.

16. ♀ *Oreo.* When Oreo was born, one of her two likely fathers, High Tail, was dead. The other, Slim, was early associated with the dyad (G. Hausfater, personal communication). Little evidence of such an association remained when I first observed Oreo, during her third month. During

the last trimester of Oval's pregnancy, all ties seemed to be at least temporarily severed between her and her previous offspring, Ozzie, a state which had not changed four months later. A close association did exist between Oval and her three-year-old daughter, Fanny, as did one between Fanny and Ozzie.

17. ♀ *Vicki*. Vee's first infant, Vicki, was not able to get on the nipple during her first day of life. Her mother carried her upside down and backward, even dragging her and bumping her on the ground much of the first day. The infant seemed normal at first sight in the morning, but the extreme mishandling she received soon made it hard to tell the source of poor coloration and signs of weakness and dehydration on that first day. In the sleeping trees the next morning, the infant was on the nipple when first seen. Vee behaved much more competently but remained fairly unresponsive and quite unprotective of the infant. She was still incompetent as compared even to the other primiparous females. Vicki's early deprivation may have led to her death at three weeks. She never regained adequate skin color or clinging ability and often looked somewhat emaciated. During these few weeks, Vee was associated with male Slim, most likely Vicki's father.

18. ♀ *Peach*. As the group foraged, Plum remained behind to give birth to her daughter, Peach. For almost two hours, Max remained about 100 yards from Plum, high on a fallen, dead tree, watching her and occasionally giving single, two-phased alarm barks. Peach's sister, Pooh, remained near her godfather Max. Max, Pooh, Plum, and Peach rejoined the group, staying near each other the rest of the afternoon. The next morning, Red kept Max away from Plum. Male Even then displaced and chased Red away, remaining in a close reciprocal relationship with Plum and Peach throughout the remainder of the study (three weeks). Plum, wary and quite protective, followed Even closely. Even was Peach's most likely father. Red, Max, and High Tail (dead by Peach's birth) were the other consorts during Plum's last estrous period.

Acknowledgments

I gratefully acknowledge the financial assistance of the American people through the National Institute of Mental Health (MH 19617) and the cooperation of the government and people of the Republic of Kenya, especially Mr. Joseph Kioko, the Warden of Amboseli National Park.

This chapter was made possible thanks to the enthusiasm and speed with which Christina Kioko, in Kenya, and Rebecca McCauley, in Chicago, performed the computer preparation and first analyses of my voluminous data. Mihaly Csikszentmihalyi has consistently helped through critical and stimulating discussion. David Stein provided critical comments on an earlier draft of this chapter. Rachel and Michael Altmann contributed patience, pleasure, and often assistance in both Amboseli and Chicago. Stuart Altmann's contributions have been invaluable to every stage of this project.

References

Altmann, J. 1974. Observational study of behavior: sampling methods. *Behaviour* 49:227–267.

Altmann, J., S. A. Altmann, G. Hausfater, and S. A. McCuskey. 1977. Life history of yellow baboons: physical development, reproductive parameters, and infant mortality. *Primates* 18:315–330.

Altmann, S. A., and J. Altmann. 1970. *Baboon ecology: African field research.* Chicago: Univ. of Chicago Press.

Deag, J. M., and J. H. Crook. 1971. Social behaviour and "agonistic buffering" in the wild barbary macaque *Macaca sylvana* L. *Folia Primatol.* 15:183–200.

DeVore, I. 1963. Mother–infant relations in free-ranging baboons. In *Maternal behavior in mammals,* ed., H. L. Rheingold, pp. 305–335. New York: Wiley.

Hausfater, G. 1975. Dominance and reproduction in baboons. *Contributions to primatology.* Vol. 7. Basel: Karger.

Hinde, R. A. and Y. Spencer-Booth. 1967. The behaviour of socially living rhesus monkeys in their first two and a half years. *Anim. Behav.* 15:169–196.

Pielou, E. C. 1969. *An introduction to mathematical ecology.* New York: Wiley-Interscience.

Post, D. 1976. Feeding and ranging behavior of the yellow baboon. *J. Am. Phys. Anthropol.* 44:199.

Ransom, T. W., and B. S. Ransom. 1971. Adult male–infant relations among baboons (*Papio anubis*). *Folia Primatol.* 16:179–195.

Rowell, T. E., N. A. Din, and A. Omar. 1968. The social development of baboons in their first three months. *J. Zool.* 155:461–483.

Slatkin, M. 1975. A report on the feeding behavior of two East African baboon species. In *Contemporary Primatology,* S. Kondo and co-eds., pp. 418–422. Basel: S. Karger.

Slatkin, M., and G. Hausfater. 1976. A note on the activities of a solitary male baboon. *Primates* 17:311–322.

Chapter Fourteen

Genes, Environment, and Development of Temperament in Young Human Twins

ROBERT PLOMIN and DAVID C. ROWE

Institute for Behavioral Genetics
Department of Psychology
University of Colorado at Boulder
Boulder, Colorado 80309

Discussions of evolution usually emphasize variability among species but do not adequately consider the range of variability within species. Behavioral genetics swings the spotlight to variability within species and determines the relative influence of genetic and environmental factors in the development of behavioral individual differences. In this chapter, a behavioral genetic study of social responsiveness in one– and two–year–old human infants will be described after a brief prelude concerning personality.

PERSONALITY

Personality is the complex of individual differences in behavior, individual differences that are stable across time and situations. Individual differences in cognitive behavior have been pulled out for extensive study. However, the myriad of other behavioral traits lumped together in the category of personality contributes to a sense of being overwhelmed by diversity and complexity. Research in human personality has centered around factor-analytic description of these behaviors (e.g., Eysenck, 1972; Cattell, 1973), but these descriptions are not completely satisfying when we remember the old saw, "You only get out of factor analysis what was put in." Although factor-analytic descriptions of personality were a useful antidote to the older theories of personality that had no empirical grounding, it may be useful to return to a theoretical framework that includes an empirical orientation with the descriptive power of techniques such as factor analysis and the etiological analyses of quantitative behavioral genetics.

A theoretical framework is needed to guide excursions into the hinterland of personality, not just for human personality, but also for nonhuman animals. Behaviors related to personality that are studied in nonhuman animals often seem selected for convenience of measurement rather than for a theoretical rationale. This has been particularly true of behavioral genetics research, which, during the 1960s, simply tried to demonstrate that behavior *can* be influenced by genetic factors. However, such demonstrations are no longer needed because it now seems safe to say that behavior can no longer be assumed innocent of genetic influences until proven guilty.

TEMPERAMENT THEORY

One way to focus personality research is to consider the temperamental core of personality. Temperament consists of those personality traits with an inherited component. Temperaments are, of course, modifiable by the environment, but they also modify the environment through the selection of environments, setting the tone of social interactions, modifying the impact of environment, and providing reinforcing feedback that shapes behavior of others.

A temperament theory of human personality development has been proposed (Buss and Plomin, 1975) that used genetic, evolutionary, and developmental criteria to isolate certain personality traits as possible temperaments. By definition, the most important criterion of a temperament is its inheritance. This criterion requires genetic evidence from behavioral genetic studies such as strain and selection studies for nonhuman animals and twin and adoption studies for humans. It also demands that the behavioral character can be measured adequately from a psychometric standpoint in terms of factor structure and reliability. However, many personality traits tend to show heritable influences; indeed, an important book by Loehlin and Nichols (1976) reported a study of 850 pairs of adolescent twins and suggested that all personality traits in humans may be equally influenced genetically. Although there are reasons to doubt this hypothesis (Horn, Plomin and Rosenman, 1976; Plomin *et al.*, 1977), it suggests that additional criteria be used to whittle down personality to its temperamental core.

An evolutionary perspective that considers adaptiveness and the phyletic history of temperaments is another useful criterion. Interest in an evolutionary perspective for human infancy has been rekindled by a book by D. G. Freedman (1974). However, this criterion is speculative and not as easily interpreted as it may seem at first glance. Temperaments are heritable and thus require genetic variability, yet directional selection (selection for a high or low extreme of a trait) *reduces* genetic variability. For example, if emotional arousal is considered to be adaptive in a directional sense (i.e., greater emotional reactivity is selected), then alleles favoring emotional arousal would tend to become locked into the species, thus reducing genetic variability and lowering heritability. One way out is to look for directional adaptiveness, but assume that selection pressure has been relaxed so that genetic variance remains as a result of mutation and migration. A more parsimonious approach is to consider traits that seem adaptive at a moderate level so that genetic variability is insured. In population genetics, this type of selection has been called stabilizing selection. For example, moderate levels of emotional reactivity seem to be crucial for survival of the individual. Very high arousability may be debilitating in the face of threat; low emotionality may not produce fight or flight. A selective edge for moderate levels of a trait balances against either extreme and maintains genetic variability. Stabilizing selection thus makes it possible to consider evolutionarily adaptive, yet heritable, temperaments.[1]

In addition to adaptiveness, the phyletic history of a trait can be used as part of an evolutionary criterion for temperament. Animals other than man, particularly those closer to man in the phylogeny, should evidence similar traits. In an interesting book, *Personality and Temperament,* Diamond (1957) traced the appearance of temperaments in rats, cats, dogs, and chimpanzees, suggesting that "the distinctive values of human cultural existence are outgrowths of this basic pattern, but they do not displace it" (p. 9). However, like many comparative analysts, he emphasized similarities and differences between species and only alluded to variability within species.

Another set of criteria is developmental: early appearance of a trait and longitudinal stability. Traits that emerge early in development and remain stable throughout development are likely to be evolutionarily significant and are, of course, more significant in a contemporaneous sense. Obviously, the converse is not true; adaptive, genetically influenced traits do not necessarily emerge early or show stability in development. For example, human infant reflexes, such as rooting, appear early but are specific to infancy; symbolic, cognitive ability is quite stable longitudinally but does not emerge until after infancy. Nonetheless, these two developmental criteria are useful as part of an initial sieve for temperaments.

FOUR POSSIBLE TEMPERAMENTS

Buss and Plomin used these genetic, evolutionary, and developmental criteria to suggest four traits as possible temperaments: emotionality, activity, sociability, and impulsivity (EASI). *Emotionality* involves intensity of reaction, including arousal and its expressive aspects. Level of *activity* is total energy output. *Sociability* refers to affiliativeness, and *impulsivity* is the tendency to respond quickly rather than inhibiting a response. They reviewed behavioral genetic studies which were, for the most part, twin studies using self-report personality questionnaires, and concluded that the data, although spotty in quality, are consistent with the hypothesis that the four EASI traits are to some extent inherited (although the evidence is not as clear for impulsivity).

The evolutionary criteria of adaptiveness and phyletic comparisons also seem to be met by these four traits. It is reasonable to hypothesize that a certain moderate level of each of these traits is most adaptive. Emotionality was mentioned earlier as an example of stabilizing selection. Too much or too little activity and impulsivity are also obviously maladaptive. The case for stabilizing selection for sociability is not as obvious because greater gregariousness is usually viewed as adaptive for social animals. However, it is likely that too much sociability is maladaptive in producing animals unable to function independently.

The second part of the evolutionary criterion involves presence of the temperaments in animals close to man. Diamond's (1957) analysis suggested four temperaments shared by man and animals close to man: fearfulness, affiliativeness, impulsiveness, and aggressiveness. Diamond's *fearfulness* is close to our temperament of emotionality; *affiliativeness* is the same as sociability; and *impulsiveness* includes activity and impulsivity. However, *aggressiveness* was not included among the EASI temperaments. Although it may be another temperament, there is no evidence for its heritability for humans, and we felt that it could be derived from the temperaments of emotionality, activity, and impulsivity.

In addition to passing the heritability and evolutionary criteria for temperaments, the EASI temperaments have also received some support for the developmental criteria of early appearance and longitudinal stability. For humans, research has isolated stable individual differences during the first few months of life and these include some of the EASI temperaments. Longitudinal studies of personality are quite rare for both human and nonhuman animals, but the few that exist lend some support to the stability hypothesis of the EASI temperaments, at least after infancy. There is, however, essentially no information on the stability of impulsivity and very little information is available for activity. Nonetheless, present data are consistent with the hypothesis that the EASI temperaments pass the developmental criteria.

Thus, in terms of our genetic, evolutionary, and developmental criteria, it seems reasonable to propose the traits of emotionality, activity, sociability, and impulsivity as temperaments. The crucial criterion is the genetic one and more research is needed to confirm the heritability of the EASI traits and components of these traits and to study the developmental interaction of genes and environment. Because most previous human behavioral genetic research on temperament consisted of self-report data from adolescent twins, our previous research focused on studies of temperament in young twins as rated by their parents. One twin and family study (Plomin, 1976) found that the EASI traits and components of these traits were consistent with the hypothesis that the EASI traits are influenced by genetic factors in the first few years of life. Another study (Rowe and Plomin, 1977) combined EASI work with information from the New York Longitudinal Study (Thomas, Chess, and Birch, 1968) to merge the two systems of temperament in early childhood. A twin study (Plomin and Rowe, 1977) based on the new rating instrument again showed genetic influence in the etiology of the EASI temperaments.

Although parental rating studies are a useful preliminary sieve for human behavioral genetic studies, the next step involved a more fine-grained and more objective analysis of the behaviors associated with the EASI temperaments. We are currently conducting a twin study of five- to nine-year-old twins using videotape observations, polygraphic recordings, and standardized tests of emotionality, activity and impulsivity. However, the remainder of this chapter will present the results from an observational study that focused on the temperament of sociability in one- and two-year-old children, and which is described in greater detail in Rowe and Plomin (1977) and Plomin and Rowe (1978).

Observational Twin Study of Social Responsiveness in Infancy

The purpose of the study was to focus on the temperament of sociability in infancy. There were four goals: (1) provide more refined and objective measures of behaviors associated with sociability and measure them in a natural context; (2) determine the relationships among different dimensions of social responsiveness; (3) examine differential responsiveness of the infant to mother and stranger; and (4) determine the relative contribution of heredity and environment to the development of these dimensions of social responsiveness using the twin methodology.

There are very few behavioral genetic studies of human personality that have used measures other than self-report of subjects. Behavioral genetic studies have measured temperament more objectively for non-human animals but often without concern for the evolutionary context of behaviors. Although the control of the laboratory is appealing, we decided that it would be unwise to measure the social responsiveness of infants in a foreign environment. We chose instead to study infants in their homes and obtained some control over their natural environment by presenting standardized situations to the infants. These situations included the appearance of a stranger in the infant's home, playing and cuddling with his (or her) mother and the stranger, and separation from the mother. The infants' social behavior during these situations was assessed by time-sampling and behavioral checklists conducted by observers who assumed unobtrusive positions in the home.

Concerning the second and third goals, we hypothesized that sociability is not a unitary trait. Research on social behavior in human infancy has been dominated by the notion of attachment, which assumes a unitary trait in which social behavior of infants is preferentially directed toward their mother. Masters and Wellman (1974) questioned the extent to which social behaviors of infants are, in fact, directed preferentially toward the mother. We assessed different dimensions of social responsiveness such as fear of strangers and cuddliness as well as more traditional measures of attachment (distal behaviors, proximity-seeking, and separation distress) in order to study their interrelationship. We were also able to compare social responsiveness to the stranger and to the mother in order to test the critical assumption of attachment, i.e., that infants respond preferentially to their mothers.

Despite the evolutionary origins of research on attachment, little thought has been given to genetic variability in individual differences in attachment. Thus, we proposed to study social behavior of infants, using twins in order to apply the twin methodology to describe individual differences in social responsiveness and to ascribe those differences to genetic and environmental influences. The twin method is a natural experiment in which pairs of genetically identical twins are compared to same-sex fraternal twins who are only half as similar genetically. If a trait is not influenced by genes, the twofold-greater genetic similarity of identical twins should not make them more similar phenotypically than fraternal twins. If, however, genes influence a trait, identical twins will be more similar phenotypically than fraternal twins. Of course, both kinds of twins are of the same sex, share the same womb, and grow up in the same family. However, the twin method is often dismissed because it is assumed that identical twins have more similar environments than do fraternal twins, and thus the greater phenotypic similarity of identical twins is due to their greater environmental similarity rather than their greater genetic similarity. Because this is such a common criticism of twin methods, a short digression is in order to address it.

One commonly reads that identical twins share more similar environments than fraternal twins, but evidence supporting this assertion is not easily found. It is reviewed in Vandenberg (1976) and in Loehlin and Nichols (1976), who find only a few such differences, such as identical twins tending to be dressed alike more often. However, the amount of variance accounted for by these different treatments cannot explain much of the phenotypic difference between identical and fraternal twin correlations. Moreover, it is not enough to ask if identical twins are treated more similarly than fraternal twins; one must also ask if such differential treatment makes a difference for the particular trait under consideration. For example, it seems unlikely at the outset that being dressed slightly more similarly than fraternal twins will make identical twins more similar in cognitive ability. Loehlin and Nichols developed a test of this second, and most crucial, question concerning the twin method. Because differences within pairs of identical twins can be caused only by environmental factors, correlating identical twin intrapair absolute differences with environmental measures purported to violate the equal environments assumption of the twin method tests the question of whether differential treatment (such as being dressed alike more) makes a difference in behavior. Loehlin and Nichols examined several such environmental variables and composites of them and found no correlations between identical twin intrapair differences on behavioral measures and differences in environmental treatment. Thus, the few variables for which identical twins tend to be treated slightly more similarly than fraternal twins do *not* make a difference in behavior.

Other types of evidence also support this conclusion. Scarr (1968, in press), for example, has shown that identical twins mistaken by their parents as fraternal twins are behaviorally as similar as other identical twins and vice versa for misdiagnosed fraternal twins. Furthermore, for personality, identical twins reared apart from early in life are *more* similar than identical twins living together, suggesting that, if anything, identical twins living together may actually be treated *less* similarly than expected (Newman, Freeman, and Holzinger, 1937; Shields, 1962). Together, these data suggest that the equal environments assumption of the twin method is a reasonable one and put the burden of proof on those who merely assert that the twin method is biased.

THE SAMPLE

The 92 children in this study were same-sex twins living in 46 Denver-area families. Their mean age was 22.2 mo, with a standard deviation of 6.3 mo. Their mothers were recruited through mothers-of-twins club meetings and came from predominately middle class families. The average education of the mothers was 14.2 yr.

PROCEDURE AND MEASURES

Each home session involved an experimenter (hereafter called "the stranger"), two undergraduates who each rated one twin (to avoid "halo" effects in rating the twins), the twins' mother, and the twins. During the "warm-up" situation, the project was explained to the mother, her questions were answered, and she was asked to sign an informed consent form. Most sessions were conducted in the late morning.

Table 1 summarizes the measures of the seven standardized situations used during each home visit.

In Situation 1, the infants had 5 min to "warm up" to the stranger. In Situation 2, the stranger enticed the infants to play with him, and in Situation 3, the stranger played with the infants using a standard toy. The mother played with her infants in the same way in Situation 4 and cuddled them in Situation 5. In Situation 6, the infants were cuddled by the stranger. Finally, the mother was briefly separated from her infants in Situation 7.

Results

INDIVIDUAL DIFFERENCES

Individual differences in social responsiveness were pronounced for all measures, a finding that tends to be masked in other studies that focus on mean comparisons among children of different ages. Table 2 lists the means and standard deviations for all variables. This variability can be assessed reliably: The average interobserver reliability (for 16 children who were observed by two raters) was .90 for the individual behaviors in Situations 1–4, and above .84 for the scale scores of Situations 3–7, with the exception of "*Quality of Play with Stranger.*" In nearly all cases, the

TABLE 1. Seven standardized situations used to assess social responsiveness.

SITUATION	MEASURES
1. Warm-up (5 min). Stranger (S) on floor, 5 f in front of Mother (M), talking to M.	Alternating 15-s interval recordings of behavior directed toward M and toward S: approaches, proximity in feet, touches, positive vocalizations, smiles, and looks.
2. Stranger Approach (1 min). S offers toy to children; if no approach after 30 s, S moves 3 f closer and offers toy.	Latency to approach S.[a]
3. Play with Stranger (2 min per child). S hands block for perforated toy to child 5 times.[b]	15-s recordings of positive vocalizations, smiles, looks, offering blocks, imitation, playing with S, and playing alone.[c]
4. Play with Mother (2 min per child). M repeats play sequence.	Same as Situation 3.[c]
5. Cuddle with Mother (30 s per child). S demonstrates cuddling with doll, holding doll to chest with doll's head resting on S's shoulder, and asks M to similarly cuddle each child.	Recording duration of hold, seconds squirming, number of times child looks away and ratings of hesitancy to hold and facial expressions.[d]
6. Cuddle with Stranger (30 s per child). Infant was directed to S by M.	Same as Situation 5.[d]
7. Separation from Mother (45 s). M leaves as if leaving house.	Recording frequency and intensity of negative vocalization, negative facial expressions, latency and speed to M's place of departure, final distance from place of M's departure.[e]

[a]Frequency and intensity of negative vocalizations and facial expressions were also recorded but the incidence was so low that they were not useful measures—although it is interesting that, in their natural environment, children are not as fearful of strangers as the laboratory research literature might suggest.

[b]The other child was permitted to stay adjacent to the stranger, but play was directed only to the chosen child.

[c]Because the last four behaviors were interrelated, they were combined in a scale called *Quality of Play with Stranger*. For the Play with Mother situation, the scale was called *Quality of Play with Mother*. Touches were also recorded but not analyzed because of their low incidence.

[d]These behaviors were intercorrelated and thus combined into a composite called *Cuddliness with Mother* for Situation 5 and *Cuddliness with Stranger* for Situation 6.

[e]The interrelationships among these measures suggested that they could be combined into a scale which was called *Separation Distress*.

range of individual variability was the full possible range. For example, in the warm-up situation, some children came running to the stranger and other children were reticent for the entire home visit. When asked how characteristic such behavior was of their children, mothers nearly always said that their behavior during the home visit was characteristic of the way the children usually react to strangers.

The coefficients of variation (SD/$\bar{x}$ × 100) for the scale scores were

TABLE 2. Means and standard deviations of all variables

SITUATION	BEHAVIOR	MEAN	STANDARD DEVIATION
1	Approaches to stranger	1.3	2.0
	Proximity to stranger	6.3	2.7
	Touching stranger	0.2	0.7
	Positive vocalizations to stranger	2.8	3.8
	Smiling at stranger	2.9	3.7
	Looking at stranger	8.4	5.3
	Approaches to mother	1.5	1.8
	Proximity to mother	4.7	4.2
	Touching mother	2.2	1.9
	Positive vocalizations to mother	2.8	3.9
	Smiling at mother	2.4	3.4
	Looking at mother	3.0	2.7
2	Latency to approach stranger	24.2	22.1
3	Quality of play with stranger	22.1	4.6
	Positive vocalizations to stranger	2.5	2.7
	Smiling at stranger	2.4	3.0
	Looking at stranger	2.6	2.6
4	Quality of play with mother	23.2	4.6
	Positive vocalizations to mother	4.0	3.7
	Smiling at mother	3.0	3.5
	Looking at mother	1.7	1.7
5	Cuddliness with mother	32.8	7.2
6	Cuddliness with stranger	31.8	6.4
7	Separation distress	9.7	6.8

20.8%, 19.8%, 22.0%, 20.1%, and 70.1% for "quality of play" with stranger and mother, "cuddliness" with mother, and stranger and "separation distress," respectively. Findings such as these make it unreasonable to focus on between-group variability (such as age or sex differences) without recognizing the wide range of within-group variability.

It should be mentioned that the results to be reported are not a function of sex, age, or experimenter effects. Only one sex difference was found: males were significantly more cuddly in Situation 6. Although the children ranged in age from 13 mo to 37 mo, there were few age effects. There were no more significant experimenter effects than expected by chance.

DIMENSIONS OF SOCIAL RESPONSIVENESS

One of the goals of this research was to determine the relationships among different dimensions of social responsiveness. The results clearly show that there is no monolithic concept like attachment or sociability. Although age did not covary with many of the social behaviors, we nonetheless partialed out age from a correlation matrix among the social behaviors. The intercorrelations indicated that social behaviors in different situations were generally independent, particularly across situations. For example, individuals who were less fearful of the stranger in the warm-up situation were not necessarily more playful in the play situations nor more cuddly. In general, at least five independent dimensions are required to assess social responsiveness: distal behaviors (vocalization, smiling, looking), proximal behaviors (approaching, proximity, touching), playfulness, cuddliness, and separation distress.

However, within situations, there was somewhat more generality among social behaviors. As indicated in Table 1, behaviors from Situations 3, 4, 5, 6, and 7, were combined into scales for each situation. Cronbach Alpha reliabilities of .54, .66, .84, .78, and .86 for these scales indicate substantial covariance among the behaviors within a situation.

ATTACHMENT

Another goal of this research was to examine the hallmark of the concept of attachment, i.e., preferential responding toward the mother. Children actually showed significantly more distal social behavior to the stranger than to the mother during the warm-up situation; they looked significantly more at the stranger than the mother in the play situations, and they cuddled as well with the stranger as the mother. These findings suggest that the notion of fear of strangers may be overplayed—it seems to be more like information-seeking or wariness as indicated by increased vocalization, smiling and looking. It should be reiterated that these data were obtained in the child's natural home environment rather than a laboratory which is the usual source of information about fear of strangers.

TWIN STUDY

The reliable individual differences in social responsiveness described above may be caused by genetic or environmental differences among the children. The twin method untangles genetic and environmental influences by comparing the phenotypic similarity for identical twins to same-sex fraternal twins. In the past, zygosity of twins was determined by analysis of genetic markers in the blood, i.e., comparing twin pairs on a number of markers. If they differ on any of these markers, the twin pair must be fraternal. However, zygosity diagnosis has been made easier, particularly for young twins from whom it is difficult to obtain blood, because it has been demonstrated that physical markers such as eye color, hair color, hair texture, skin complexion, and height can be used in the same way to diagnose zygosity with greater than 90% accuracy as validated against blood analysis. For example, Cohen *et al.* (1973) found 98% accuracy for physical similarity zygosity diagnosis for young twins. Analysis of physical markers, using a method similar to that of Cohen *et al.*, yielded 21 pairs of identical twins and 25 pairs of fraternal twins. It should be noted that misdiagnosis of zygosity, while unlikely, is a conservative bias in terms of a genetic hypothesis because if identical twins are misdiagnosed as fraternal and vice versa, the difference between the identical and fraternal twin correlations will be reduced.

Intraclass correlations were computed separately for identical and fraternal twins for all of the behaviors listed in Table 1. Because the sample sizes were not very large, the most conservative analysis of heritable influence is to test for identical twin correlations that are significantly greater than fraternal twin correlations. Table 3 lists the social behaviors with a substantial heritable component.

Many of the social behaviors toward the stranger were heritable, but only one of the behaviors toward the mother was genetically influenced. In the warm-up situation, for example, both distal and proximal social

TABLE 3. Heritable social behaviors [a]

		r mz	r dz
Situation 1:	Positive vocalization to stranger;	(.58,	.34)
	Looking at stranger;	(.67,	.08)
	Approaching stranger;	(.50,	−.05)
	Proximity to stranger;	(.40,	−.03)
	Touching mother.	(.47,	.22)
Situation 2:	Latency to approach stranger	(.51,	.30)
Situation 3:	Positive vocalization to stranger;	(.49,	−.03)
	Smiling at stranger.	(.58,	.03)
Situation 5:	Cuddliness with mother.	(.15,	−.30)

[a]Social behaviors for which the identical correlation was substantially greater than the fraternal twin correlation

behaviors toward the stranger were heritable but only touching the mother showed genetic influence. In the play situation, both positive vocalization and smiling at the stranger indicated the influence of genes, but none of the social behaviors toward the mother was heritable. Although cuddliness with the mother showed significant heritability, the insignificant identical twin correlation and the negative fraternal twin correlation deny the importance of genetic influence. This finding of genetic influence for social interactions with strangers but not with mother may have important ramifications for understanding social behavior, and we will return to it in the discussion section.

In the past, behavioral genetic analyses have estimated the influence of genetic differences and ascribed all the rest of phenotypic differences to environmental differences and error. Obviously, all of the behaviors not listed in Table 3 are social behaviors for which individual differences are primarily determined by environmental factors. Even the heritable behaviors in Table 3 also show environmental influences. However, it is possible to separate the influence of the environment into more distinct categories: environmental influences that operate "between" families to make family members similar to one another and different from other families and those environmental influences that operate "within" families to make family members different from one another. In the social sciences, most research has focused on between-family environmental factors such as childrearing practices. However, Loehlin and Nichols (1976) have suggested that, for personality, most of the environmental variance lies within, not between, families.

Within-family environmental factors can be estimated by the differences within pairs of identical twins. Because they are identical genetically, differences within pairs of identical twins can be caused only by environmental factors and these are environmental factors making family members (in this case, siblings of a twin pair) different from one another, i.e., within-family environment. Between-family environmental variance can be estimated by subtracting the within-family environmental component from the total environmental variance. However, such subtractions are hazardous with sample sizes numbering less than hundreds of pairs, and thus we have again been conservative by listing in Table 4 those social behaviors that showed significant correlations for both identical and fraternal twins (suggesting between-family influences) but with no significant difference between their correlations (suggesting that the between-family influences are environmental rather than genetic). Because these behaviors show between-family similarity but no heritable influence, we tentatively list them as between-family environmental variables (although they show within-family environmental influence as well).

The most noticeable feature about Table 4 is the paucity of social

TABLE 4. Social behaviors that are influenced by between-family environmental factors[a]

		r_{mz}	r_{dz}
Situation 1:	Positive vocalization to mother	(.56,	.46)
Situation 4:	Smiling at mother	(.60,	.53)
Situation 6:	Cuddliness with stranger	(.44,	.42)

[a]Identical and fraternal twin correlations are significantly greater than zero but there is no significant difference between the identical and fraternal twin correlations

behavior influenced by between-family environmental factors, in line with Loehlin and Nichols' hypothesis. They suggest that serious attention be paid to the possibility of a major role for environmental factors that make family members different from (rather than similar to) one another.

Most of the environmental action lies within families, as indicated in Table 5.

Discussion

There appear to be several dimensions of social responsiveness in human infancy and they show wide-ranging individual differences. These dimensions can be reliably assessed, they are largely independent, and they seem to be unrelated to the concept of attachment. The behavioral genetic twin data suggest the intriguing possibility that genetic influences are important in the development of social responsiveness to strangers but not to the mother.

This finding is particularly exciting because it fits with other behavioral genetic data on adult personality. Although sociability appears on every self-report personality questionnaire and the evidence for its inheritance is well known (Scarr, 1969), there have been few attempts to study components of sociability. Buss and Plomin (1975) suggested two components: quantity and quality. *Quantity* refers to superficial relationships implied by the term "gregariousness," and *quality* denotes "warmth" or "affectionateness." When these two components were assessed, they were correlated only .30, and the quantity rather than the quality component showed greater genetic influence in a family study and to some extent, in a twin study (Plomin, 1974). Social responsiveness to a familiar person may fall in the category of quality or warmth of a relationship; social responsiveness to an unfamiliar person may involve the quantity component of sociability. Thus, in the present study, social responsive-

TABLE 5. Social behaviors that are influenced predominantly by within-family environment factors[a]

		r mz	r dz
Situation 1:	Smiling at mother	(.19,	.19)
	Looking at mother	(−.01,	.11)
	Approaching mother	(.14,	−.03)
	Proximity to mother	(.23,	.11)
	Smiling at stranger	(.08,	.25)
	Touching stranger	(−.07,	−.03)
Situation 3:	Looking at stranger	(.17,	.17)
	Quality of play with stranger	(−.02,	.21)
Situation 4:	Positive vocalization to mother	(.10,	.37)
	Looking at mother	(−.05,	.23)
	Quality of play with mother	(.32,	.21)
Situation 7:	Separation distress	(.26,	.40)

[a]Identical twin correlations are not significant

ness to the mother (which indicated no heritability) may reflect the quality component and social responsiveness to the stranger (which was significantly heritable) may tap into the quantity component.

These conclusions are supported by an independent line of research. Rather than using phenotypically defined traits to measure personality, it is possible through the use of multivariate techniques and factor analysis to derive genetically and environmentally defined traits (Plomin et al., 1977). The simplest method for self-report personality data is to use *items* rather than phenotypically-defined scales which are based on many items with an unknown mix of genetically and environmentally influenced items. Items high in heritability can be factor analyzed separately from items low in heritability. Horn, Plomin, and Rosenman (1976) used the 180 personality items of the California Psychological Inventory to conduct such an analysis with data from 198 pairs of adult male twins. In a factor analysis of 41 highly heritable items, one factor accounted for 28% of the variance. All the highly loading items on this factor involved interacting with strangers. For example, the highest-loading item on this factor was: *It is hard for me to find anything to talk about when I meet a new person.* This analysis again suggests that one of the most heritable aspects of personality is social responsiveness to strangers.

Evolutionarily, it may be that the social bonds involved in making man a social animal are limited to the cohesiveness necessary for interaction in groups. The quality, warmth, or affection of close dyadic interaction may not be part of our evolutionary heritage. Although the present data suggest that too much has been made of the concept of attachment, it is interesting to note how our theory corresponds to Bowlby's (1969) formulation concerning the evolution of attachment. In the present study, the typical indices of attachment such as separation distress were not heritable. However, Bowlby felt that the evolutionary underpinnings of attachment were not intrinsic to the relationship between child and mother but rather to predator pressure favoring wariness toward novel stimuli. It may be that this wariness is the other side of the coin of the quantity component of sociability that leads to group cohesion. This component, rather than attachment per se, is heritable.

In summary, we suggest that social responsiveness is not a unitary trait and that it is situation specific. Social responsiveness to a familiar figure such as mother is unrelated to social responsiveness to strangers and it is the latter that has an inherited component.

Summary

A temperament theory of personality development was discussed and one aspect of that theory, social responsiveness, was studied in detail. The social behavior of 92 one- and two-year-old children was assessed towards their mother and a stranger using seven standardized situations: Warm-up, stranger approach, play with stranger, play with mother, cuddle with mother, cuddle with stranger, and separation from the mother. Individual differences in social behavior were considerable. Correlational analyses among these behaviors indicated that there are several independent dimensions'of social behavior, i.e., no single concept can capture the full flavor of social behavior in infancy. A behavioral genetic analysis of twin data led to the conclusion that significant genetic influences exist for social behaviors directed toward a stranger, but *not* toward mother. This finding seems to fit with other data that suggest that social responsiveness to strangers rather than familiar figures is heritable. Environmental influences predominate in social responding, but contrary to common belief, the environmental influences on social behavior in infancy operate primarily within families rather than between families.

Acknowledgments

Supported in part by NIMH grant 28076 and a grant from the William T. Grant Foundation.

Note

1. This problem and this approach to the problem were suggested by J. C. DeFries of the Institute for Behavioral Genetics.

References

Bowlby, J. 1969. *Attachment.* New York: Basic Books.

Cohen, D., E. Dibble, J. M. Grawe, and W. Pollin. 1973. Separating identical from fraternal twins. *Arch. General Psychiatry,* 29:465–469.

Cohen, L. J. 1974. The operational definition of human attachment. *Psychol. Bull.* 81:207–217.

Masters, J. C. and H. M. Wellman. 1974. The study of human infant attachment: a procedural critique. *Psychol. Bull.* 81:218–237.

Buss, A. H., and R. Plomin. 1975. *A temperament theory of personality development.* New York: Wiley-Interscience.

Cattell, R. B. 1973. *Personality and mood by questionnaire.* San Francisco: Jossey-Bass Publisher.

Diamond, S. 1957. *Personality and temperament.* New York: Harper and Brothers.

Eysenck, H. J. 1972. Primaries or second-order factors: a critical consideration of Cattell's 16 PF Battery. *Br. J. Soc. Clin. Psychol.* 11:265–269.

Freedman, D. G. 1974. *Human infancy: an evolutionary perspective.* Hillsdale, N.J.: Lawrence Erlbaum Associates.

Horn, J., R. Plomin, and R. Rosenman. 1976. Heritability of personality traits in adult male twins. *Behav. Genet.* 6:17–30.

Loehlin, J. C., and R. C. Nichols. 1976. *Heredity, environment, and personality: a study of 850 twins.* Austin, Texas: Univ. of Texas Press.

Mischel, W. 1968. *Personality and assessment.* New York: Wiley.

Newman, J., F. Freeman, and K. Holzinger. 1937. *Twins: a study of heredity and environment.* Chicago: Univ. of Chicago Press.

Plomin, R. 1974. *A temperament theory of personality development: parent–child interactions.* (Unpublished doctoral dissertation, Univ. of Texas.)

Plomin, R. 1976. A twin and family study of personality in young children. *J. Psychol,* 94:233–235.

Plomin, R., J. C. DeFries, D. C. Rowe, and R. Rosenman. Genetic and environmental influences on human behavior: multivariate analysis. *Paper presented at the Seventh Annual Meeting of the Behavior Genetics Association, Louisville, April 29, 1977.*

Plomin, R., and D. C. Rowe. 1977. A twin study of temperament in young children. *J. Psychol.* 97:107–113.

Plomin, R., and D. C. Rowe. An observational twin study of social responsiveness in infancy. *Dev. Psychol.* (In press.)

Rowe, D. C. and R. Plomin. 1977. Temperament in early childhood. *J. Personality Assessment* 41:150–156.

Rowe, D. C. and R. Plomin. An observational study of individual differences in social responsiveness in infancy. (Submitted, 1977.)

Scarr, S. 1968. Environmental bias in twin studies. *Eugenics Q.* 15:34–40.

Scarr, S. Twin method: defense of a critical assumption. (In press.)

Shields, J. 1962. *Monozygotic twins brought up together and apart.* Oxford: Oxford Univ. Press.

Thomas, A., S. Chess, and H. Birch. 1968. *Temperament and behavior disorders in children.* New York: New York Univ. Press.
Vandenberg, S. G. 1976. Twin studies. *In Human behavior genetics,* A. R. Kaplan, ed., pp. 164–197. Springfield, Ill.: Thomas.

Chapter Fifteen

The Development of Sex Differences in Human Behavior: Cultural Impositions, or a Convergence of Evolved Response-Tendencies and Cultural Adaptations?

LAWRENCE V. HARPER and KAREN SANDERS HUIE

Department of Applied Behavioral Sciences
University of California at Davis
Davis, California 95616

Traditionally, animal behaviorists have been concerned with the analysis of at least three aspects of behavior: function, underlying mechanisms (including ontogeny), and evolution. We will concentrate upon the analysis of the latter two aspects and will argue that the early development of sexually dimorphic behavior patterns in humans can be understood as biologically based predispositions which evolved as a result of hominids' adopting a hunting-gathering way of life.

Evolutionary Origins

From an evolutionary standpoint, differences between the sexes can be traced back to the specialization of sex cells (Ghiselin, 1974; Trivers, 1972; Williams, 1975). The "male" form is small and mobile whereas the "female" type is large and immobile. Presumably, once reproduction by gametic fusion had evolved, the production of undifferentiated sex cells could not be maintained because, beyond the basic hereditary instructions, there were inevitable differences in cytoplasmic stores. Assuming that a rich cytoplasm favors zygote survival, but reduces motility, selection should have favored specialization. The smaller, more mobile (sperm) cells would be more likely than intermediates to contact and combine with the most abundantly endowed (egg) cells. Heavily laden egg cells, being minimally mobile, would be least likely to combine with each other and less likely to combine with intermediates than with sperm. Intermediates would be at a disadvantage both ways: when combining with motile, but nutrient-poor sperm cells, fewer zygotes would survive relative to sperm-egg combinations, and because they were less mobile than the sperm cells, intermediates also would be less likely to combine with each other or the richly endowed egg cells. Thus, gamete specialization probably evolved relatively early (cf. Dawkins, 1976; Trivers, 1972).

The factors that determine whether an individual produces eggs or sperm, or both, vary across species. Among many forms, the fixity of gamete production (i.e., specialization for egg or sperm production) may be a function of the degree to which the fitness of the sexes depends upon environmental conditions, and whether or not the individual has control over the conditions it will encounter (cf. Charnov and Bull, 1977). At any rate, there exists a range of strategies from almost purely environmental through strict genetic determination of reproductive sex.

The latter appears to have evolved from the specialization of one pair of chromosomes into two distinctive forms—a highly modified "Y" or "Z" and a more basic "X" or "W"—without alteration in the remaining hereditary materials (Mittwoch, 1973). Although a number of animals such as fish or amphibians maintain considerable developmental lability despite the possession of this chromosomal dimorphism, among birds and mammals the hereditary determination of sex is relatively absolute; it depends upon whether the individual is hetero- or homogametic (Mittwoch, 1973; Silvers and Wachtel, 1977). In birds, the male is the homogametic sex (WW), whereas, in mammals, it is the female (XX). Which genotype produces eggs (is female) may depend upon the degree to which the zygote is insulated from the maternal milieu (Mittwoch, 1973). In any case, the mechanisms underlying the development of sexual *dimorphisms* seem to have evolved as a unit, the same for both sexes (Ghiselin, 1974) and common across all vertebrates (Mittwoch, 1973). It appears that the homogamete's developmental creode is the "basic" one, and the phenotype of the heterogamete is "induced" as the fertilized egg develops.

DEVELOPMENT OF SEXUAL DIMORPHISMS IN MAMMALS

In mammals, the male is heterogametic and the female homogametic; reproductive sex is thus determined at conception. Generally speaking, the embryo has the potential to develop either a male or a female phenotype until the gonads begin to differentiate. The latter event is largely, if not exclusively, controlled by the Y chromosome. Although it is not entirely clear whether the Y chromosome specifies testicular development directly (cf. Silvers and Wachtel, 1977) or indirectly, by altering the rate of gonadal growth (Mittwoch, 1973; Ounsted and Taylor, 1972), in the absence of the Y, the gonad develops as an ovary. Gonadal *differentiation* is thus the first step in sexual differentiation.

The second step depends upon gonadal *function*. The female phenotype appears to be the "basic" creode in mammals, and the induction of a male phenotype depends upon the presence of testicular secretions. The primordia of the fetal reproductive organs are essentially bisexual to this point. In the absence of testicular secretions, the specifically masculine components regress and a feminine phenotype develops. In the presence of testicular hormones, the reproductive tract differentiates as a male and the primordia of the uniquely feminine organs regress (Diamond, 1976; Mittwoch, 1973).

In addition, at some point after the reproductive tract has begun to differentiate, gonadal secretions also influence neural development. Here, too, the "basic" creode appears to be largely feminine, and testicular hormones more or less permanently "bias" the CNS to develop and function in a characteristically "masculine" fashion (Diamond, 1976; Ounsted and Taylor, 1972). This biasing presumably involves a variety of processes, including the lowering of thresholds for male-typical behavior without necessarily raising the thresholds for (concurrently developing) female-typical activities (cf. Beach, 1976b). In addition, it may be that certain female-typical behaviors are potentiated by ovarian secretions and that sensitivity to gonadal output depends upon genotype. In any case, some of these early hormonal events may affect behavior in the absence of further hormonal stimulation (Beach, 1976b; Diamond, 1976).

Later on, corresponding to the labile dimorphisms of lower forms, additional sex-related differences in anatomy and behavior will become manifest in response to, and depending upon, further gonadal secretions. A number of these hormonally induced changes will have been conditioned by earlier hormonal secretions (Beach, 1976b).

A final mechanism that may account for some of the sex differences observed in mammals has to do with the fact that the male's X chromosome has no "backup" like that of the female. Thus, although the X chromosome is common to both sexes, it will be subject to more intense selection in the male (Mittwoch, 1973; Ounsted and Taylor, 1972).[1] In addition, it has been suggested that the regulatory "message" of the Y chromosome affects the developmental rate of the male such that a wider range of genetic potential is expressed than in the female (Ounsted and

Taylor, 1972). In any event, it seems that the Y chromosome message is largely, if not exclusively, regulatory; we are dealing with differential expression of shared genetic potentials.

The latter point is crucial to the understanding of the development of behavioral sex differences. It now seems clear that the substrata for most masculine and feminine activities are not physiologically mutually exclusive. Rather, they represent the functioning of independent, semiautonomous neural systems, both of which develop in all members of the species. Hence, sex differences in behavior simply represent the relative dominance or unequal development of one or the other of these shared systems (Beach, 1976a,b; Whalen, 1974). As we have pointed out, although there may be some structural information coded by the Y chromosome, its message seems to be primarily regulatory. Indeed, one of the genes involved in determining mammals' responsiveness to androgens is probably located on the X chromosome (Mittwoch, 1973; Silvers and Wachtel, 1977). In short, here too, we are dealing primarily with differential activation of information contained in the rest of the genome as opposed to a unique, structural message from the Y.

Evolution and Development of Sex Differences in Humans

The precise nature of the sex differences characteristic of any form depends upon the adaptive strategies that the species has adopted. However, among most animals there seem to be common dimorphisms corresponding to behaviors that have presumably arisen in response to the fundamental differences in gamete production. Relative to sperm, eggs are in short supply, and, reproductively speaking, egg-producers (females) tend to be sought after more often than seekers, whereas sperm-producers (males) tend to seek out females and compete for them—or commodities that will provide them with access to females. Thus, on the average, males tend to be better "equipped" for search and conflict whereas females, who are sought after and also responsible for care of the young, are less likely to expose themselves to danger (Campbell, 1972; Williams, 1975). These differences are neither absolute nor universal; other factors, such as the relative investment in the care of young—and the nature of the care provided—can profoundly affect the form and degree of sexual dimorphism (Trivers, 1972; Ghiselin, 1974; Wilson, 1975). However, the fact that the female can be more certain who her offspring are may tend to favor the development of different functional roles even where care-giving is a shared endeavor (Ghiselin, 1974).

This specialization seems to have been true of our own species throughout most of its evolutionary history. Although the earliest known representatives of the hominid lineage were not as markedly dimorphic as contemporary ground-dwelling primates, the males were larger and physically stronger than females (Campbell, 1974; Leaky, 1973). We may presume that, like most primates, they functioned to protect the group from predators and to regulate intra- and intergroup conflict, while the females were relatively more concerned with the immediate care of young (Eisenberg, Muckenhirn, and Rudran, 1972). It is very likely that when hominids adopted a hunting-gathering way of life, further specializations occurred. Building upon the already-existing division of labor (reinforced by the constraints imposed by pregnancy and nursing), hunting then became a typically male activity while gathering was largely, although by no means exclusively, a female function (Lee and DeVore, 1968). Such a division of labor among the sexes has been typical of most human societies (including agricultural ones) until very recent times. A clear majority of nonindustrial cultures prescribe mobility, physical exertion, and group cooperation—all required for hunting large game with primitive weapons—for males, while female roles involve more sedentary, solitary activity—appropriate for gathering (cf. D'Andrade, 1966).

Insofar as selection might favor individuals who were capable of fulfilling their societal roles, one would expect that males and females would become predisposed to develop the appropriate behaviors (cf. Beach, 1974, 1976a; Harper and Sanders, 1978). Specifically, one would expect that, relative to girls, boys should be physically more active, and more physically "aggressive"; as potential hunters, they should also spend more time away from the center of the group, be more adept at visualizing spatial relationships, especially as they relate to locating one's position in space and guiding projectiles, and they should engage in more mutually regulated activity, especially while moving about at a distance from the larger group (as in pursuit of large game). As future gatherers, girls should be more likely to engage in activities which, while similar to those of others, do not require mutual coordination;[2] and, as care-givers for the young, even when not threatened, they should remain closer to the main body of the group and to those who could protect them.

With regard to our first two expectations there is abundant evidence that males are more given than females to gross muscle activity—even from birth—and that across a number of cultures they are more physically aggressive (cf. Block, 1976; Maccoby and Jacklin, 1974).[3] In accord with the third, we have found that, between the ages of 3 and 5, preschool boys spend more time outdoors and use more space during free play than do girls (Harper and Sanders, 1977). The latter sex differences may be influenced by early hormones: genetic females who are exposed to androgenic stimulation *in utero* spend more time in vigorous outdoor activity than do untreated girls (Money and Ehrhardt, 1972) or their sisters (Ehrhardt and Baker, 1974)—despite the fact that they were reared as girls without further exposure to androgens.[4] These data are consistent with the notion that little boys may be congenitally predisposed to develop behavior patterns appropriate to a hunting role. Moreover, they suggest that the differences occur as a result of the same kinds of hormonal mechanisms as those involved in the ontogeny of gender-differentiated behavior in other mammals.

With respect to our fourth attribute, it is now generally accepted that among most human groups males are more adept at visualizing spatial relations than are females (cf. Maccoby and Jacklin, 1974). Although the genetic and/or hormonal mechanisms underlying these differences are unclear (cf. DeFries *et al.,* 1976), there is evidence that the right cerebral hemisphere (which normally is dominant when analyzing spatial relations) develops more rapidly in boys than in girls. This seems to be true for the deposition of myelin, observable as early as age 4 (Buffery and Gray, 1972), and the onset of specialized function in spatial analysis, which is apparent by age 6 (Witelson, 1976). Thus, although there is some debate concerning the developmental dynamics involved (cf. Waber, 1976), this sex difference has clear biological roots and is in accord with our evolutionary model.

Our investigations of the social interactions of preschool children also support the view that males would be more likely than females to engage in mutually regulated activity while away from the "main group" and that females would indulge in more individual pursuits, closer to sources of protection.

SEX DIFFERENCES IN SOCIAL CONTACTS
AMONG PRESCHOOLERS

Our data are drawn primarily from 51 girls and 56 boys who were enrolled at the University Laboratory preschool between fall 1971 and spring 1975. Enrollment varied between 25 and 30 children per quarter; between 15 and 25 children remained in the program throughout every quarter of each of the 4 academic years. Additional observations were made at a parent cooperative preschool in the neighboring community of Dixon on 6 girls and 8 boys in winter and spring of 1973–74, and on 13 boys and 11 girls over all three academic quarters of the 1975–76 year. The Davis facility is very large, affording 186 m[2] indoor play area and 1780 m[2] outdoor space. The Dixon facility had 141 m[2] indoor area available for play in both years. In 1973–74, the children had access to 375 m[2] outdoors; in 1975–76 this area was expanded to about 550 m[2].

For the Davis samples, there were between 4 and 9 h of observation per child per quarter. The Dixon samples were observed for shorter periods of time, between 1 and 4 h per quarter. At Davis, each child was observed during free play once each week for 35–50 min and his or her behavior was recorded continuously, every 15 sec. At Dixon, we could not adhere to a weekly schedule and the observation periods varied from 15 to 35 min due to shorter free-play sessions. Otherwise, the observational procedures were the same.

In each 15-sec period we recorded the following data relevant to this report: the child's: (1) location; (2) play activity, which was categorized as (a) solitary, apart from and independent of the activities of others, (b) parallel, within 2 m of, and similar to the activity of another child, but not directly regulated by it, (c) interactive, involving action on or reciprocal response from another child; and (d) cooperative, involving a mutually agreed-upon goal or theme. These four categories were mutually exclusive. (In 1971–72 we scored all activity involving rapid changes of location simply as "in transit" without further modification; in all subsequent years we further categorized the quality of transit behavior.) (3) In every year we also recorded contacts with the teaching staff, which were categorized as either (a) "disciplinary" or (b) "general." From fall 1971 to spring 1973 general adult contacts were subclassified as adult- or child-initiated whenever it was obvious who initiated a bout. (In 1975–76 we revised our protocols and reinstituted these categories to cover *all* general adult contacts.) Furthermore, we also noted whether adult contact involved a male teacher. Unfortunately there were no males on the Dixon staff in 1975–76, and there were too few instances of such behavior at Davis to warrant subclassification according to the identity of the initiator. Interobserver agreement on all categories reported here ranged from 70 to 98%.

Free-Play Peer Relationships

Despite the facts that the arrangement of the play areas at Davis changed twice over the years and that the enrollment and teaching staff varied both within and across years, a consistent trend in peer play emerged. The girls engaged in more parallel play while the boys spent more time interacting with their peers. The same trends appeared at Dixon, especially in the 1975–76 school year when there was greater outdoor space available (see Table 1). In addition, when one examines the locations in which these activities occurred, it becomes clear that, relative to total play time, less parallel play occurred when the children were outdoors and there was a consistent trend for them to engage in more interactive play when outside. These data are also in accord with our expectations:

TABLE 1. Free play

Davis Laboratory School

	TIME IN SOLITARY ÷ TOTAL TIME	TIME IN PARALLEL ÷ TOTAL TIME	PARALLEL OUTDOORS ÷ TIME OUTDOORS	TIME IN INTERACTION ÷ TOTAL TIME	INTERACTIVE OUTDOORS ÷ TIME OUTDOORS	TIME IN COOPERATIVE ÷ TOTAL TIME
1971-72						
FALL[b]						
♂ N[a]=15	.29	.37	.40	.20	.25	.02
♀ N=12	.24	.46*	.48	.18	.19	.03
♂ N=13	.20	.46	.39	.26	.27	<.01
WINTER[b]						
♀ N=14	.24	.54*	.41	.15*	.18	<.01
♂ N=13	.25	.33	.34	.31	.37	<.01
SPRING[b]						
♀ N=15	.26	.44**	.45	.21*	.23**	.01*
♂ N=12	.35	.41	.40	.22	.23	.01
1972-73						
FALL						
♀ N=11	.41	.45	.42	.13	.11*	<.01
♂ N=13	.27	.37	.31	.35	.40	<.01
WINTER						
♀ N=12	.40*	.45*	.39	.14***	.14***	<.01

	.28	.33	.30	.39	.43	< .01
♂ N = 14	.28	.33	.30	.39	.43	< .01
SPRING						
♀ N = 12	.39	.40	.39	.20**	.24**	.01
♂ N = 12	.44	.38	.34	.15	.20	.03
FALL 1973–74						
♀ N = 12	.35*	.47**	.43	.15	.18	.03
♂ N = 13	.35	.39	.28	.22	.29	.04
WINTER						
♀ N = 14	.38	.40	.30	.19	.20	.03
♂ N = 16	.29	.39	.38	.27	.30	.05
SPRING						
♀ N = 14	.32	.43	.44	.20*	.22	.05
♂ N = 11	.27	.41	.35	.25	.27	.07
FALL						
♀ N = 12	.33	.50	.41	.13*	.17*	.04
♂ N = 12	.26	.43	.33	.26	.27	.05
WINTER 1974–75						
♀ N = 12	.25	.53**	.41	.18	.18	.04
♂ N = 11	.23	.43	.40	.30	.32	.04
SPRING						
♀ N = 12	.26	.54*	.53*	.17*	.19*	.03

TABLE 1. *(cont.)*

	TIME IN SOLITARY ÷ TOTAL TIME	TIME IN PARALLEL ÷ TOTAL	PARALLEL OUTDOORS ÷ TIME OUTDOORS	TIME IN INTERACTION ÷ TOTAL TIME	INTERACTIVE OUTDOORS ÷ TIME OUTDOORS	TIME IN COOPERATIVE ÷ TOTAL TIME
Dixon Parent Cooperative						
1973–74						
WINTER ♂ N = 8	.37	.38	.30	.16	.28	.08
♀ N = 6	.33	.38	.29	.20	.16*	.08
SPRING ♂ N = 8	.36	.41	.42	.18	.25	.04
♀ N = 6	.28	.48	.49	.20	.22	.03
FALL ♂ N = 12	.25	.49	.35	.20	.23	.06
♀ N = 9	.26	.52	.41	.16	.21	.06
1975–76						
WINTER ♂ N = 12	.20	.45	.39	.28	.31	.07
♀ N = 12	.21	.55	.47	.20	.23	.05
SPRING ♂ N = 11	.19	.49	.46	.27	.30	.05
♀ N = 11	.19	.64**	.61*	.14**	.18*	.02*

[a] Sample size varied from season to season.
[b] Did not include activity "in transit."
*Difference significant at .05 level; ** at .01 level; *** at .001 level.

gathering and domestic tasks are relatively solitary; they require less mutual regulation and far-ranging activity than does hunting. Moreover, a comparison of the children's behavior at Dixon across the two years raises the interesting possibility that the boys' greater interactive play depends upon the availability of a minimum amount of outdoor space. (Their behavior was otherwise much the same; as in 1973–74, the boys in the 1975–76 group spent more time outdoors than did the girls.) This too would be compatible with the hypothesis of a congenital predisposition to engage in behaviors appropriate to a hunting role—the development of any behavior requires appropriate environmental support.

Moreover, reasoning from the premise that hunting involves interaction while in *pursuit* of game, we predicted that boys would display more interactive play than girls while in transit and that, for boys, time in transit should correlate positively with interactive play. Although the results were only suggestive, they were generally in accord with our hypothesis (Harper and Sanders, 1978).

So far then, our findings are consistent with the picture of the male-as-hunter. But what about the girls? We have already noted that they engage in more indoor and parallel play than boys as we would expect from the demands of tending the young and gathering. Moreover, in accord with the view that the female should be more prone to remain in protected areas, we find that, overall, girls spend more time interacting with adults and that they initiate more contacts with teachers than do boys. Although the differences are not great, they are remarkably stable for both programs (see Table 2).

In accord with our hypothesis that such sex differences may have a congenital basis, there are data suggesting that "social style" may be influenced by early hormonal events. Reinisch (1977) studied the male and female offspring of otherwise normal mothers who had received progestin- or estrogen-dominant medications while they were pregnant. On a standardized, paper-and-pencil test of "personality," the children who were exposed to estrogens were more "group-oriented" and "group-dependent," i.e., looking to others for authority, than their siblings or the progestin-exposed subjects. Here, too, the evidence is compatible with the notion that early hormonal events may "bias" certain response tendencies—in this case, that prenatally occurring ovarian activity may cause girls to be more responsive to adults, and to seek their company more often than boys while at free play.

However, when we look at time spent with male teachers we note that the trend is the opposite: boys spend more time with them than do girls (see Table 3). These findings could be interpreted as a congenital predisposition to be more attentive to actions typical of appropriate

TABLE 2. Contacts with teachers

		DAVIS												DIXON				
		1971–72			1972–73			1973–74			1974–75			1973–74		1974–75		
		F	W	S	F	W	S	F	W	S	F	W	S	F	W	F	W	S
General ÷ Total time	♂	.17	.10	.09	.17	.19	.22	.20	.22	.16	.13	.17	.18	.23	.20	.22	.21	.23
	♀	.17	.12	.12*	.24	.24	.31	.27*	.25	.22*	.25	.22	.24	.28	.25	.21	.23	.30
Disciplinary ÷ Total time	♂	.006	.006	.006	.004	.007	.004	.004	.006	.009	.002	.003	.003	.005	.010	.008	.004	.004
	♀	.003	.002**	.003**	.002	.002	.001*	.003	.004	.002*	.001	.001**	.001**	.008	.011	.004	.002	.002

*Difference significant at .05 level; **Difference significant at .01 level.

TABLE 3. Contact with male teachers[a]

SEX	DAVIS						DIXON				
	1973–74			1974–75			1973–74		1974–75		
	F	W	S	F	W	S	W	S	F	W	S
♂	.06	.07	.07	.03	.05	.08	.014	.007	No		
♀	.11	.04	.05*	.02	.04	.03	.003	.006	male teachers		

[a]Decimal numbers represent the fraction "Time with teacher ÷ Total time."
*Difference significant at .05 level.
 F = fall; W = winter; S = spring.

role-models, but they also can be taken as evidence that the children are "identifying" with like-sexed individuals (cf. Mussen, Conger, and Kagan, 1974).

From the latter viewpoint, the teachers might be selectively influencing the children's behavior according to cultural stereotypes of gender-appropriate activities (Birns, 1976). Indeed, there exists quite unambiguous evidence that adults' responses to infants' behavior vary solely according to whether the child is labeled a "boy" or a "girl" (Condry and Condry, 1976), and that preschool teachers respond differently according to the child's gender (Cherry, 1975). On the other hand, the evidence for *direct* adult-influence on sex-stereotyped behavior is ambiguous (Etaugh *et al.,* 1975; Harper and Sanders, 1978).

Effects of Contacts with Teachers

In order to see whether there was any indication of subtle but differential adult responses to the children in the larger Davis sample, we looked at the locations in which adult-child contacts occurred. Corresponding to the fact that boys played outdoors more, there was a general tendency for them to interact with adults outdoors more often than did girls. Likewise, girls spent more time in contact with adults indoors than did boys; and the latter differences tended to be larger. The boys received more discipline than girls and disproportionately more of it outdoors (Table 4). Analysis of where adults initiated general contacts indicated the same trends as overall general contact. Sex differences in location of child-initiated contacts were less marked, especially for outdoor contacts (Table 5).

The fact that boys received disproportionately more discipline while outdoors is inconsistent with the idea that adult pressures caused them to

TABLE 4. Location of contacts with teachers[a]

CATEGORY	1971-72			1972-73			1973-74			1974-75			1975-76		
(TOTAL TIME[a])	F	W	S	F	W	S	F	W	S	F	W	S	F	W	S
							At Davis								
General outdoors															
♂	.082	.018	.041	.100	.058	.122	.085	.036	.082	.072	.038	.094	—	—	—
♀	.074	.007	.030	.096	.048	.137	.089	.035	.099	.073	.035	.101	—	—	—
General indoors															
♂	.089	.083	.048	.064	.128	.100	.115	.184	.078	.059	.130	.085	—	—	—
♀	.097	.117	.090**	.147*	.193*	.170*	.177	.216	.124	.173*	.187*	.137*	—	—	—
Discipline outdoors															
♂	.004	.002	.005	.002	.003	.003	.003	.001	.006	.002	.001	.002	—	—	—
♀	.000*	.000	.000***	.001	.000*	.001**	.000	.000**	.001***	.000*	.000*	.000	—	—	—
Discipline indoors															
♂	.003	.004	.002	.002	.003	.001	.001	.005	.003	.001	.002	.001	—	—	—
♀	.002	.002*	.002	.001	.002	.001	.002	.004	.001	.001	.001*	.000	—	—	—

At Dixon

Category													
General outdoors													
♂	—	—	—	—	—	.020	.051	—	—	—	—	.058	.049	.109
♀	—	—	—	—	—	.016	.060	—	—	—	—	.041	.034	.099
General indoors													
♂	—	—	—	—	—	.167	.098	—	—	—	—	.159	.155	.111
♀	—	—	—	—	—	.215	.136	—	—	—	—	.161	.191	.194
Discipline outdoors													
♂	—	—	—	—	—	.002	.003	—	—	—	—	.004	.002	.002
♀	—	—	—	—	—	.010	.007	—	—	—	—	.002	.001	.002
Discipline indoors													
♂	—	—	—	—	—	.003	.007	—	—	—	—	.004	.002	.002
♀	—	—	—	—	—	.007	.003	—	—	—	—	.002	.001	.001

*Difference significant at the .05 level; ** .01 level; *** .001 level.

F = fall; W = winter; S = spring.

[a]Each decimal number represents the ratio of time for the item in the first column ("Category") to total time.

TABLE 5. Teacher- and child-initiated contact[a]

CATEGORY ÷ (TOTAL TIME)[a]	1973–74			1974–75			1975–76		
	F	W	S	F	W	S	F	W	S

At Davis

Teacher-initiated outdoors

♂	.024	.012	.031	.026	.014	.039	—	—	—
♀	.024	.009	.028	.012**	.012	.036	—	—	—

Teacher-initiated indoors

♂	.051	.090	.029	.023	.057	.033	—	—	—
♀	.066	.090	.051	.082**	.096*	.068**	—	—	—

Child-initiated outdoors

♂	.005	.003	.011	.008	.004	.016	—	—	—
♀	.007	.003	.011	.005	.003	.021	—	—	—

Child-initiated indoors

♂	.007	.029	.013	.008	.026	.012	—	—	—
♀	.023**	.033	.020	.014	.024	.023	—	—	—

At Dixon

Teacher-initiated outdoors

♂	—	.004	.023	—	—	—	.036	.034	.083
♀	—	.002	.021	—	—	—	.029	.024	.076

Teacher-initiated indoors

♂	—	.070	.043	—	—	—	.124	.112	.074
♀	—	.080	.076	—	—	—	.114	.133	.135*

Child-initiated outdoors

♂	—	.001	.006	—	—	—	.022	.015	.026
♀	—	.001	.013	—	—	—	.012	.010	.022

Child-initiated indoors

♂	—	.017	.018	—	—	—	.035	.048	.037
♀	—	.032	.040*	—	—	—	.047	.058	.059

*Difference significant at the .05 level; ** .01 level.

F = fall; W = winter; S = spring.

[a] Each decimal number represents the ratio of time for the item in the first column ("Category") to total time.

stay outside. On the other hand, one might argue that, because girls tended to receive more adult contacts indoors, they stayed indoors more often. To clarify the issue, we selected all the subjects from the Davis samples who had attended the program regularly through an entire year. This procedure yielded a subsample of 34 boys and 34 girls. A preliminary analysis of variance revealed no significant yearly differences and indicated that all the foregoing sex differences characterized this group as well. Given that the behavior of these children was the same as that of the larger group, we attempted to see if early (e.g., fall) contacts with the staff could predict later child behavior (e.g., time out in spring).[5]

An extensive analysis of the predictive relationships between child and teacher behaviors indicated that, while early contacts with adults were clearly related to some later child behaviors, the relationships were complex. Many of those adult behaviors which predicted later child activity were themselves related to antecedent *child* behaviors, and some of the cross-season relationships that at first glance appeared to be predictive proved to be attributable to the stability of (stronger) concurrent relationships. Where adult behavior seemed clearly related to subsequent child activity, the patterns of prediction often were not the same for both sexes. Moreover, a number of these sex-differentiated predictive relationships should have led to *reductions* in the observed sex differences.

Although such cross-lagged correlations can never prove the existence of "causal" relationships, our data do suggest that the teacher-child interactions at Davis were reciprocal, resulting in modifications of the behavior of the adults as well as that of their charges. Because the pattern of predictive relationships did not indicate any simple or direct adult influence upon the ways in which the behavior of the boys differed from that of the girls, we conclude that the relationship was not only bidirectional, but also that the nature of the behavioral modifications varied according to the gender of the children. Although we cannot rule out the *possibility* that there was some subtle, qualitative difference in the ways in which the adults treated the children (we did not observe any),[6] the data are consistent with the hypothesis that the sexes were predisposed to respond differently to environmental conditions. (For a consideration of the argument that the home environment shapes differences such as those found here see Chapter 14.)

Summary

We have proposed that, as a result of at least 50,000 years of subsisting by hunting and gathering, hominids evolved biological predispositions toward gender-dimorphic behavior patterns appropriate to that mode of existence. We have further suggested that because they share the same structural genes, both males and females probably develop the substrates necessary for performing behavior "typical" of the opposite sex. However, as a result of early (often embryonic) hormonal events associated with the development of anatomical/physiological dimorphisms, the behavioral thresholds of boys and girls are "biased" so that, on the average, they will engage in different activities.

We have argued that a number of these presumably congenitally determined predispositions arc observable in the free-play behavior of young children. In particular, we suggest that the facts that preschool boys engage in more outdoor interactive peer play involving large expenditures of energy whereas girls spend more time in parallel play and in the company of adults can be understood as reflecting interactions among gender-differentiated biological predispositions (thresholds) and environmental (cultural) variables. With respect to the questions of vigorous, outdoor play and proximity to adults, we have reviewed studies suggesting that these sex differences are mediated in part by prenatal hormones.

In conclusion, given the foregoing, the cross-cultural commonality of sex differences in physical aggressiveness, and solid evidence for biologically based gender dimorphisms in the development of spatial abilities, it is quite possible that traditional "sex-role socialization" represents, in part, a convergence of congenital response biases and culturally prescribed roles.

Acknowledgments

Supported by the Agricultural Experiment Station, University of California.

Notes

1. An analogous phenomenon occurs in birds where the female is heterogametic (WZ). Like the male mammal (XY), the avian female phenotype is "induced" and the embryonic mortality of hens is higher than that of cocks (Landauer, 1967), suggesting that the "W" chromosome is also subject to more intense selection in the heterogamete.

2. Women are sometimes considered to be more "sociable" than men; however, "sociability" can be manifested in several ways. Whereas females seek closer, more intense relationships with individual friends and family members, males show greater positive sociability toward a wider range of individuals, especially peers (Block, 1976; Waldrop and Halverson, 1975). Indeed, such qualitative difference might have been predicted from the fact that females can be more

certain of who their offspring are than can males. Ghiselin (1974) points out that females would benefit more from stronger loyalties to individuals whereas males would gain more from loyalty to the group as a whole.

3. It seems that male children assigned household duties and care of younger siblings display less physical aggressiveness than their peers who are allowed to pursue more "typically masculine" pursuits (Ember, 1973). However, this fact cannot, by itself, be taken as proof that "socialization" *suffices* to explain the sex differences observed in most cultures. In the first place, these boys were still more aggressive than girls. In the second place, although "maternal" responsiveness can be induced in male mammals by continuously exposing them to infants (cf. Bell and Harper, 1977), few would use this fact to argue that male or female behavior is "learned." As we have pointed out, it is likely that in man, as in other mammals, both sexes are *capable* of performing masculine and feminine activities; whether they *do* so depends upon both organismic and environmental variables.

4. These data are not unequivocal; they are based upon maternal and self-report. Moreover, they have been criticized because prenatally administered exogenous progestins or excessive adrenal androgens masculinize the genitalia of genetic females. Thus it is possible that the girls' anomalous anatomy could have colored parental expectations and thereby influenced maternal reports directly. Parental expectations also could have accounted for the data indirectly by affecting the androgenized girls' self-images and/or behavior (Quadagno *et al.,* 1977). The foregoing criticisms are serious; however, we feel that the Money and Ehrhardt and the Ehrhardt and Baker data cannot be dismissed so readily because (1) they are consistent with findings from well-controlled studies on primates (cf. Goy, 1968) and (2) there are few unequivocal demonstrations of parents' influencing gender-dimorphic behavior in human children (cf. Harper and Sanders, 1978). Quadagno, Briscoe, and Quadagno (1977) also expressed concern about the possibility that exogenous cortisone could affect the behavior of some of the subjects in the Ehrhardt and Money studies. However, this concern applies only to the girls who were virilized as a result of the adrenogenital syndrome; those who were masculinized *in utero* because of maternal medication did not require—or receive—cortisone postnatally. Nevertheless, they too were reported to engage in "tomboyish," vigorous, outdoor activity.

5. These data will be presented in detail elsewhere.

6. Our observers varied over the years and included several "liberated" young women who presumably might have noticed any obvious signs of differential treatment. Furthermore, when advised of our finding sex differences in outdoor play, the (predominantly female) nursery school staffs expended considerable ingenuity in attempting to induce the girls to play outside. For example, in spring of one year, the girls' favorite playthings were placed outdoors at the beginning of the day. (The girls then went outside to get the toys and took them back in to play with them.)

References

Beach, F. A. 1974. Human sexuality and evolution. *In Reproductive behavior,* eds., W. Montagna and W. A. Sadler, pp. 333–365. New York: Plenum.

Beach, F. A. 1976a. Human sexuality in four perspectives. *In Human sexuality in four perspectives,* ed., F. A. Beach, pp. 1–21. Baltimore: Johns Hopkins Univ.

Beach, F. A. 1976b. Hormonal control of sex-related behavior. *In Human sexuality in four perspectives,* ed., F. A. Beach, pp. 247–267. Baltimore: Johns Hopkins Univ.

Bell, R. Q., and L. V. Harper. 1977. *Child effects on adults.* Hillsdale, N.J.: Erlbaum.

Birns, B. 1976. The emergence and socialization of sex differences in the earliest years. *Merrill-Palmer Q.* 22:229–254.

Block, J. H. 1976. Issues, problems and pitfalls in assessing sex differences: a critical review of *The Psychology of Sex Differences. Merrill-Palmer Q.* 22:283–308.

Buffrey, A. W. H., and J. A. Gray. 1972. Sex differences in the development of spatial and linguistic skills. *In Gender differences: their ontogeny and significance,* eds., C. Ounsted and D. C. Taylor, pp. 123–157. London: Churchill.

Campbell, B. G. 1972. *Sexual selection and the descent of man.* Chicago: Aldine.

Campbell, B. G. 1974. *Human evolution,* 2d ed. Chicago: Aldine.

Charnov, E. L., and J. Bull. 1977. When is sex environmentally determined? *Nature* 266:828–830.

Cherry, L. 1975. The preschool teacher-child dyad: sex differences in verbal interaction. *Child Devel.* 46:532–535.

Condry, J., and S. Condry. 1976. Sex differences: a study in the eye of the beholder. *Child Devel.* 47:812–819.

D'Andrade, R. G. 1966. Sex differences and cultural institutions. *In The development of sex differences,* ed., E. E. Maccoby, pp. 174–204. Stanford Calif.: Stanford Univ. Press.

Dawkins, R. 1976. *The selfish gene.* New York: Oxford Univ. Press.

DeFries, J. C., G. C. Ashton, R. C. Johnson, A. B. Kuse, G. E. McClearn, M. P. Mi, M. N. Rashad, S. G. Vandenbergh, and J. R. Wilson. 1976. Parent-offspring resemblance for specific cognitive abilities in two ethnic groups. *Nature* 261:131–133.

Diamond, M. 1976. Human sexual development: biological foundations for social development. *In Human sexuality in four perspectives,* ed., F. A. Beach, pp. 22–61. Baltimore: Johns Hopkins Univ.

Ehrhardt, A. and S. W. Baker. 1974. Fetal androgens, human central nervous system differentiation and behavioral sex differences. *In Sex differences in behavior,* eds., R. C. Friedman, R. M. Richart, and R. L. Vande Wiele, pp. 33–51. New York: Wiley.

Eisenberg, J. F., N. A. Muckenhirn, and R. Rudran. 1972. The relation between ecology and social structure in primates. *Science* 176:863–874.

Ember, C. R. 1973. Feminine task assignment and the social behavior of boys. *Ethos* 1:424–439.

Etaugh, C., G. Collins, and A. Gerson. 1975. Reinforcement of sex typed behavior of two-year-old children in a nursery school setting. *Dev. Psychol.* 11:255.

Ghiselin, M. T. 1974. *The economy of nature and the evolution of sex.* Berkeley and Los Angeles: Univ. of California Press.

Goy, R. W. 1968. Organizing effects of androgen on the behaviour of rhesus monkeys. *In Endocrinology and human behaviour,* ed., R. P. Michael, pp. 12–31. New York: Oxford Univ. Press.

Harper, L. V., and K. M. Sanders. 1977b. Preschool children's use of space: sex differences in outdoor play. *In Readings in child development and relationships,* 2d ed., eds., R. C. Smart and M. S. Smart, pp. 99–111. New York: Macmillan.

Harper, L. V., and K. M. Sanders. 1978. Sex differences in preschool children's social interactions and use of space: an evolutionary perspective. *In Sex and behavior: retrospect and prospect,* eds. T. E. McGill, B. Sachs, and D. A. Dewsbury, New York: Plenum.

Landauer, W. 1967. The hatchability of chicken eggs as influenced by environment and heredity, revised ed. *Storrs Agriculture Experiment Station Monograph No. 1.* Storrs, Conn: Univ. of Connecticut.

Leaky, R. E. F. 1973. Further evidence of lower Pleistocene hominids from East Rudolf, No. Kenya, 1972. *Nature* 242:170–173.

Lee, R. B., and I. DeVore. 1968. *Man the hunter.* Chicago: Aldine.

Maccoby, E. E., and C. N. Jacklin. 1974. *The psychology of sex differences.* Stanford, Calif.: Stanford Univ. Press.

Mittwoch, U. 1973. *Genetics of sex differentiation.* New York: Academic Press.

Money, J., and A. A. Ehrhardt. 1972. *Man and woman, boy and girl.* Baltimore: Johns Hopkins Univ.

Mussen, P. H., J. J. Conger, and J. A. Kagan. 1974. *Child development and personality,* 4th ed. New York: Harper & Row.

Ounsted, C., and D. C. Taylor. 1972. The Y chromosome message: a point of view. *In Gender differences: their ontogeny and significance,* eds., C. Ounsted and D. C. Taylor, pp. 241–262. London: Churchill.

Quadagno, D. M., R. Briscoe, and J. S. Quadagno. 1977. Effect of perinatal gonadal hormones on selected nonsexual behavior patterns: a critical assessment of the nonhuman and human literature. *Psychol. Bull.* 84:62–80.

Reinisch, J. M. 1977. Prenatal exposure of human fetuses to synthetic progestin and estrogen affects personality. *Nature* 266:561–562.

Silvers, W. K., and S. S. Wachtel. 1977. H-Y antigen: behavior and function. *Science* 195:956–960.

Trivers, R. L. 1972. Parental investment and sexual selection. *In Sexual selection and the descent of man 1871–1971,* ed. B. G. Campbell, pp. 136–179. Chicago: Aldine.

Waber, D. P. 1976. Sex differences in cognition: a function of maturation rate? *Science* 192:572–574.

Waldrop, M. F., and C. F. Halverson, Jr. 1975. Intensive and extensive peer behavior: longitudinal and cross-sectional analyses. *Child. Dev.* 46:19–26.

Whalen, R. E. 1974. Sexual differentiation: models, methods, and mechanisms. *In Sex differences in behavior,* eds., R. C. Friedman, R. M. Richart, and R. L. Vande Wiele, pp. 467–541. New York: Wiley.

Williams, G. C. 1975. *Sex and evolution.* Princeton, N.J.: Princeton Univ. Press.

Wilson, E. O. 1975. *Sociobiology the new synthesis.* Cambridge, Mass.: Belknap Press.

Witelson, S. F. 1976. Sex and the single hemisphere: specialization of the right hemisphere for spatial processing. *Science* 193:425–426.

Section Six

MUSIC, PLAY, AND TOOLS

This section of the book deals with some particular areas of development, rather than emphasizing aspects of ontogeny within a group of closely related animals. John Fentress considers the development of grooming and swimming in mice, particularly the ways in which some stereotyped movement patterns become integrated with one another during ontogeny. He develops a musical analogy that is very useful in helping to elucidate some of the on-going development processes. Like Anne Bekoff, he discusses the importance of endogenous processes in behavioral development. While many of us are interested in studying nonhuman sensory cues that humans also perceive, Victor DeGhett, in his analysis of the development of ultrasonic communication in rodents, stresses the importance of "getting into" the *umwelt* of animals that inhabit perceptual worlds different from our own. Also, he shows that there is a tight link between mothers and offspring, based on ultrasonic communication, that has both evolutionary and ontogenetic implications. Marc Bekoff and Robert Fagen in their respective chapters consider play behavior in light of recent sociobiological thought. Both stress that play is not at all the "garbage pail" category of behavior to which it is frequently relegated and that detailed and question-directed analyses of play can shed considerable light on classical theories of behavioral evolution and ontogeny and those currently in vogue. Finally, Benjamin Beck discusses the ontogeny of tool use by nonhuman animals. As with play behavior, speculation far outweighs data. Beck considers questions of origin and questions of dissemination. He concludes that the study of the ontogeny of tool use requires no unique terminology, concepts, or strategies. Considering the importance of such information for understanding human evolution, it is to be hoped that more data will be forthcoming.

Chapter Sixteen

Mus Musicus: The Developmental Orchestration of Selected Movement Patterns in Mice

JOHN C. FENTRESS

Department of Psychology
Dalhousie University
Halifax, Nova Scotia
Canada
B3H 4J1

The orchestration of integrated behavior can be viewed both in terms of ephemeral performance characteristics and in terms of deeper underlying "score" (cf. Lorenz, 1957; Chomsky, 1968). A main task of developmental analyses is to bring these two perspectives together (e.g., Fentress, 1976b, 1977a).

In this chapter I shall concentrate upon aspects of the developmental integration of some *relatively* stereotyped movement patterns in mice as model systems for inquiry. Within this context I find the analogy between integrated behavior and a musical composition to be useful in guiding initial questions. As in a musical performance, for example, we can view problems of temporal patterning in behavior as *necessarily* dependent upon *rules of relationship* between individually abstracted behavioral "notes" and "pauses" through time. Similarly, we can think of the "score" that stabilizes and sets the constraints for individual performances as emerging from the organism's phenotypic structure. In some sense we are seeking to find the conductor who translates the score into performance at different stages in the animal's life. We are also concerned of course with the composer, but must remind ourselves that both the writing of the score *and* its translation into performance contribute to the final product we observe.

I. Grooming

In adult mice it is possible to divide facial grooming sequences into component stroke "types" and to determine rules of relationship among these stroke "types" through time in the construction of higher-order melodies of expression (Fentress, 1972; Fentress and Stilwell, 1973) (Fig. 1). At finer levels of analyses each stroke "type" can be decomposed into various kinematic details with the stroke as a unit representing a convenient abstract representation of these details (e.g., Woolridge, 1975; Fentress and Golani, in preparation). Or, one can take all facial grooming "strokes" as distinct, for example, from body grooming at a still higher level of abstraction, or all grooming as distinct, say, from walking or sitting still. Obviously, for behavioral analysis, we can divide the "stream of events" (Hinde, 1970) in a variety of ways and at a variety of levels. It is the question of relations between these levels that pose some of the more fundamental issues for developmental analyses.

Here I shall confine most of my attention to descriptions at two basic levels. First, I shall accept the identity of stroke "types" and ask how these individually defined strokes and their relations to one another are formed during development. Secondly, I shall survey changes in kinematic details in the development of grooming in comparison with contact pathways between the forepaws and face which in turn are primary defining characteristics of stroke "types." I shall then conclude this section by noting briefly the relations between facial grooming and other aspects of ongoing behavior within a developmental context. I acknowledge with gratitude that my insights from these descriptions are drawn heavily from active collaboration first with Ms. Frances Stilwell and subsequently with Dr. Ilan Golani, who have worked in my laboratory. Without their inspiration these studies would not have reached their present stage, and I accept full responsibility for this highly abstracted summary of my colleagues' many contributions.

A. DEVELOPMENT OF STROKE TYPES AND THEIR INTEGRATION[1]

A question raised by the description of stroke types and their combination as represented in Figure 1 is whether there is a differential development of individually classified strokes and the combination of these strokes through time.

STROKES

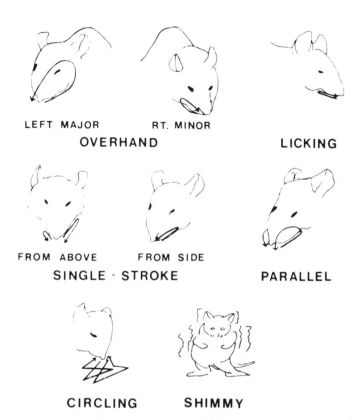

FIG. 1. A. *Grooming stroke categories defined for adult mice. (From Fentress, 1972.) B. (next page) Sequencing of stroke categories and statistically defined higher-order units or "melodies" (1–5): B = body grooming, C = circling, L = licking (L < 1/3 sec; L· = 1/3-2/3 sec; L: = > 2/3 sec), N = pause, O = overhand, P = parallel, S = single stroke. H_0 = information measures of numbers of stroke categories; H_1 = relative stroke probabilities; and H_2 = paried dependencies. (From Fentress and Stilwell, 1973.)*

1. Stroke Categories

Clearly defined (i.e., separable) stroke types were first observed in mice at approximately ten days of age. Prior to this time (below) a wide variety of forepaw movements around the face could be observed, but their precise

GROOMING SERIES

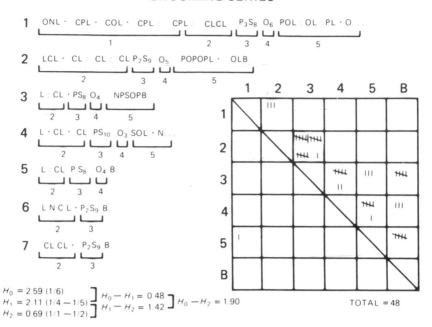

1 ONL · CPL · COL · CPL : CPL : CLCL P_3S_8 O_6 POL : OL : PL · O . . .

2 LCL · CL : CL : CL P_2S_9 O_5 POPOPL · OLB .

3 L : CL · PS_8 O_4 NPSOPB .

4 L · CL : CL PS_{10} O_3 SOL · N . . .

5 L : CL P S_8 O_4 B .

6 L N C L · P_2S_9 B .

7 CL CL · P_2S_9 B .

$H_0 = 2.59 \ (1/6)$
$H_1 = 2.11 \ (1/4 - 1/5)$
$H_2 = 0.69 \ (1/1 - 1/2)$

$H_0 - H_1 = 0.48$
$H_1 - H_2 = 1.42$

$H_0 - H_2 = 1.90$

TOTAL = 48

relation to categories initially defined for adults (Fentress, 1972; Fentress and Stilwell, 1973) was difficult to determine. This, of course, represents a major problem in developmental analyses of behavior; when, for example, does one define a given behavior as an "immature form" of adult behavior X as opposed to a different behavior Y? In this context we found ourselves using a variety of cues such as form of movement, position of movement in the overall sequences, etc., and made no attempt to segregate each possible cue precisely.

The data were gathered from films taken at 32 f.p.s. of individual mice enclosed in a special filming chamber. From frame by frame analyses more than 1500 illustrations of individual strokes were drawn, and a variety of quantitative determinations of sequential relations among stroke types was made (cf. Fentress, 1972; Fentress and Stilwell, 1973). Detailed evaluations of three DBA/2J mice filmed on days 10, 11, 12, 13, 14, 16 and 19 make up the data reported here.

As a general practical point we found it useful to examine our films at a variety of projection speeds, normally in the range of 1 f.p.s. through

16 f.p.s. The basic reason for this is that slower speeds permitted more detailed evaluations of individual limb segment movements while higher projection speeds permitted a more satisfactory perceptual assimilation of overall stroke categories and their integration through time.

Several general comments concerning stroke types are useful here. (1) *Circling (C)*. High-speed circling (cf. Fentress, 1972; Fentress and Stilwell, 1973) movements were observed only once in a single 19 d old mouse, while this is a common component in adult grooming sequences (ibid). (2) *Licking (L)*. Until day 19 the mice were more prone to show extended periods of inactivity during forepaw licking, with the forepaws held immobile beneath the snout, than were adults. (3) *Overhands (O)*. A progressive increase in the size and smoothness of these large amplitude facial grooming strokes was seen through day 19, accompanied by pro- gressively reduced inactivity periods (with forepaws held motionless just below the nostrils) following individual strokes. The velocity of overhand movements also showed a progressive increase. (4) *Parallels (P)*. These "half moon"-shaped synchronous strokes as defined in adult mice could be discerned in young mice, but they were much more variable than in adults. (5) *Singlestrokes (S)*. While in adult animals these strokes along the side of the snout alternate strictly between large- and small-amplitude movements at ten per second, the young animals through day 19 showed greater variability in timing and placement, a decreased tendency to alternate strictly between large- and small-amplitude strokes for a given limb, and longer pauses between successive strokes (which in adults is essentially negligible). (6) *X-Strokes*. Often in the young animals in- dividual strokes combined defining characteristics of more than one previously classified adult stroke, which led us to utilize a new, and necessarily ambiguous, category "X." Some examples of immature forms of large-amplitude "overhand" strokes and of "X" strokes are shown in Figures 2 and 3, respectively.

With respect to individual strokes, defined from criteria originally employed for adult animals (Fentress, 1972; Fentress and Stilwell, 1973), it was not possible to state unambiguously that one stroke "type" matured more rapidly than did others. For example, while the *form* of single- strokes approached a basically adult trajectory profile more quickly than did the form of overhands, detailed timing characteristics of each con- tinued to show selected deviations from the adult form through day 16. However, one's overall impression is that relatively simple and rapid stroke types attain their basic adult *form* more quickly than do more complex and/or less rapid stroke types. This general conclusion is based largely upon comparisons between rapid and "simple" singlestrokes, and slower and more complex overhands, but the persistent low probability of

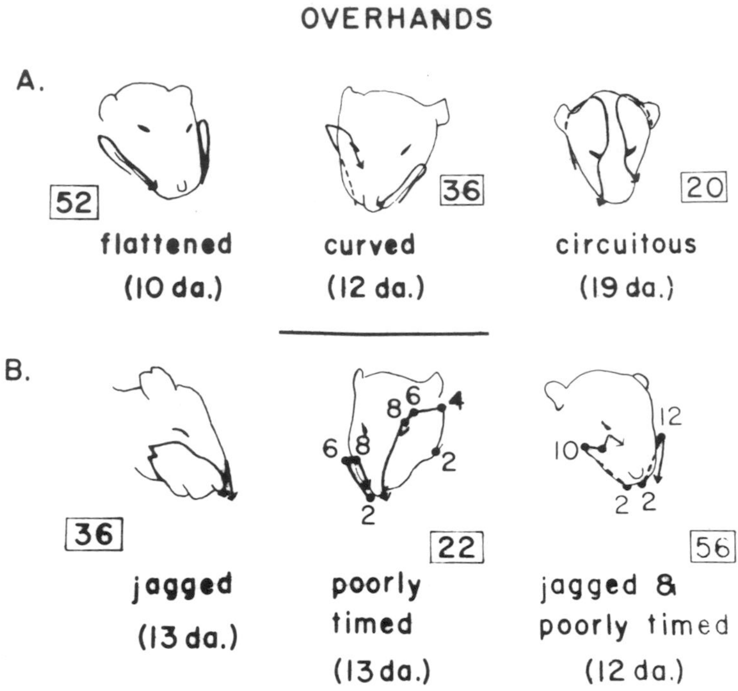

OVERHANDS

A.

52

flattened
(10 da.)

36

curved
(12 da.)

20

circuitous
(19 da.)

B.

36

jagged
(13 da.)

22

poorly
timed
(13 da.)

56

jagged &
poorly timed
(12 da.)

FIG. 2. **A.** *Examples of smooth trajectories for overhand type strokes in young mice, which may be subdivided into broad categories of flattened, curved, and circuitous in comparison to adults.* **B.** *Examples of jagged and/or poorly timed overhand type strokes in young mice. The numbers on the face indicate limb positions for consecutively counted film frames. Numbers in squares indicate total duration of illustrated grooming sequence standardized to 64 f.p.s. (From Stilwell and Fentress, in preparation.)*

extremely rapid yet complex circling movements through day 19 illustrates the necessity for caution in making broad conclusions at the present time.

2. Linkage Between Strokes

If we view stroke "types" as analogous to individual notes in a musical performance, we must next seek rules of relationship between these "notes" as composed and orchestrated into higher-order "melodies." Stil-

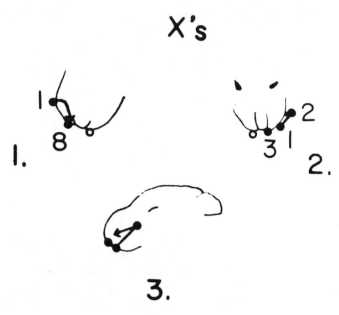

FIG. 3. *In young mice stroke types were often observed that did not fit unambiguously into any single category previously defined for adults, in which case they were designated "X-Strokes." Three illustrations are given. (1) Resembles a parallel stroke (adult definition) except only one paw is involved and its looping trajectory is exaggerated in length, and resembles adult overhands except that "major limb" excursion is too small (see Fig. 1). (2) Resembles a left-handed singlestroke (but right forelimb did not participate), or a licking stroke, except one limb was not on a line below the snout. (3) Resembles parallel stroke except the loop is too high on the head, or an overhand except its duration is truncated. "X" strokes declined progressively in probability from day 10, and were absent by day 19. Numbers on faces refer to consecutive film frames. (From Stilwell and Fentress, in preparation.)*

well and I have previously described five such melodies in mice (initially labeled as "units," e.g., Fig. 1B).

The first of these melodies was never observed in mice up to the age of 19 days, largely because of the absence of circling and shimmying movements which were included in our definitions for adult animals. An interesting point here is that with the absence of circling and shimmying, Melody 1, which in adult animals frequently initiated a grooming sequence, is indistinguishable from Melody 5, which often concludes a facial grooming sequence. Our second melody in infant mice never reached the prolonged uninterrupted licking observed for adults, although relatively constant progression in licking duration with age was observed. The third melody was defined for adults largely in terms of uninterrupted sequences of singlestrokes in which there is a strict alternation between the limbs in terms of which limb makes the larger amplitude movement. Up until day 19 progressively decreasing pauses between singlestrokes in series were observed, with certain strokes in the series (often *not* interrupted by pauses) frequently failing to alternate strictly in terms of "major" and "minor" movements (Fig. 4). The number of singlestrokes in series also increased with age, while the presence of licking during this series declined. It is important to note that while fewer singlestrokes per time unit were observed in the young mice, this is *not* primarily due to stroke speed but to the pauses interspersed between successive strokes. Melody 4 in adults is defined by multiple overhand strokes with occasional short licking bouts. Serially connected overhand strokes were observed only twice in our developmental study, and these in a single mouse on day 19. Melody 5 in adults is a loosely constructed "jam session" which may contain all "notes" except circling, shimmying, and singlestrokes. This melody was observed in infant mice from day 10, although its sequential connection to other melodies was much less precise than in audlts.

This leads directly to questions of sequential connections between "melodies" into, dare I say "movements"? Melodies 1–5 follow one another in quite strict sequence in adult animals. In young animals we never observed Melody 1 (above). Melody 2 was recognizable in abbreviated form by day 19, and occurred in appropriate sequential position in the overall grooming sequence. The more simple Melody 3 was observed from day 10. Melody 4 is also recognizable early (i.e., day 10), and is connected temporally to Melody 3 from the beginning (but often with interspersed intervals in young animals; Fig. 4). In contrast while Melody 5 was also recognizable from day 10 its temporal location in the overall movement sequence is irregular (i.e., unpredictable) until day 16. That is, one observes Melody 5 in places where it would not be expected on the basis of data obtained with adult animals. In this context it should be emphasized that "jam session" Melody 5 is particularly loose (i.e., heterogeneous) in its definition, which indicates the practical difficulty in clearly separating properties of behavior at different circumscribed levels. That is, Melody 5 is almost a "nonmelody" (for classical purists, at least), and thus its presence at inappropriate temporal locations in young mice might in part indicate the failure of young animals to form clearly defined

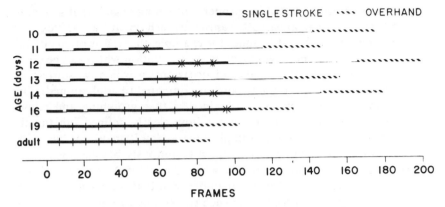

FIG. 4. *Sequencing of singlestroke series (Melody 3) and sub-sequent overhand (Melody 4) averaged separately for observations on postnatal days 10–19 and for adults. Length of bars indicates mean duration of individual strokes. Gaps between singlestrokes indicate pauses. X's on singlestrokes indicate failure to alternate between major and minor limb movements. Gaps between singlestroke series as a whole and first overhand most frequently indicate interspersing of stroke combinations not expected from adult observations (e.g., licks, "X's", parallels and Melody 5). Note that gaps and failures to alternate were each spaced randomly throughout the sequence; the segregated representation here is for schematic convenience. (From Stilwell and Fentress, in preparation.)*

melodies rather than temporal looseness in the sequential staging of well defined melodies.

In this context it is interesting to note that Melody 5 is frequently interspersed between Melody 3 (repeated singlestrokes) and Melody 4 (repeated overhands) until serially connected singlestrokes (3) are directly connected to overhand strokes (4). This occurs suddenly on day 16. This *direct* sequential connection between Melodies 3 and 4 coincides with the internal perfection of each. A simplified representation of the main points raised in this section is provided in Figure 4.

3. Context of Grooming

In adult rodents facial grooming commonly occurs between protracted periods of locomotion and stationary behavior (e.g., Fentress, 1968a, 1968b, 1972). The trained observer can quite readily predict when grooming is likely to occur. A striking property of young mice is that grooming may be initiated in an apparently capricious fashion, at least as judged by adult criteria. Stilwell and I found, for example, that it was often difficult to anticipate grooming in young animals, a situation which can be determined objectively by the number of occasions in which we started our cameras too late when filming young as compared to adult animals. One clear impression that we have obtained is that grooming and related activities in young animals do not follow clearly defined functionally integrated "goals". Thus, for example, mice at day 10 may "unexpectedly" groom during locomotion if a forepaw fortuitously brushes against the face. This is clearly analogous to what Lind (1959) has defined as a "transitional action," and it is our impression that ontogenetic analyses may reveal important processes in such transitions with particular clarity (see Fentress, 1972). Subsequent observations by Golani and myself with even younger animals confirm this general expectation (Fentress and Golani, in preparation).

B. KINEMATIC ANALYSIS OF TEMPORAL FLOW[2]

The difficulty in establishing clear stroke types ("notes") and clearly distinguishable higher-order combinations of grooming movements ("melodies") in mice less than ten days of age led to a subsequent and complementary analysis of temporal flow in infant animals (Fentress and Golani, in preparation). Here explicit comparisons were made between the kinematic details of facial grooming movements and the consequent contact pathways between forepaws and the face.

We found that by placing mice in an upright sitting posture it was possible readily to produce and observe grooming-type movements from postnatal day 1. This led to the construction of a special filming chamber in which the infant mice were supported in a mirror apparatus that provided orthogonal views of participating limb segments. Films were taken via a specially designed stroboscopic unit that permitted filming rates of 100 f.p.s. (LoCam camera) (Fig. 5). It was very striking that once the infant animals were supported in an appropriate "grooming posture" a variety of moderate intensity stimuli, such as light pinches of the tail, could elicit grooming-type movements in a highly reliable manner (cf. Fentress, 1976a, 1977b). Films were analyzed through frame by frame examination. The Eshkol-Wachmann notation system (Eshkol and Wachmann, 1958; Golani, 1976) was employed to depict details of kinematic articulation, and a variety of illustrative methods were used to summarize contact pathways between the forepaws and the face (e.g., Fig. 6). In this way the emergence of order in facial grooming could be

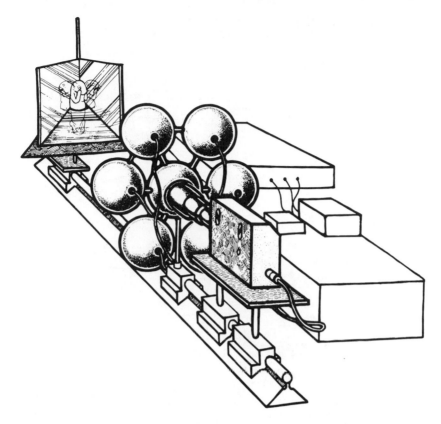

FIG. 5. *Stroboscopic filming apparatus and mirror chamber used to document ontogeny of facial grooming movements in mice from postnatal day 1. (From Fentress and Golani, in preparation.)*

documented from a variety of explicitly designed perspectives from postnatal day 1. Analyses were also made of grooming behavior of mice in the home cage for the purpose of comparison to behavior in the mirror filming chamber.

In broad terms we found it possible to define three distinct phases of facial grooming in the mirror chamber from birth to postnatal day 10. An abbreviated summary taken from single frame analyses of more than 6000 ft of cine film is provided below. A more detailed account is in preparation in collaboration with I. Golani.

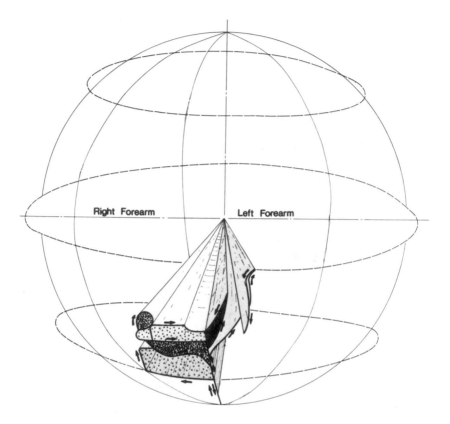

FIG. 6. *Schematic representation of facial grooming strokes in a three-day-old mouse, derived from the Eshkol-Wachmann movement notation system. Each forelimb is viewed as tracing movement patterns along the surface of a sphere from within, thereby depicting the relative continuity of ipsilateral limb movements and basic asymmetry between the two forelimbs at this ontogenetic phase. The figure is drawn as if the two shoulders are superimposed on a single spot at the center of the sphere for purposes of schematic simplification. (From Fentress and Golani, in preparation.)*

Phase I: Birth Through Postnatal Days 5–6

During postnatal days 1 and 2 kinematic similarities to adult grooming are evident, but actual contact with the face on descending strokes is rare and apparently fortuitous. From day 3 most grooming strokes involve facial contacts, which range from stationary contacts to trajectories that begin at the forehead and move to the end of the snout. These contact pathways vary widely in area and amplitude, with short-term regularities apparent between strokes of a single sequence. Up to ten strokes were observed in sequence with the contact pathways being either symmetrical or asymmetrical, and traced either simultaneously or in alternating fashion. Often the pathways are irregular with up to six stops (pauses) observed in an individual stroke.

The head is either held at the midline (day 1) or moved to this position at the initiation of grooming. Head movements during grooming are essentially absent at this phase; thus the details of contact pathways are largely attributable to limb segment articulations of the forelimbs. Coupling between both contralateral and ipsilateral forelimb segments is loose during this phase. For example, the two forelimbs may move essentially independently of one another, and movements of the lower arm which in older animals occurs in concert with upper arm rotations during ascending strokes are frequently absent in the young animals. Coordinated licking is also absent. Thus, while a rich variety of individual component movements are observable from day 1 they are not orchestrated precisely with one another either in their simultaneous expression ("chords") or sequential distribution ("melodies"). The rhythmicity of movement is also irregular in comparison to older animals. In many respects grooming appears nonfunctional (as defined by systematic removal of peripheral irritants) throughout this phase. Toward the end of Phase I there is a dramatic simplification of grooming movements, leading to Phase II.

Phase II: Days 5–6 Through 9–10

Grooming movements in the mirror chamber become progressively restricted from day 5. By day 7 individual strokes were never observed to start above the mystacial vibrissae, paw contacts are established on each side of the face simultaneously, and the resulting bilateral trajectories follow a symmetrical and nearly straight path toward the midline of the snout where forepaw contact is released. Individual strokes are separated by pauses of varying duration rather than occurring in uninterrupted sequences.

Forelimb kinematics also become simplified and restricted during this phase, with a resulting high congruence with contact pathways. The details of movement in one forelimb segment are tightly linked to movements in other segments, and movements of one limb can be "predicted" through observations of the contralateral limb, indicating tight bilateral coupling. Head position remains fixed at the midline during grooming as in Phase I. Functional licking of the forepaws is rarely if ever present.

Phase III: Days 9–10 and Beyond

Contact pathways gradually become more elaborate, extended, and smooth, and fall within relatively distinct classes, as previously described for adult mice (Fentress, 1972; Fentress and Stilwell, 1973; Woolridge, 1975). Several strokes may occur in uninterrupted sequence with both simultaneous and alternating contact pathways on the two sides of the face. Contact pathways similar to those observed in Phase I reappear, but they are smoother and more clearly defined. Phase II pathways are still observed occasionally, but their bout frequency declines, they may be repeated without release of contact, and they may be elaborated. Distinguishable contact pathways converge to and diverge from nodal points on the snout.

The kinematics incorporate more body segments which are orchestrated into various elaborate but repeatable combinations. Head and neck movements become involved. Together they serve to increase the extent of contact trajectories (e.g., if paws move downward the head moves upward). In most cases there is a high congruence between these elaborated kinematic details and resulting diverse contact pathways. Functionally coordinated licking movements are now observed. In general terms, Phase III captures much of the richness and variation of Phase I and the regularity of Phase II, as if providing a functionally integrated synthesis of each.

C. ISSUES OF INSTRUMENTATION AND CONTROL

1. Descriptive Overview

If we view contact pathways as the functional consequence of kinematic orchestration, several developmental themes emerge from these observations. First, the *tuning* of stroke types as measured by spatial configurations through time is initially loose, then simple but precise, then elaborate. Secondly, the *timing* and *rhythmicity* of grooming strokes and their connections also changes from an irregular pattern, to simple serial discontinuities, to elaborate temporal patterns. Thirdly, the *chord* structure of kinematic components similarly becomes both more precise then elaborate during ontogeny. Fourthly, the construction of higher-order *melodies* from individual grooming *notes* becomes refined and elaborated through the first 19 days of life.

There is a fifth property, which might be called *transposition,* in which alterations in one parameter of grooming are accompanied by alterations in other parameters, with the resulting maintenance of the overall chord structure and melodic composition. For example, during Phase I and Phase II one of the lower arms may get caught beneath the chin, the digits of the two forepaws may become entangled, or one paw may become stuck on the face (e.g., at the eye or mystacial vibrissae). Under these conditions a predictable variety of kinematic adjustments is made which preserves important aspects of the total grooming composition. Such *transpositions* and resulting melodic constancies become even more pronounced in Phase III. For example, Golani (Fentress and Golani, in preparation) has documented in detail adjustments made by a ten-day-old mouse in the nest which was leaning with its left elbow on a sibling. While only its right forearm could thus be employed normally in grooming, contact pathway symmetries on each side of the face were maintained through appropriate adjustments of rear legs, shoulder, neck, and head.

This latter observation in particular suggests that routes (priorities) of control can be dissected at different stages of ontogeny. Careful descriptive studies are of obvious value in clarifying questions at this level.

2. Routes of Control

A major question in ontogeny concerns the degree to which the developmental orchestration of movement depends upon intrinsic versus extrinsic sources (e.g., Chapter 2; Hinde, 1970; Bateson, 1976; Fentress, 1976b, 1977a). Fentress (1973) demonstrated that many of the developmental changes in grooming, including coordination of shoulder movements, head movements, and even tongue and eye movements are preserved in mice which had their forelimbs painlessly amputated at birth (Fig. 7). This indicates that endogenous maturational factors may play a more important role than would initially be expected. However, Fentress (*ibid*) also noted that the animals articulated certain well orchestrated adjustments (e.g., exaggerated tucking of the head) which indicates the need for more refined analyses.

An experimental approach to the question of control pathways and priorities is to perturb a part of the orchestrated movement complex, and to observe adjustments in other parts of the system. The above-mentioned observation of grooming in the nest by a mouse that was leaning on its left arm, for example, suggests that at ten days the animal maintains bilateral symmetry of contact pathways and adjusts limb segment kinematics to do so. This could in part be due to the management of tactile input on the face. During the early stages of Phase I, on the other hand, management of facial contact is not a sufficient explanation since the forepaws rarely contact the face. This suggests, then, that proprioceptive input or central

FIG. 7. *Line drawing representation of facial grooming movements in ten-day-old mouse whose forelimbs were amputated painlessly at birth. Note tongue movements coordinated with upper arm and head excursions. (From Fentress, 1973.)*

regulation is of primary importance. Further, the similarities of kinematic details in facial grooming of mice with amputated forelimbs (Fentress, 1973) indicates control by central or proprioceptive routes. However, it is important here to note that some adjustments were made, including the licking of objects after a grooming sequence *as if* to provide "expected" tactile input on the tongue. Deafferentation studies (dorsal root and trigeminal nerve lesions) in adult animals also suggest central programming mechanisms (Fentress, 1972). This may relate to observations by Northup (1977) that mice with genetically produced atrophy of cerebellar Purkinje cells ("nervous strain," autosomal recessive) fail to adjust as readily to weights attached to the forelimbs as do their normal littermates. Golani and I have recently found that when one forelimb in a grooming adult animal is pulled gently away from the face with a string, the contralateral limb may join it in the production of several strokes *in front of* the face, indicating a high priority of bilateral symmetry of limb movements. Further application of similar techniques to developing animals could be most revealing. It may be possible, for example, to test in a more critical fashion the common ethological observation that the form of movement (described in terms of selected kinematic details) frequently appears to mature prior to its orientational component (measured either in terms of an environmental locus [cf. Fentress, 1967] or orientation to other body segments (here measured in terms of contact pathways).

II. Locomotory and Swimming Movements

In the space available I can only mention briefly some recent observations on the development of locomotory and swimming movements which compliment the themes discussed previously in the context of grooming. I shall concentrate here on questions of coupling and timing of movement episodes.

A. EARLY LOCOMOTORY MOVEMENTS

During ontogeny coupling between the forelegs in locomotion achieves a clearly orchestrated form prior to that observed for the hind legs. Two observations are of particular interest. First, in mice 4–7 days of age tactile input to one limb may generate rhythmical "locomotory" movements in that limb without apparent effects on the other limbs. During the early part of the period (e.g., day 4) these observations can be made after tapping any one of the four legs. Later (e.g., day 7) taps on one forepaw tend to generate "locomotory" movements in each forepaw while taps on a hindleg may continue to elicit rhythmic movements in that hindleg only. The normal coordination between foreleg and hindleg movements during locomotion remains imperfect throughout this period. If a 4–7-day mouse is placed on its back it may flail three legs in a relatively coordinated fashion while one leg (usually a hindlimb) remains stationary. Thus, in early locomotion there is a deficit in interlimb coordination, with the hindlegs achieving a mature form of expression after the forelegs.

The second observation concerns transitions between early locomotory movements and face grooming. Mice 2–4 days of age, for example, may fortuitously bring one forelimb in close proximity to the face during apparent attempts at locomotion, at which time one or more face grooming movements may follow (cf. Lind, 1959, on transitional behavior). Similarly when mice up to 4–5 days of age are placed on their backs initial flailing movements of the forepaws may lead to recognizable grooming kinematics when one forepaw fortuitously passes closely to the face. Thus it is not simply a case of progressive coupling of limb segments during development with which one must be concerned, but also maintenance of one functionally oriented sequence (e.g., locomotion, righting) without fortuitous transitions.

Observations on locomotion as well as grooming suggest that part of the difficulty in making detailed developmental analyses is that young animals may have the neural capability of performing integrated sequences of behavior prior to the time they are normally recorded, but that more general problems such as balance, body weight, and muscular strength prevent the occurrence of these behavior patterns. For example, it was not until Golani and I supported infant mice in an upright sitting posture that we observed reliable Phase I grooming sequences. An additional observation of interest in this context is that once the infant animals were placed in an appropriate body posture for grooming a variety of mild inputs, such as moderate intensity pinches of the tail, could elicit grooming reliably (cf. Fentress, 1976a, 1977b).

B. ONTOGENY OF SWIMMING[3]

Limb movements during swimming provide a potentially useful assay for orchestrated movements during ontogeny (cf. Fox, 1965; Shapiro *et al.,* 1970). Here, for example, the body is suspended in water which removes problems of muscular strength and balance noted above.

In collaboration with T. L. Templeman the ontogeny of swimming behavior has been examined in ten litters of C57/6J mice that were raised with their respective mothers until 32 days of age. Swimming was tested by placing the animals in 35°C water and filming samples of behavior at 64 f.p.s. Test trials ranged from just under 30 s to three minutes, timed by the observed welfare of the animals. Limb movements were quantified by counting the number of frames per limb rotation, pauses between rotations, and coupling between the limbs.

Mice can perform swimming movements from postnatal day 1, but the orchestration of these movements changes progressively with age. On day 1 each of the four limbs moves in an irregular pattern, with little regularity in their coupling. By day 2 the coupling of the forelimbs improves, and the resultant alternating patterns propel the animals through the water, usually in circles. During these early trials the animals would occasionally perform Phase 1 grooming strokes (above) if their forepaws fortuitously passed in front of their faces. Progressive coordination of all four limbs was observed through day 9, with an increase in mean rotations per second for both the forelegs and hindlegs from 1.5 rotations/sec on day 1 to 2.5 rotations/sec on day 8 (Fig. 8). From the beginning the foreleg rotations, as defined between successive limb extensions, were accompanied by pauses between extensions, while the hindleg rotations were more continuous. Thus the speed of foreleg movements was actually higher than the speed of hindleg rotations. By day 9 the mice would extend their forepaws in front of them for varying intervals (mean = 0.13 seconds), and by day 14 most mice showed well coordinated swimming with the hindlegs only. Hindleg rotation speed

ONTOGENY OF SWIMMING LIMB ROTATIONS

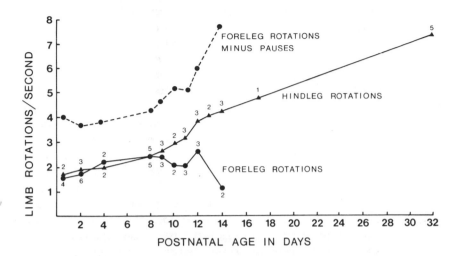

FIG. 8. *Ontogeny of forelimb and hindlimb rotations/second in mice during swimming. Note overall progressive similarity in forelimb and hindlimb rotations/second through day 8 (bottom two curves) plus shorter actual kicking duration in forelimbs, as calculated by substraction of inter-rotation pauses for forelimbs (top curve). Similar pauses were not seen in hindlimb rotations. Numbers of animals observed per day for both forelimb and hindlimb rotations are indicated on the lower two graphs; differences between forelimb and hindlimb observations on days 1 and 2 reflect earlier development of forelimb movements, while differences on days 13 and 14 indicate progressive forelimb disuse. (From Fentress and Templeman, unpublished.)*

increased to a value of around 4 rotations/sec by day 14, and around 7 rotations/sec by day 32. (Here it is interesting to note that the forelimb rotations minus pauses on day 14 yield a similar figure.) Thus there is rostral-caudal progression in limb coordination together with subsequent increase in forepaw pauses, and finally forelimb disuse when the hindlimb movements approached their adult form. It was possible to restore forelimb use through the addition of weights or surgical paralysis of the hindlegs through day 16. Templeman and I also obtained suggestive evidence that the rate of maturation of swimming movements could be accelerated slightly by giving animals daily swimming practice and comparing their performance to "first-time" swimmers through day 32. However, while these results are statistically significant for individual animal comparisons ($p < .01$), we prefer to remain cautious in our interpretation at this time.

III. Conclusions

As with grooming the ontogeny of locomotory and swimming movement patterns indicates the value of examining progressive changes in such features of behavior as *tuning, note structure, chord structure, melodic flow,* and *rhythmicity*; hence the value of the analogy between behavior and music. Several broad parallels emerge among these diverse observations, such as progressive tightness, elaboration, and refinement of the orchestrated complex, and the considerable role of endogenous maturational factors. Equally important distinctions are also apparent, such as in the progressive decrease in pauses between individual grooming strokes in mice between postnatal week one and two, and the parallel *increase* in forelimb stroke pauses during swimming over a similar age span. This indicates clearly the value of examining several forms of movement development as a test of the scope to which particular generalizations may be accurately applied (cf. Chapter 2).

It is evident that for many aspects of vertebrate behavior endogenous processes are achieving renewed emphasis (e.g., Chapter 2; Bateson, 1976; Fentress, 1977a). It is also evident that we can no longer think of experience in behavior as simply *adding* new information to the developing animal. For example, external stimuli can also act to *preserve* behavioral mechanisms that are already formed, and to *select* (as in unlocking a file drawer) potentialities for development expression that would otherwise be unavailable to the animal (e.g., Fentress, 1977a). Finally, I argue here that more detailed descriptive analysis along operationally definable dimensions, such as suggested in the analogy between behavior and a musical performance plus its underlying score, may provide important new data from which we can establish a more adequate basis for our understanding of the final orchestrated complex of behavioral expression that we observe in diverse species.

Acknowledgments

Reported research and manuscript preparation were aided by USPHS grant MH16995 and National Research Council of Canada Grant A9787.

Notes

1. This section represents research done in collaboration with F. P. Stilwell.
2. Research in collaboration with I. Golani.
3. Research in collaboration with T. L. Templeman.

References

Bateson, P. P. G. 1976. Specificity and the origins of behavior. *Adv. Stud. Behav.* 6:1–20.

Chomsky, N. 1968. *Language and Mind.* New York: Harcourt, Brace.

Eshkol, N., and A. Wachmann. 1958. *Movement notation.* Israel: Wiedenfeld and Nicholson. Movement Notation Society.

Fentress, J. C. 1967. Observations on the behavioral development of a hand-reared male timber wolf. *Am. Zool.* 7:339–351.

Fentress, J. C. 1968a. Interrupted ongoing behavior in two species of vole (*Microtus agrestis* and *Clethrionomys britannicus*): I. Response as a function of preceding activity and the context of an apparently "irrelevant" motor pattern. *Anim. Behav.* 16:135–153.

Fentress, J. C. 1968b. Interrupted ongoing behavior in two species of vole (*Microtus agrestis* and *Clethrionomys britannicus*): II. Extended analysis of motivational variables underlying fleeing and grooming behavior. *Anim. Behav.* 16:154–167.

Fentress, J. C. 1972. Development and patterning of movement sequences in inbred mice. *In The biology of behavior,* ed., J. A. Kiger. pp. 83–131. Corvallis, Oregon: Oregon State Univ. Press.

Fentress, J. C. 1973. Development of grooming in mice with amputated forelimbs. *Science* 179:704–705.

Fentress, J. C. 1976a. Dynamic boundaries of patterned behaviour: interaction and self-organization. *In Growing points in ethology* eds., P. P. G. Bateson and R. A. Hinde, pp. 135–169. London: Cambridge Univ. Press.

Fentress, J. C., ed. 1976b. *Simpler networks and behavior.* Sunderland, Mass.: Sinauer Associates.

Fentress, J. C. 1977a. Opening remarks: constructing and potentialities of phenotype. *N. Y. Acad. Sci.* 290:220–225.

Fentress, J. C. 1977b. The tonic hypothesis and the patterning of behavior. *N. Y. Acad. Sci.* 290:370–395.

Fentress, J. C., and F. P. Stilwell. 1973. Grammar of a movement sequence in inbred mice. *Nature* 244:52–53.

Fox, W. M. 1965. Reflex ontogeny and behavioral development of the mouse. *Anim. Behav.* 13:234–241.

Golani, I. 1976. Homeostatic motor processes in mammalian interactions: a choreography of display. *In Perspectives in ethology,* Vol. 2, eds., P. P. G. Bateson and P. H. Klopfer. pp. 69–134. New York: Plenum.

Hinde, R. A. 1970. *Animal behavior. A synthesis of ethology and comparative physiology,* 2d ed. New York: McGraw-Hill.

Lind, H. 1958. The activation of an instinct caused by a "transitional action." *Behaviour* 14:123–135.

Lorenz, K. Z. 1957. The past twelve years in the comparative study of behaviour. *In Instinctive behaviour: the development of a modern concept.* ed., C. H. Schiller. pp. 288–310. London: Methuen.

Northup, L. R. 1977. Temporal patterning of grooming in three lines of mice: some factors influencing control levels of a complex behaviour. *Behaviour* LX1:1–25.

Shapiro, S., M. Salas, and K. Vukovich. 1968. "Hormonal effects on ontogeny of swimming ability in the rat: assessment of central nervous development." *Science* 168:147–150.

Woolridge, M. W. 1975. *A quantitative analysis of short-term rhythmical behaviour in rodents.* Doctoral dissertation, Wolfson College, Oxford, England.

Chapter Seventeen

The Ontogeny of Ultrasound Production in Rodents

V. J. DE GHETT

Department of Psychology
Ethology Laboratory
State University of New York
Potsdam, New York 13676

The pioneering work in the area of neonatal ultrasound production in rodents was done by Zippelius and Schleidt (1956). They reported that young house mice (*Mus musculus*), European field mice (*Apodemus flavicollis*), and voles (*Microtus arvalis*) emitted both sonic and ultrasonic vocalizations. The ultrasonic calls were quite frequent when the pups were handled or when they cooled off. Zippelius and Schleidt (1956) considered these vocalizations to be distress or discomfort (*Unbehagen*) calls, similar to the *Pfeifen des Verlassenseins* that Lorenz (1935) had described in birds. At the time, their work seemed to have had little impact on ethology. Later Noirot (1966a) published her work on the ontogeny of the rate of ultrasound production in the albino house mouse and Hart and King (1966) published their developmental work on sonic and ultrasonic vocalization in *Peromyscus maniculatus*. Two years later Noirot (1968) published a developmental study of ultrasound production in the albino rat.

Since these studies of Noirot and of Hart and King, considerable information about ultrasonic vocalization in infant rodent species has been developed. Much of this work has been reviewed by Noirot (1972), by Sales and Pye (1974), and by Brown (1976). Table 1 shows the species of murid and cricetid rodents known to emit ultrasounds as neonates.

These early studies have given us a glimpse at the *Umwelt* of the

TABLE 1. List of murid and cricetid rodents known to emit ultrasounds as neonates

TAXON	AUTHORITY
Muridae	
Mus musculus	Noirot (1966a) and others
Mus minutoides	Sewell (1970a)
Rattus norvegicus	Noirot (1968); Allin and Banks (1971)
Rattus exulans	De Ghett and McCartney (unpublished); Sales (personal communication)
Rattus rattus	De Ghett and McCartney (unpublished)
Apodemus sylvaticus	Sewell (1970a)
Apodemus flavicolis	Zippelius and Schleidt (1956)
Acomys cahirinus	Sewell (1970a); Sales and Smith (in press)
Praomys natalensis	Sewell (1970a)
Thamnomys sp.	Sewell (1970a)
Arvicanthis niloticus	Sales (personal communication)
Notomys alexis	Watts (1975)
Notomys cervinus	Watts (1975)
Notomys mitchellii	Watts (1975)
Notomys fuscus	Watts (1975)
Cricetidae	
Microtus arvalis	Zippelius and Schleidt (1956)
Microtus pennsylvanicus	Colvin (1973)
Microtus montanus	Colvin (1973); De Ghett (1977)
Microtus californicus	Colvin (1973)
Microtus longicaudus	Colvin (1973)
Microtus ochrogaster	Colvin (1973)
Microtus agrestis	Sewell (1970a)
Clethrionomys glareolus	Sewell (1970a)
Dicrostonyx groenlandicus	Brooks and Banks (1973)
Mesocricetus auratus	Okon (1971)
Meriones unguiculatus	Sewell (1970a); De Ghett (1974)
Meriones libycus	Sewell (1970a)
Meriones shawi	Sewell (1969 cited in Sales and Pye, 1974)
Gerbillus sp.	Sewell (1970a)
Onychomys leucogaster	De Ghett (unpublished)
Calomys callosus	Smith (1972)
Lagurus lagurus	Sales and Pye (1974)
Peromyscus maniculatus	Hart and King (1966); Smith (1972); Huff (1973)
Peromyscus leucopus	De Ghett and McCartney (unpublished); Huff (1973)
Peromyscus polionotus	Huff (1973)

Peromyscus californicus	Huff (1973)
Peromyscus melanophrys	Huff (1973)
Peromyscus difficulis	Huff (1973)
Reithrodontomys megalotis	De Ghett (unpublished)
Cricetulus griseus	Sales (personal communication)
Phodopus sp.	Sales (personal communication)
Sungorus sp.	Sales (personal communication)

maternal rodent: The seemingly helpless and passive little rodent pup, under certain conditions, emits ultrasonic vocalizations; and these vocalizations may be an important link in the mother-young relationship. Certainly these ultrasonic distress vocalizations must be good for something. As Dawkins (1976) has said: "Natural selection entitles us to expect with confidence that good, though not necessarily optimal, design principles will pervade the internal organization of animals down to the smallest levels" (page 15).

These vocalizations might be signals sent from an infant to an adult. For the moment, let us assume that the adult is a maternal female. Otte (1974) presents a useful model of a communication network (Fig. 1). Using Otte's (1974) model we would label the "functional system" as the infant rodent pup (signaler) and the maternal female rodent (receiver).

Functional System

The typical frequency range of ultrasonic vocalization from rodent pups is between 30 and 80 kHz. Figure 2 shows an ultrasound sonographic record from a one-day-old mouse pup, *M. musculus* (from Noirot, 1966b).

Adult rodents appear to be able to hear these ultrasounds (Ralls, 1967; Brown, 1973a, b; Brown and Pye, 1975). Figure 3 (from Brown, 1973b) shows that *M. musculus* can hear well into the ultrasonic range with an average peak high-frequency response at about 50 kHz when measured at the level of the inferior colliculus. Using behavioral techniques with mice, Markl and Ehret (1973) and Ehret (1974) have found a region of maximum high-frequency responsiveness between 40 and 60 kHz.

The relationship between peak high-frequency response at the level of the inferior colliculus in the adult and the typical range of neonatal ultrasound production in certain selected rodent species is shown in Figure 4. These data indicate that the adult rodent does possess the

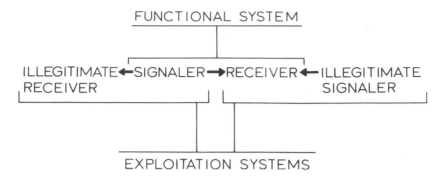

FIG. 1. *Communication model redrawn from Otte* (1974). (*Reprinted by permission of Annual Reviews, Inc.*)

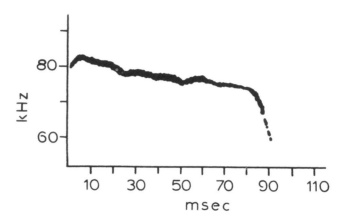

FIG. 2. *Sonographic record of an ultrasonic vocalization of a one-day-old mouse* (Mus musculus) *pup.* (*Redrawn from Noirot [1966b]; reprinted by permission of the Société Royale Zoologique de Belgique.*)

acoustic design features to permit the reception of neonatal ultrasonic vocalizations. The signaler-receiver functional system appears to be intact.

The communication link between the pup as a signaler and the mother as a receiver should not be considered as purposeful. The pup is emitting ultrasounds that are associated with a presumed state of affect (Scott and De Ghett, 1972). Verhaeghe and Noirot (in press) have shown that pups being retrieved by deaf and hearing mouse mothers emit the same number of ultrasounds even though deaf mothers are known to be

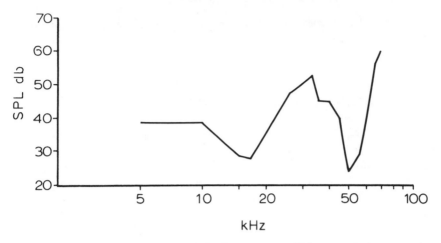

FIG. 3. *An audiogram of the house mouse* (M. musculus). *Measurements were taken at the inferior colliculus.* (*Redrawn from Brown* [*1973b*]; *reprinted by permission of Springer-Verlag.*)

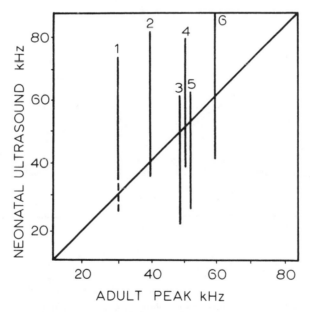

FIG. 4. *The relationship between the range of neonatal ultrasounds and the typical peak high frequency response* (*inferior colliculus*) *in several rodent species:* 1, Rattus norvegicus; 2, Meriones unguiculatus; 3, Clethrionomys glareolus; 4, Mus musculus; 5, Microtus agrestis; 6, Apodemus sylvaticus.

less efficient as retrievers of displaced pups. The number of ultrasounds emitted is independent of the consequences of ultrasound emission.

Exploitation Systems: Illegitimate Receivers

If these ultrasonic signals can be detected by an adult rodent as a legitimate receiver, then can they be detected by some illegitimate receiver such as a predator? According to Peterson, Heaton, and Wruble (1969), many carnivores are equipped to hear into the ultrasonic range. For example, the coyote, *Canis latrans,* has an upper frequency limit of about 80 kHz (cochlear microphonic potential). The frequency of the ultrasonic vocalizations of many North American rodent species is near this upper frequency limit of the coyote's hearing. Figure 5 shows the auditory sensitivity (as measured by the cochlear microphonic potential) for the coyote (from Peterson *et al.,* 1969) and for the house mouse (from Brown, 1973a) and, as a rectangular insert, the approximate frequency and intensity of a neonatal mouse

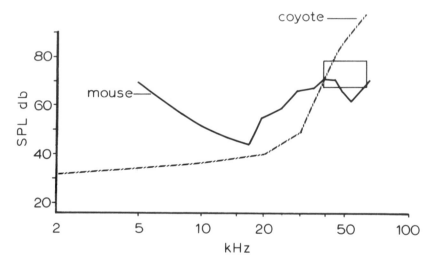

FIG. 5. *Audiograms from a coyote (redrawn from Peterson et al. [1969]) and a house mouse (redrawn from Brown [1973a]) as measured by cochlear microphonic potentials. The rectangular insert represents the typical frequency and intensity of house mouse neonatal ultrasounds. (Reprinted by permission of the American Society of Mammalogists and Springer-Verlag.)*

ultrasonic vocalization. Notice that the ultrasonic vocalization is at and below the coyote threshold but above the mouse threshold.

The relationships depicted in Figure 5 may not really be as precise as the auditory sensitivity functions indicate. But the general relationship should hold. A possibility exists that a coyote might hear a neonatal rodent ultrasonic vocalization; so might a red fox (*Vulpes vulpes*). The red fox appears to have a threshold function beyond 50 kHz similar to the coyote's (Peterson *et al.*, 1969). Isley and Gysel (1975) suggest that ultrasounds from young rodents would be difficult for the red fox to locate by phase-difference localization techniques. Thus while a red fox might occasionally be able to hear such an ultrasound, it would have a difficult time discovering the location of the pup. No doubt, this conclusion could be extended to the coyote and other carnivores with similar auditory sensitivity functions and a distance between the tympanic membranes similar or greater than that of the red fox. It is, therefore, highly doubtful that moderate- to large-size carnivore predators could be illegitimate receivers.

The tayra (*Eira barbara*), a mustelid, seems to be more sensitive beyond 50 kHz than either the red fox or the coyote (Peterson *et al.*, 1969). If the data for the tayra are somewhat representative of Mustelidae, then one might expect that weasels, ferrets, stoats, polecats, mink (*Mustela* sp.), and skunks might be troublesome as illegitimate receivers. Because of the narrow heads (less distance between the tympanic membranes) of the small mustelids, phase-difference localization would be moderate to good. Not only could the small mustelids hear some of the neonatal ultrasounds but they probably could also locate the vocalizing young.

According to Gould (1969), shrews (*Suncus murinus, Blarina brevicauda,* and *Cryptotis parva*) emit ultrasounds between 50 and 107 kHz that might be used for echolocation. Buchler (1976) has shown that *Sorex vagrans* does echolocate. If we assume that shrews can hear throughout most of the frequency range over which they are vocalizing, then they should be able to detect many neonatal rodent ultrasounds. In the laboratory, the least shrew (*C. parva*) will kill and consume two- to three-day-old house mice and can be maintained quite effectively on a steady diet of house mice (Barrett, 1969). In the field, it is not clear to what extent shrews prey upon live rodents. Many species of shrews use *Microtus* runways and are sufficiently small so that they could easily get in or near rodent nests containing young. Hamilton (1939) believed that shrews might occasionally "blunder onto a nest of young mice." That might be the extent of the shrew's interest in detecting and locating young rodents in the field. Theoretically, shrews could be illegitimate receivers; whether they actually are must await further data on food habits and auditory sensitivity.

Onychomys leucogaster young emit ultrasounds in the same 30- to 80-kHz range as most other rodents. The adults in my laboratory appear to detect these ultrasounds and locate displaced young that are emitting ultrasounds. An adult *O. leucogaster* should be able to detect and locate young of many other rodent species. In the laboratory *O. leucogaster* will readily capture, kill, and consume house mouse pups. Often, the adult's first response appears to be to the ultrasonic vocalization of the young mouse pup. Grasshopper mice (*Onychomys* sp.) should be as efficient as illegitimate receivers as they are as legitimate receivers.

Ultrasonic vocalizations will not be effective signals for a predator if the predator is not fairly close to the source of the ultrasonic vocalization. Let us assume that there is a rodent pup emitting ultrasounds on a flat surface. The vocalizations are at 50 kHz and 80 db when measured at 0.1 m from the pup. The ambient temperature is 20°C and the relative humidity is 50%. The predator is 10 m from the vocalizing pup and has no olfactory or visual clues to suggest the presence of the pup. The inverse-square law would predict that the intensity of the vocalization would drop about 20 db over a distance of 10 m (-20 db $= 10 \log_{10} 0.1$ m$/10$m). There will also be an additional attenuation of the ultrasonic vocalization caused by the atmospheric conditions (20°C and 50% relative humidity). Griffin (1971) discusses this atmospheric attenuation in detail and gives the necessary formulas and graphs to calculate the attenuation. Using this material, we would estimate an additional 35-db loss at the predator. The predator would now have to be able to detect a 50-kHz vocalization at about 25 db to be considered as an illegitimate receiver. I think that would be an unlikely event.

The effect of burrow systems on the transmission of ultrasounds is difficult to assess. In a personal communication, Griffin wrote as follows:

> I have often wondered about the transmission of ultrasonic distress calls of infant rodents through burrows and tunnels. The pups are normally in nests and sound transmission from them to places [where] other small mammals are likely to be listening is very different from the propagation of sounds through open air. Transmission might be either worse or better. Sound waves travel very well along pipes, partly because there is no spreading loss. But natural burrows have rough and irregular walls. The net result of these and other factors is very difficult to predict, but direct measurements should be relatively easy. One would need only to locate a small high-frequency loud-speaker in or near the nest and measure sounds similar to the calls of the pups at various locations along burrows, tunnels, or runways connecting to the nest.

Probably the most likely illegitimate receivers would be the small predators. These are the predators that might be spending a fair portion of their time in close proximity to rodent burrows and nests. Under these conditions the small predators might be receiving an ultrasonic signal that has dropped 10 or 20 db from the source. Although the necessary auditory sensitivity functions for the small predators seem lacking, I think it would be safe to assume that a 10- or 20-db reduction in intensity at or beyond 50 kHz should not be severely handicapping to an individual that may already have some visual or olfactory cues.

Exploitation Systems: Illegitimate Signalers

Illegitimate signalers could theoretically be bats and certain insects, as well as conspecific and heterospecific rodents. While rodents could make good bat detectors, they probably do not actually hear too many ultrasonic bat calls. The echolocating bat cruising over a field at dusk is probably not directing much ultrasonic energy through the field grasses and into nests and burrows. Katydids emit ultrasonic calls. For example, *Conocephalus strictus,* found in the United States, has in its stridulations ultrasonic components at 20 to 25 kHz and at 40 kHz (Pielemeier, 1946). Many species of rodent mothers should be able to detect these signals. The Little Brown Cricket (*Nemobius fasciautus*) does stridulate with a peak frequency near 28 kHz (Pielemeier, 1946). This should be easily detectable by *Rattus norvegicus*. Stridulating insects near burrows and nests may just be troublesome illegitimate signalers. Conspecific and heterospecific rodents cohabiting burrows or in nearby nests are also a potential source of illegitimate signals. However, the level of maternal aggression and defense should be sufficiently high to keep intruders at a distance (Noirot, 1969; Gandelman, 1972; St. John and Corning, 1973). The maternal female is probably not distracted too much by ultrasounds coming from other than her own young. The functional system in Otte's (1974) model would appear to have a reasonably high signal/noise ratio. "Good, though not necessarily optimal, design principles" (Dawkins, 1976) seem to exist in the communication system.

Ultrasounds and Maternal Behavior

A number of studies have demonstrated the importance of neonatal ultrasounds in the process of finding and retrieving displaced young, in nest building, and in the general care and maintenance of infant rodents during the days and weeks after birth. Zippelius and Schleidt (1956) believed that these ultrasounds serve as signals that alert and guide the parent to displaced young. They also provide some empirical verification of the value of these ultrasounds.

Allin and Banks (1972) played recorded seven-day-old rat ultrasounds to lactating females, males, and virgin females. Lactating females left the nest box and moved in the direction of the recorded signals. Males and virgin females only oriented to the source but did not leave the nest. Smith (1976) used electronically generated ultrasounds to test the approach responses of lactating female mice, primed (24-hr exposure to a pup) females and males, and naive females and males. Lactating females and primed females showed a greater number of approach responses than did other groups of mice. Sewell (1970b) has shown that lactating *Apodemus sylvaticus* will leave a nest of pups when recorded ultrasounds are present and choose the side of a T-partition containing the speaker. Colvin (1973) has shown that *Microtus montanus, M. californicus,* and *M. ochrogaster* males and females respond equally well to pups emitting ultrasounds. Calling pups were more effective for bringing either parent than pups that did not vocalize. Huff (1973) has shown that lactating *Peromyscus maniculatus* mothers will leave a nest and search for a displaced pup when recorded *P. maniculatus* ultrasounds are played. The female also shows a distinct preference for recorded ultrasounds over recorded control sounds. Both *P. maniculatus* and *P. leucopus* mothers show a slight preference for recorded ultrasounds from conspecifics. *P. maniculatus* mothers are as efficient at locating the source of recorded ultrasounds as they are at locating a pup emitting ultrasounds.

Smotherman *et al.* (1974) used lactating female rats in a Y-maze-type choice situation. The choices were a rat pup emitting ultrasounds, a tape of neonatal rat ultrasounds, a chilled pup that did not vocalize, or an empty maze arm. The pup emitting ultrasounds provided both ultrasonic and olfactory cues. The taped ultrasounds were ultrasonic cues only. The chilled pup provided olfactory cues only (extremely cold pups become comatose and do not vocalize when they are very young). The most effective stimulus to initiate retrieval was a combination of olfactory and ultrasonic cues. Smotherman, *et al.* (in press) have shown that pup odor determines the speed at which the female initiates her search for a displaced rat pup. They have also shown that olfactory cues seem to increase the female rat's tendency to approach a source of ultrasounds. Noirot (1970) demonstrated that olfactory cues exert a positive influence on pup cleaning and ultrasonic cues exert a positive influence on nest building in mice. It is reasonable to suppose that segments of maternal care would not have evolved as a dependency on a single cue system. Olfactory stimulus gradients would provide little directional information while ultrasounds would provide little information persistence.

When all studies are combined, it appears that adults do respond to neonatal ultrasounds and that these ultrasounds do aid in the organization of various aspects of maternal care. Many aspects of maternal care decline as the young get older. Age of the young (an exogenous factor) and time since parturition (an endogenous factor) are naturally confounded variables in the life history of mammals. For years, the main focus of the maternal behavior research has been the changing endogenous hormonal state of the mother. The influence of neonatal ultrasounds on the behavior of the mother seems to be an extremely important aspect of an exogenous control system that organizes maternal behavior. So that the pendulum does not swing too far in the direction of exogenous control, recall that Allin and Banks (1972) have demonstrated that it is the lactating female rat that responds best to recorded ultrasounds. Endogenous elements in the system cannot be ignored even when examining the exogenous elements.

Ontogenetic Changes

Changes in the ontogeny of ultrasound production have been investigated in the following species; mice, *M. musculus* (Noirot, 1966a; Noirot and Pye, 1969; Okon, 1970a, b; Nitschke, Bell, and Zachman, 1972; Sales and Smith, in press); rats, *R. norvegicus* (Noirot, 1968; Allin and Banks, 1971; Okon, 1971; Sales and Smith, in press); golden hamsters, *Mesocricetus auratus* (Okon, 1971); deer mice, *P. maniculatus* (Hart and King, 1966; Smith, 1972; Huff, 1973), *P. polionotus* and *P. leucopus* (Huff, 1973); collared lemmings, *Dicrostonyx groenlandicus* (Brooks and Banks, 1973); Mongolian gerbils, *Meriones unguiculatus* (De Ghett, 1974); montane voles, *Microtus montanus* (Colvin, 1973; De Ghett, 1977) and other voles *M. pennsylvanicus, M. californicus, M. longicaudus,* and *M. ochrogaster* (Colvin, 1973); woodmice, *Apodemus sylvaticus* and spiny mice, *Acomys cahirinus* (Sales and Smith, in press).

The developmental changes in the rate of ultrasound production are quite dramatic. No species studied to date vocalizes at a uniform rate during development. Very early in development, the rate of calling is moderate, but increases daily. It then reaches a maximum or peak rate, and gradually declines. The form of the Rate X Age function is that of an inverted J.

In house mice, ontogenetic changes can be found in ultrasonic call duration, total bandwidth, and intensity (Noirot and Pye, 1969; Okon,

1970a, and Sales and Smith, in press, for intensity) in addition to the changes in the rate of ultrasound production. Ontogenetic changes in intensity have also been reported in rats and woodmice by Sales and Smith (in press) and in rats and golden hamsters by Okon (1971).

Although the literature is still sketchy, there is some evidence for two kinds of neonatal vocalizations. The labels vary with different authors, but the phrases *isolation call* and *handling call* seem to be accepted by most. Isolation calls come from rodent pups that are allowed to vocalize undisturbed by anything except the temperature of the air around them and the temperature of the surface they are lying on. This condition is similar to that of a rodent pup that has been abandoned and separated from a warm nest and/or mother. Handling calls come from rodent pups that are receiving some form of tactile, vestibular, or painful stimulation. This condition resembles that of a rodent pup that is being retrieved, cleaned, and/or manipulated by its mother.

The Ontogeny of Handling Calls

The work of Okon (1970a, b) suggests two kinds of ultrasonic vocalization in the house mouse. Maximum intensity ultrasounds (2–3 μbars peak) are found between days 6 and 16 when the pups are isolated at ambient temperatures of 2°–3°C, 12°C, and 22°C (Okon, 1970a). Each ambient temperature produces a slightly different developmental profile. Maximum intensity ultrasounds (4–13 μbars peak) are produced between birth and day 9 or 10 when the pups receive various types of handling, including maternal retrieval (Okon, 1970b). The developmental profiles for the handling calls and the isolation calls are very different. The peak intensities (μbars) differ and the ages at which the maximum intensity is reached also differ. Handled pups produce very intense ultrasounds at an age when nonhandled pups vocalize softly and infrequently. Noirot (personal communication) notes that mice do not appear to emit ultrasounds when they are handled at nest temperatures, presumably in the 35°–40°C range. This fact suggests that the handling calls may be partially temperature dependent.

Bell, *et al.* (1971) found an average ultrasonic call duration of 108 msec for handled *P. maniculatus* pups and an average call duration of 45 msec for nonhandled pups. Smith (1972), working with *P. maniculatus,* found that an intense audible vocalization was characteristic of young pups under cold stress (isolation) and that a less intense ultrasonic call was characteristic of older pups that were being handled. Sewell (1969, cited in Sales and Pye, 1974) studied a variety of rodent species and concluded that more intense calls were characteristic of handled pups rather than of nonhandled (isolated) pups.

I have observed the intense calls from *M. musculus* C57BL/6 during the process of retrieval. Often the retrieving mother will halt and release the pup, apparently in response to the intense ultrasonic call, and then pick up the pup again. I have also observed that intense calls often occur immediately before the pup is dropped. The assumption is that the pup was being bitten too hard by the retrieving mother.

The handling calls are believed to have an inhibitory effect on parental aggression and excessive manipulation by a parent. Definitive studies are yet to be done. Sales and Pye (1974) reviewed the handling call literature and concluded: "There is no direct experimental evidence that the calls produced by pups on handling have a negative or inhibitory effect on maternal behavior although there are several indications of such an effect" (page 170). And Brown (1976) has said: "It may be . . . that researchers have indeed tried to derive more information from the cries of infants than does the rodent mother" (page 315).

The Ontogeny of Isolation Calls

When young cricetid and murid rodents are born they do not possess the ability to thermoregulate; they are ectothermic. They gradually develop the ability to regulate their internal body temperature; when they have developed this ability they are endothermic. Small to medium-size rodents appear to develop thermoregulatory ability in three stages: (1) ectothermic stage—birth to about day 5, when the core body temperature falls to near the ambient temperature; (2) partial endothermic stage—day 6 to about day 14, when endothermy can be achieved at ambient temperatures of about 25°C; and (3) final endothermic stage—day 15 to about day 20, when adult-type temperature regulation is possible across a fairly broad range of temperatures.

From the early ultrasound production work of Zippelius and Schleidt (1956), Hart and King (1966), and Noirot (1966a) to the present, these distress vocalizations have been linked to a decline in core body temperature. When the newborn rodent is maintained at the nest temperature (35° to 40° C), it will not vocalize. As the ambient temperature falls below the nest temperature, young rodents will start to vocalize and, if they are very young, when the ambient temperature approaches 0°C they will become comatose and stop vocalizing. In the laboratory it is possible to drive the rate of ultrasonic vocalization by changing the ambient temperature. The ultrasonic vocalizations emitted under these conditions are the isolation calls.

That young rodents emit more ultrasounds at low ambient temperatures has been shown in laboratory rats (Allin and Banks, 1971; Okon, 1971; Ostwalt and Meier, 1975), laboratory mice (Okon, 1970a; Nitschke,

Bell, and Zachman, 1972) and in golden hamsters (Okon, 1971). Certainly, more species should be studied but so far the results are quite uniform.

McManus (1971) studied the ontogeny of temperature regulation in the Mongolian gerbil (*M. unguiculatus*). Figure 6 shows McManus's results (1971). Figure 7 shows the development of the rate of ultrasonic vocalization in the Mongolian gerbil (De Ghett, 1974). The functions are almost mirror images of each other. They correlate significantly at $r = -.91$ ($p < .01$). An examination of McManus's data reveals that the young gerbils actually lost more body temperature on day 4 than they did on days 2, 3, and 5. The core body temperature is significantly lower ($p < .05$) on day 4 than on any other day of postnatal ontogeny. The rate of ultrasonic vocalization for the Mongolian gerbil is at its highest (peak) rate on day 4 (De Ghett, 1974). The rate is significantly greater ($p < .05$) on day 4 than on any other day of postnatal ontogeny.

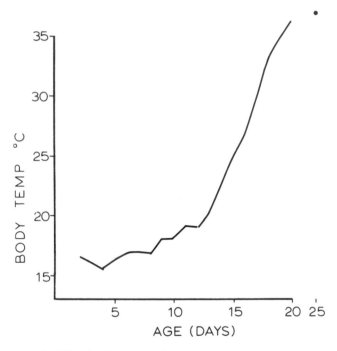

FIG. 6. *The development of temperature regulation in the Mongolian gerbil. Core body temperatures were measured after one hour at 14–15° C. The day 25 value is within the normal range of adult gerbils. Notice the low value for day 4. (Redrawn from McManus [1971]; reprinted by permission of the American Society of Mammalogists.)*

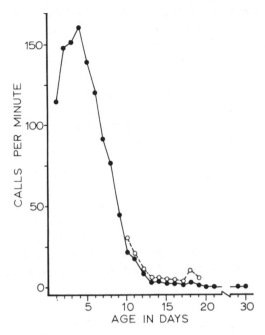

FIG. 7. *Developmental changes in the rate of ultrasound production in the Mongolian gerbil. Note the peak rate on day 4. (Ignore the open circles with the dashed lines.) (From De Ghett [1974]; reprinted by permission of John Wiley & Sons, Inc.)*

This drop in body temperature is convenient for the thesis that the development of the isolation call rate is related to the development of endothermy. It does provide an additional point for the argument and it does tie the development of ultrasound production to the developing physiology of the animal. But, this drop in body temperature is a curious phenomenon in itself.

After one hour at an ambient temperature of 14°–15°C, the core body temperature of the Mongolian gerbil was significantly greater ($p < .05$) than the ambient temperature (McManus, 1971). This was true even on day 4. Some minute degree of endothermy appears to be present very early in development. But why is there a decline in this ability from birth to day 4? An improvement would be expected. Perhaps there is a change in the surface/volume ratio of the young gerbil during the early days that makes heat retention less likely on day 4 than on previous days. The growth rate for weight in the Mongolian gerbil is very high and it increases daily from birth until day 5 and then declines (De Ghett, 1972).

Between day 4 and day 5 there is a 15.6% increase in the weight of the young gerbil. A very high growth rate could also indicate that more of the neonate's metabolism is being directed toward tissue production and that little is available to use on day 4 for primitive thermoregulation. After day 4, the young gerbil loses less and less core body temperature. At this time the young gerbil is able to assume certain postures that will reduce the amount of ventral surface in contact with the substrate. This should aid in the reduction of conductive heat loss. Convective heat loss will also be reduced as more and more dorsal and ventral fur develops.

Some litters of house mice (*M. musculus* C57BL/6) lose more body temperature on day 5 and vocalize more on day 5 than on any other day. The data in Figure 8 concern a single litter of five pups. While the relationship between body temperature loss and rate of ultrasonic vocalization is not as nice as for the previous data on the Mongolian gerbil, the general relationship still appears to hold. Following day 5, the C57BL/6 mouse is capable of assuming postures that will reduce conductive heat loss, and fur development is sufficient to provide some insulative value.

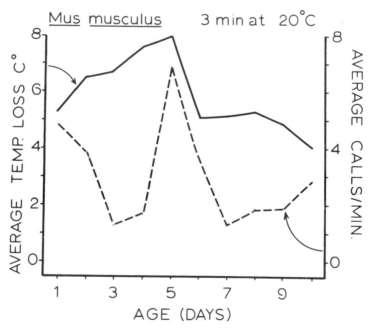

FIG. 8. *Data on the development of ultrasound production* (dashed lines) *and the development of temperature regulation as measured by body temperature lost* (solid lines) *for* M. musculus. *Data are from a litter of five.*

Some litters lose more body temperature on day 3 or day 4, and these litters have average peak rates of ultrasound production on these days (see Fig. 9).

The ontogeny of the rate of ultrasound production in the montane vole (*M. montanus*) shows the same changes that are seen in the Mongolian gerbil and the C57BL/6 house mouse (De Ghett, 1977) (see Fig. 10). As in the mouse and the gerbil, these changes are in the ontogeny of the isolation calls. The peak rate of ultrasonic vocalization occurs on day 2 of postnatal ontogeny. The peak rate occurs very early in ontogeny. The peak rate is also very high (average calls/min = 160.88). No other rodent species studied to date appears to have so early and so high a peak rate of ultrasonic vocalization. I have no information on the development of temperature regulation in this species. However, I would predict that the young pups have a high requirement for warmth very early in development. This is partially confirmed by the intense level of maternal care

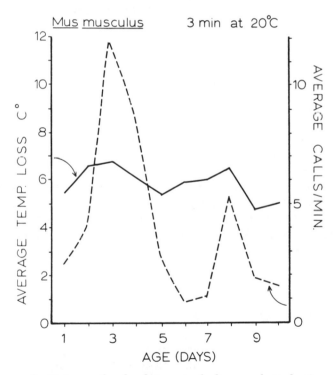

FIG. 9. *Data on the development of ultrasound production* (dashed lines) *and the development of temperature regulation as measured by body temperature lost* (solid lines) *for* M. musculus. *Data are from a litter of five.*

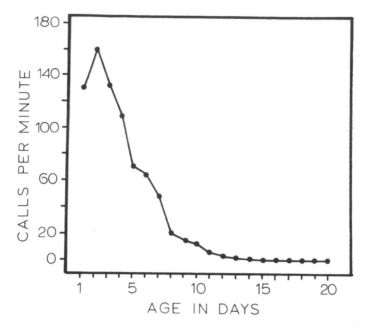

FIG. 10. *The developmental changes in the rate of ultrasound production in* Microtus montanus. (*From De Ghett* [*1977*]; *reprinted by permission of E. J. Brill, Publishers.*)

supplied by the female. When maintained in large cages with about 15 cm of bedding, the female would build a nest under the surface of the bedding and line the nest with nest material and food. Compared to nests built by a number of other rodent species in my laboratory, the *M. montanus* nests seemed to be an extremely efficient temperature-insulating system. The early and high peak rate of ultrasound production is also consistent with the precocious level of development of *M. montanus* and other microtine species.

Evolutionary Implications

The young rodent faces a problem at birth. How do you develop endo-thermy when you need an optimal internal temperature to develop the necessary physiological endothermic mechanisms? The young rodent takes the mammalian way out of the problem: let mother do it. To make sure that the warmth is supplied and that heat loss is reduced, the young rodent emits ultrasounds as its core body temperature falls below some optimum. The behavioral signaling system is conveniently tied to developing physiology. This is also the mammalian way. We would not expect the evolution of ultrasonic vocalization to have occurred in any other way in a mammal. In discussing mammalian evolution, Olson (1959) wrote: "In cases in which characters not primarily related to the developing physiology take ascendency in the selective hierarchy there will not be fruition of the total suite of ostensive characters of the skeleton which we call mammalian" (page 352). Olson (1959) is saying that basic mammalian characteristics are related to developing physiology.

Ultrasonic vocalization is the neonate's way of achieving behavioral thermoregulation before it develops physiological thermoregulation. Richards (1973) considers thermoregulatory ability to be the achievement of "freedom from the effects of the physical laws of heat flow" (page 5). The ultrasonic signals from the pup are an effective cue to the maternal female to search for a pup, retrieve it, and be generally maternal to it. As Noirot (1970) has shown, these ultrasounds also appear to stimulate nest building. The mother provides the heat and the nest provides the insulation which reduces heat loss.

I would surmise that the evolution of the ultrasonic call in response to a drop in the core body temperature evolved before the call in response to handling. Of course, this assumes that they are truly different calls. The handling call does not appear to be tied to developing physiology. It may have become "emancipated" from the original isolation call. We still know very little about the ontogeny of these handling calls, so it is premature to draw conclusions.

Ultrasound probably evolved very rapidly and early in murid and cricetid rodents. There was probably heavy selection pressure favoring neonatal ultrasound production because of its link to the process of behavioral thermoregulation. Olson (1959) said: "With continuing selection favoring the physiological process of homoeothermy any feature which appeared and was favorable to this process would tend to have a selective advantage" (page 351). Behavioral thermoregulation could be considered as a feature favorable to the process of physiological (internal) thermoregulation.

I doubt that we will find a murid or cricetid rodent that does not emit ultrasounds as a neonate; even the precocious spiny mouse emits ultrasounds. I would consider that neonatal ultrasound production represents a conservative evolutionary trend in these rodent forms.

Dobzhansky (1951), Romer (1967), and Mayr (1970) agree that maternal care has played an important role in mammalian evolution, its function being educational. Romer (1967) said: ". . . in the nursing habit we see the establishment of the world's first educational institution" (page 1635). The educational role played by the mother is usually considered to be like that of a teacher. In addition to the active role, there is an important passive role like that of the person who keeps the furnaces running in Romer's educational institution. This role is also played by the maternal female. She supplies the warmth in response to the isolation calls of her offspring. Her warmth, however, supplies more than just comfort. Misanin *et al.* (1971) have shown that a drop in core body temperature causes a loss of long-term memory in 9-day-old rats, and Nagy *et al.* (1976) have shown an identical effect in nine-day-old mice. The memories gained by the young through the active efforts of the mother are preserved in a passive way by the warmth she supplies in response to the ultrasonic isolation calls. Her rapid response will prevent a severe loss of core body temperature and will also prevent the disruptive effects of forgetting. This preserved memory organization might be a key link in the early experiences and subsequent survival of young rodents.

Acknowledgments

I would like to thank Dr. J. I. Dalland, Dr. D. R. Griffin, Dr. G. D. Sales, and Dr. E. Noirot for their helpful comments on an earlier version of this chapter. Part of the research reported in this paper was supported by a State University of New York Grant-in-Aid and Faculty Research Fellowship (#30-7137-A).

References

Allin, J. T., and E. M. Banks. 1971. Effects of temperature on ultrasound production by infant albino rats. *Dev. Psychobiol.* 4:149–156.

Allin, J. T., and E. M. Banks. 1972. Functional aspects of ultrasound production by infant albino rats (*Rattus norvegicus*). *Anim. Behav.* 20:175–185.

Barrett, G. W. 1969. Bioenergetics of a captive least shrew, *Cryptotis parva. J. Mammal.* 50:629–630.

Bell, R. W., W. Nitschke, T. H. Gorry, and T. A. Zachman. 1971. Infantile stimulation and ultrasonic signaling: a possible mediator of early handling phenomena. *Dev. Psychobiol.* 4:181–191.

Brooks, R. J., and E. M. Banks. 1973. Behavioural biology of the collared lemming (*Dicrostonyx groenlandicus* (Traill)): an analysis of acoustic communication. *Anim. Behav. Monogr.* 6:1–83.

Brown, A. M. 1973a. High frequency peaks in the cochlear microphonic response of rodents. *J. Comp. Physiol.* 83:377–392.

Brown, A. M. 1973b. High levels of responsiveness from the inferior colliculus of rodents at ultrasonic frequencies. *J. Comp. Physiol.* 83:393–406.

Brown, A. M. 1976. Ultrasound communication in rodents. *Comp. Biochem. Physiol.* 53(A):313–317.

Buchler, E. R. 1976. The use of echolocation by the wandering shrew (*Sorex vagrans*). *Anim. Behav.* 24:858–873.

Colvin, M. A. 1973. Analysis of acoustic structure and function in ultrasounds of neonatal *Microtus. Behaviour* 44:234–263.

Dawkins, R. 1976. Hierarchical organization: a candidate principle for ethology. *In Growing points in ethology,* eds., P. P. G. Bateson and R. A. Hinde, pp. 7–54. London: Cambridge Univ. Press.

De Ghett, V. J. 1972. The behavioral and morphological development of the Mongolian gerbil (*Meriones unguiculatus*) from birth until thirty days of age. Ph.D. dissertation. Bowling Green State Univ. University Microfilms, Ann Arbor, Mich. (Diss. Abstr. 73-12022.) 273 pp.

De Ghett, V. J. 1974. Developmental changes in the rate of ultrasonic vocalization in the Mongolian gerbil. *Dev. Psychobiol.* 7:267–272.

De Ghett, V. J. 1977. The ontogeny of ultrasonic vocalization in *Microtus montanus. Behaviour* 60:115–121.

Dobzhansky, T. 1951. *Genetics and the origin of the species.* New York: Columbia Univ. Press.

Ehret, G. 1974. Age-dependent hearing loss in normal hearing mice. *Naturwissenschaften* 11:506.

Gandleman, R. 1972. Mice: postpartum aggression elicited by the presence of an intruder. *Horm. Behav.* 3:23–28.

Gould, E. 1969. Communication in three genera of shrews (Soricidae): *Suncus, Blarina,* and *Cryptotis. Commun. Behav. Biol. Part A,* 3:11–31.

Griffin, D. R. 1971. The importance atmospheric attenuation for the echolation of bats (Chiroptera). *Anim. Behav.* 19:55–61.

Hamilton, W. J. 1939. *American mammals.* New York: McGraw-Hill.

Hart, F. H., and J. A. King. 1966. Distress vocalizations of young in two subspecies of *Peromyscus maniculatus. J. Mammal.* 47:287–293.

Huff, J. N. 1973. The "distress" cry of infant deermice, *Peromyscus:* physical characteristics, specific differences, social function. Ph.D. dissertation. Michigan State Univ. University Microfilms, Ann Arbor, Mich. (Diss. Abstr. 74-13909.) 206 pp.

Isley, T. E., and L. W. Gysel. 1975. Sound-source localization by the red fox. *J. Mammal.* 56:397–404.

Lorenz, K. 1935. Der Kumpan in der Umwelt des Vogels. *J. Ornithol.* 83:137–213, 289–413.

Markl, H., and G. Ehret. 1973. Die Hörschwelle der Maus (*Mus musculus*). Eine kritische Wertung der Methoden zur Bestimmung der Hörschwelle eines Saugetiers. *Z. Tierpsychol.* 33:274–284.

Mayr. E. 1970. *Populations, species and evolution.* Cambridge, Mass.: Harvard Univ. Press.

McManus, J. J. 1971. Early postnatal growth and development of temperature regulation in the Mongolian gerbil, *Meriones unguiculatus. J. Mammal.* 42:782–792.

Misanin, J. R., Z. M. Nagy, E. F. Keiser, and W. Bowen. 1971. Emergence of long-term memory in the neonatal rat. *J. Comp. Physiol. Psychol.* 77:188–199.

Nagy, Z. M., J. A. Anderson, and T. A. Mazzaferri. 1976. Hypothermia causes adult-like retention deficits of prior learning in infant mice. *Dev. Psychobiol.* 9:447–458.

Nitschke, W., R. W. Bell, and T. Zachman. 1972. Distress vocalizations of young in three inbred strains of mice. *Dev. Psychobiol.* 5:363–370.

Noirot, E. 1966a. Ultrasounds in young rodents. I. Changes with age in albino mice. *Anim. Behav.* 14:459–462.

Noirot, E. 1966b. Ultrasons et comportements maternels chez les petits rongeurs. *Ann. Soc. Roy. Zool. Belg.* 95:47–56.

Noirot, E. 1968. Ultrasounds in young rodents. II. Changes with age in albino rats. *Anim. Behav.* 16:129–134.

Noirot, E. 1969. Interactions between reproductive and territorial behavior in female mice. *Int. Mental Health Res. Newsletter* 2:10–11.

Noirot, E. 1970. Selective priming of maternal responses by auditory and olfactory cues from mouse pups. *Dev. Psychobiol.* 2:273–276.

Noirot, E. 1972. Ultrasounds and maternal behavior in small rodents. *Dev. Psychobiol.* 5:371–387.

Noirot, E., and D. Pye. 1969. Sound analysis of ultrasonic distress calls of mouse pups as a function of their age. *Anim. Behav.* 17:340–349.

Okon, E. E. 1970a. The effect of environmental temperature on the production of ultrasounds by isolated nonhandled albino mouse pups. *J. Zool.* 162:71–83.

Okon, E. E. 1970b. The ultrasonic responses of albino mouse pups to tactile stimuli. *J. Zool.* 162:485–492.

Okon, E. E. 1971. The temperature relations of vocalization in infant Golden hamsters and Wistar rats. *J. Zool.* 164:227–237.

Ostwalt, G. L., and G. W. Meier. 1975. Olfactory, thermal, and tactual influences on infantile ultrasonic vocalization in rats. *Dev. Psychobiol.* 8:129–135.

Otte, D. 1974. Effects and functions in the evolution of signaling systems. *Annu. Rev. Ecol. Systematics* 5:385–417.

Peterson, E. A., W. C. Heaton, and S. Wruble. 1969. Levels of auditory response in fissiped carnivores. *J. Mammal.* 50:566–578.

Pielemeier, W. H. 1946. Supersonic insects. *J. Acoust. Soc. Am.* 17:337–338.

Ralls, K. 1967. Auditory sensitivity in mice: *Peromyscus* and *Mus musculus.* *Anim. Behav.* 15:123–128.

Richards, S. A. 1973. *Temperature regulation.* New York: Springer-Verlag.

Romer, A. S. 1967. Major steps in vertebrate evolution. *Science* 158:1629–1637.

Sales, G., and D. Pye. 1974. *Ultrasonic communication by animals.* London: Chapman and Hall.

Sales, G. D., and J. C. Smith. Comparative studies of the ultrasonic calls of infant Murid rodents. (In press.)

Scott, J. P., and V. J. De Ghett. 1972. Development of affect in dogs and rodents. *In Communication and affect: a comparative approach,* eds., T. Alloway, L. Krames, and P. Pliner, pp. 129–150. New York: Academic Press.

Sewell, G. D. 1969. Ultrasound in small mammals. (Unpublished Ph.D. dissertation. University of London.)

Sewell, G. D. 1970a. Ultrasonic signals from rodents. *Ultrasonics* 8:26–30.

Sewell, G. D. 1970b. Ultrasonic communication in rodents. *Nature* 277:410.

Smith, J. C. 1972. Sound production by infant *Peromyscus maniculatus* (Rodentia: Myomorpha). *J. Zool.* 168:369–379.

Smith, J. C. 1976. Responses of adult mice to models of infant calls. *J. Comp. Physiol. Psychol.* 90:1105–1115.

Smotherman, W. P., R. W. Bell, W. A. Hershberger, and G. D. Coover. Orientation to pup cues: effects of maternal experiential history. *Anim. Behav.* (In press.)

Smotherman, W. P., R. W. Bell, J. Starzec, J. Elias, and T. A. Zachman. 1974. Maternal responses to infant vocalizations and olfactory cues in rats and mice. *Behav. Biol.* 12:55–66.

St. John, R. D., and P. A. Corning. 1973. Maternal aggression in mice. *Behav. Biol.* 9:635–639.

Verhaeghe, A., and E. Noirot. Ultrasound by mouse pups from deaf and hearing strains. *Dev. Psychobiol.* (In Press.)

Watts, C. H. S. 1975. Vocalizations of Australian hopping mice (Rodentia: *Notomys*). *J. Zool.* 177:247–263.

Zippelius, H.-M., and W. M. Schleidt. 1956. Ultraschall-Laute bei jungen Mäusen. *Naturwissenschaften* 43:502.

Chapter Eighteen

Social Play: Structure, Function, and the Evolution of a Cooperative Social Behavior

MARC BEKOFF

University of Colorado
Department of Environmental,
Population and Organismic Biology
Boulder, Colorado 80309

Introduction

In recent years, there has been a focus of attention on the evolution of altruistic, selfish, and spiteful behavior as well as of parent-offspring conflict (Hamilton, 1964; Williams, 1966; Trivers, 1971, 1974; Alexander, 1974, 1975; Wilson, 1975; Dawkins, 1976; Barash, 1977; Evans, 1977; Hartung, 1977). Unfortunately, there has been less emphasis on cooperative behavior (Kropotkin, 1902; Crook, 1971; Wickler, 1976). Dawkins (1976) has recently written a lucid account of the evolution of selfish behavior, advancing strong and convincing arguments against group selection and favoring selection for genes instead. Along these lines, he (and others) discusses ideas about conflicts of interest with respect to individual reproductive fitness and social interaction. At one point, Dawkins writes: "It may be that animal communication contains an element of deception right from the start, because all animal interactions involve at least some conflict of interest." (p. 70) However, he later points out (p. 93) that the model of individuals as selfish machines may break down when interacting individuals are close relatives.

In this chapter I will discuss a few aspects of social play behavior that may have some bearing on a number of current theories concerned with the evolution and ontogeny of social behavior. Specifically, it will be suggested that social play may be a behavior in which there is cooperation among the participants, especially related individuals, with cheating being selected against. I shall use Crook's definition (1971, p. 238) of cooperation as the collaborative behavior of two or more individuals in the production of some common behavioral effect. The importance of social play in the behavioral development of the limited number of mammals in which it has been carefully studied may be inferred from the following common observations: (1) There has been selection for "play signals" in most mammals in which play has been observed (but see Wemmer and Fleming, 1974). (2) Animals will handicap themselves in order to play. (3) Young individuals typically play more than do older conspecifics. (4) Animals seem to enjoy playing (see Barash, 1977, p. 179, for a discussion of enjoyment as an evolutionary strategy) and will seek it out despite parental disciplining and the risks involved (Fagen, 1977). (5) Social play is characteristic of neotenous mammals that have a protracted period of infancy and caretaker dependency (Fagen, 1977; Gould, 1977). In some mammals, lactation and parental feeding may be important in the evolution of mutualistic social behavior (Pond, 1977).

This chapter will be divided into two sections. I shall first discuss current ideas about some of the functions of social play and then consider the ways in which individuals communicate "play intention." Since young individuals tend to engage in more social play than adults of the same species, the discussion below may have more relevance to young individuals, although in some cases the argument will be extended to include older, reproductively active individuals as well. Also, since available data indicate that early social interactions may disproportionately involve related individuals (littermates, siblings of various ages, parent(s) and young, "aunts" and "uncles" and related young), especially for mammals that play socially (the data are best for primates: van Lawick-Goodall, 1968; Fady, 1969; Loy and Loy, 1974; Konner, 1975; Breuggeman, 1976; Pitcairn, 1976; E. O. Smith, 1977), one might expect, as Dawkins points out, that there will be exceptions to all-inclusive statements about the involvement of cheating and deception in social interaction and communication. Alexander (1974) suggests that we might expect to find extremes of cooperation (and competition) in the interactions of siblings (see also Maynard Smith, 1976a).

Structure, Function, Relatedness of Participants, and Social Play

The most recent reviews of social play behavior (Bekoff, 1976a; Fagen, 1976, 1977, this volume; Breuggeman, 1976; Weisler and McCall, 1976;

Ficken, 1977; Symons, 1978 in press) strongly make the point that the benefits (immediate and delayed) of participation in social play still remain unknown for almost all species in which the behavior has been described. Only rarely are the risks (injury, attraction of predators) considered (but see Hornaday, 1922; van Lawick-Goodall, 1968; Barry and Roberts, 1972; Fagen, 1977; Byers, 1977; Symons, 1978). Three functions of social play (nonsocial play could be functionally analogous in these respects) include providing exercise (Brownlee, 1954; van Lawick-Goodall, 1968; Fagen, 1976; but see McDonald, 1977), reducing excess endogenous energy, or maintaining an "optimal" arousal level (frequently tautologically argued; see Bekoff, 1976a). With respect to the exercise hypothesis, one could argue that competition between players that could result in the differential representation of an individual's genes in future generations would be present *if* (1) individuals do benefit from the exercise that is provided by play *and* (2) individuals prevent one another from engaging in the activity. I know of no observations that unambiguously show that the second situation occurs, and indeed, quantitative supportive data for the first are sorely lacking (see Fagen and George, 1977 and McDonald, 1977). However, in various ungulates (Altman, 1963; Rudge, 1970; Lent, 1974; Stringham, 1974; McDougall, 1975; Estes, 1976), some nonhuman primates (Bingham, 1927; Hinde *et al.,* 1964; Goodall, 1968; Harlow, 1969; Fedigan, 1972; Jolly, 1972), and some human cultures (Ainsworth *et al.,* 1974; for exception see Sbrzesny, 1976), mothers actively interfere in social play between their own and other youngsters (in most nonhuman cases, degrees of relatedness are not known or are not provided). A number of questions arise: (1) Are the mothers attempting to conserve energy in their young during the period in which they are dependent on her for food and care (Trivers, 1974)? (2) Do they stop play in order to provide more care (Fagen, personal communication)? (3) Does the mother interfere similarly in play between individuals that are closely related to her or does she encourage play or allow it to occur both among her own offspring (littermates, different-aged siblings) and other individuals with whom she is related, although the strenuous motor activity may increase her own energy debt to her young? (4) Or, might she be protecting her young from engaging in interactions in which there is a chance of injury (e.g., play escalating into fighting)? Fedigan (1972) noted that some mothers of vervet monkeys, *Cercopithecus aethiops,* protect their offspring from individuals who frighten or injure the infant. Similar observations have been reported by Hinde *et al.,* (1964) and Owens (1975). Of course, in these cases the mother may still be acting selfishly, not necessarily to conserve energy, but rather to protect her investment. Furthermore, one must account for large individual differences among mothers (see Chapter 13) and instances in which mothers (or other adults) are unresponsive to play invitations (e.g., Poole and Fish, 1976).

The concept of protection by caretakers should not be taken to be a one-sided affair. Infants may also discriminate among adults. Frank (1952) observed that young voles, *Microtus agrestis* Pall, resisted retrieval by females other than their mother. Savage, Temerlin, and Lemmon (1973) reported that *if* an infant permitted, mother chimpanzees, *Pan troglodytes,* responded to their own and other infants in a similar fashion. Are the infants protecting themselves from selfish adults?

In social groups or in other instances in which older males are accessible for play (and other social activities), it is also necessary to determine the role(s) that they assume during play and other types of social interactions (Burton, 1972; Redican and Mitchell, 1974; Redican, 1976; Breuggeman, 1976; Barash, 1977; see also Lamb, 1976). Steiner (1971), while studying the play of Columbian ground squirrels, *Spermophilus columbianus columbianus,* noted that dominant males would disrupt the play of all juveniles by chasing them. However, if the dominant male(s) was from an area other than that of the juveniles, the disruption was much more intense (the male would "terrorize" them). Is the intensity of the disruption a function of relatedness, the young in other areas being unrelated? Is the male treating young from other areas differently simply because he has not previously established social bonds during prior interaction? Detailed data on male-infant interactions still are forthcoming and more rigorous assessments of geneologies are necessary.

The other postulated functions of social play are more difficult to deal with, but I suspect that careful study will shed some light on how they may be related to individual fitness. For example, if dominance relationships are formed during social play *and* the relationships that are established have some bearing on later priority(ies) of access to resources that could increase one's fitness, then it would behoove an individual to engage selfishly in play and also to make certain that it emerges "dominant." If the individual cannot emerge dominant, then it would be in its best interests either to stop playing or to injure its opponent. However, if the opponent is a close relative (and in many mammals this would be the case, especially during early ontogeny), it might be in the best interests of an individual to "accept" a subordinate role and not to injure its sib (or close cousin). Injuring a relative so that it is unable to breed will result in a reduction of the subordinate's inclusive fitness as well.[1]

The relationship of social play to the ontogeny of adaptive behavior remains an unsettled issue (Bekoff, 1976a, b). If social skills and the socialization process in general are benefited by engagement in social play, then once again it would be in the best interests of a selfish individual to play and also to stop others from playing or to try to injure its playmate(s) without being injured in return. In this case, assuming that participation in play *is* beneficial, an individual should distribute its play among a variety of other individuals so that it gets "enough" and the others get "less." That is, the latter individuals either gain fewer benefits, none at all, or wind up "in the red." However, the situation probably is not this simple. If all individuals adopt the strategy of distributing play to affect negatively the development of other individuals, then the result will probably be that everyone plays the same amount (Symons, personal communication). In this case, it may be the quality and not the quantity of play that is of importance. Furthermore, if dominance relations are formed during play and the "acquisition" of various social skills is facilitated through play, then an individual must somehow weigh the relative "importance" (with respect to its fitness) of these functions and "select" the appropriate strategy. If an individual stops playing because it is unable to emerge dominant, then there must be alternate routes for socialization that can be exploited differentially. And, if he or she continues to play and acquires certain skills due to this experience, then these benefits must outweigh the debits that may accrue due to its being subordinate. In general, the best way to terminate a play bout, especially one involving unrelated individuals, would be to injure the other individual(s) or at least decrease the possibility that they will engage in any subsequent play with any other individuals. It would also be a good strategy to "use" an unrelated individual for play, gain whatever advantages possible, and then have a relative take over where the first individual left off. Here, all individuals might benefit, though the related animals would benefit more (in terms of inclusive fitness), especially if the unrelated individual's fitness is somehow reduced during the last set of interactions.

Along these lines, an individual may also engage in play to teach another individual "bad habits" (R. Dawkins, personal communication). In this way, it may learn to manipulate (see also Breuggeman, 1976) another individual but still gain whatever other benefits play provides.

However interesting the above speculations, there are virtually no supportive data available either for interactions involving unrelated or related individuals. Fagen (1977) and others have found no evidence that motor patterns that inflict serious injury are incorporated into play (though play may occasionally grade into aggression). In rats, there is no evidence that play has any ill or deterrent effects (Barnett, 1969). In pigs, retreat and defeat are seldom seen in play (McBride, 1964). Symons (1974) noted the absence of true aggression in the play of rhesus monkeys (*Macaca mulatta*) and suggests that play and aggression are mutually exclusive (Symons, 1978, in press).

In addition to the fact that play rarely results in serious injury of one participant(s) due to the behavior of the other interactant(s), there also is evidence that individuals will decrease the asymmetry of a play bout (make it a negotiated game?) by handicapping themselves (Altmann, 1962; Bertrand, 1969; Wemmer and Fleming, 1974; Owen, 1975; Le-Resche, 1976; Plooij, in press). The nature of the play bout may also change as the young age (Hinde et al., 1964; Breuggeman, 1976). Poole (1966) observed that in polecats (Mustela spp.), play is adapted to the relative strength of the interacting animals. Both pain and fear-producing situations are avoided.

Dominant individuals may also allow themselves to be dominated during play (reversals of dominance) or may behave like subordinate individuals. For example, Loizos (1969) observed that dominant chimpanzees have to behave like subordinates during chase-flee play. That is, the dominant individual(s) has to be the one who is chased. Furthermore, adults may tolerate brief "play attacks" or retaliate by inviting the infant(s) to play. However, the response of the adult may be determined by whether or not the infant is related to it. McDougall (1975) noted that if a nanny goat (Capra hircus L.) was disturbed by the play of another nanny's kid, she would lower her neck in threat at the offending kid and the interaction would end.

SOCIAL PLAY AND KIN RECOGNITION

If engagement in social play early in life is mutually beneficial to the participants, then playmates, young and old, should discriminate between kin and non-kin when given the choice. In some social carnivores and nonhuman and human primates, for example, there is a high probability that many group members are genetically related (Mech, 1970; Bertram, 1975, 1976; Bulger, 1975; Peters and Mech, 1975; Barash, 1977). Furthermore, many of these (and other) social groups are closed to outsiders (King, 1954; Bernstein, 1964; Schaller, 1972; Hamilton, 1975; Marler, 1976; Zimen, 1976) and there is little emigration from the family group early in life. Also, as mentioned above, social play between unrelated (or at best distantly related) infants frequently is disrupted by a caretaker. The closed nature of many social groups may, in the long run, favor "indiscriminate" within group play that varies in quality and quantity with the age and sex of the participants, since an individual's inclusive fitness could be increased by playing with just about any other group member. And, it is entirely plausible that kin recognition itself may be fostered by the formation of strong social bonds through repeated exposure (see Zajonc, 1971) to related individuals during early life. In fact, I would like to suggest that one delayed benefit (à la Fagen, 1977) of social play for animals that engage in the activity as youngsters may be an increased ability of an individual to discriminate (later) closely related kin from more distantly or unrelated individuals via the establishment (and maintenance) of social bonds (see Zajonc, 1971, and Wickler, 1976 for general discussions of social attraction and attachment and also Merrick, 1977). It also remains possible that genetic relatedness is not the major factor, but rather familiarity is a key determinant of choice, and in certain types of social groups, the most familiar and accessible individuals happen to be the most closely related (see also Massey, 1977). For species in which social preferences can be modified by early experience or for species in which the young are born in a precocial state and are highly mobile, closed groups may be necessary to insure that bonding is to kin, i.e., "familiar" is equivalent to closely related kin and "strange" to distantly related or unrelated individuals. The ability to discriminate strange and familiar individuals (e.g., offspring, caretakers, siblings) very soon after birth would also favor kin-bonding.

In summary, it *appears* that social play, especially between related individuals, is predominantly a cooperative social behavior (see also Kropotkin, 1902) that is of mutual benefit to the participants. With respect to self-handicapping and dominance reversals, there are no data that indicate that either strategy has evolved as a ploy to get another individual to play so that it can be dominated and/or injured. Also, reversals of dominance infrequently lead to injurious fighting. Theoretically, at least, it is possible for cooperation among individuals to be more rewarding than individual selfish behavior (Hamilton, 1975, p. 152). More detailed, systematic analyses are needed in order to generate more conclusive answers. For example, does play between unrelated young differ functionally and/or structurally from social play between littermates, different aged siblings, or distantly related individuals? Do siblings have higher thresholds for aggression (Symons, personal communication) possibly due to early bonding? Is agonistic behavior "practiced" during play between unrelated young (Konner, 1975)? Is kin recognition facilitated through early play experience? Unrelated individuals do play with one another, the amount depending on the social structure of the group or on litter size (Konner, 1975; E. O. Smith, 1977). However, I could not find any detailed analyses in which play between related individuals was compared to play between unrelated individuals. Also, we must not overlook the "contagious" nature of play. While it remains possible that when an individual joins a play group it is doing so to assert itself in one way or another or to terminate the interaction, it also could be the case that the individual(s) who is doing the "luring" will gain an advantage over the newcomer!

The most parsimonious and perhaps correct assumption that follows from a review of available literature is that within some species, individuals do gain immediate and delayed benefits (as yet, not clearly defined) from engaging in social play and will take advantage of every opportunity to partake in the activity (even with unrelated individuals). Is there any evidence for this statement? I believe that there is. First, as mentioned above, the young of many species in which play has been observed are very persistent in their efforts to play, although the strenuous motor activity may involve risks in negotiating the environment or may attract predators (Hornaday, 1922; Fagen, 1977; Byers, 1977). Second, caretakers frequently encourage play although this may make their young more dependent upon them for care, especially during early life. (Parent-offspring conflict may develop if the demands on parental resources become excessive [Trivers, 1974].) It would be interesting to know whether or not mothers encourage or allow more play if they receive aid in providing for their offspring. Third, play is an important component of the courtship of many mammals and may help to strengthen the pair bond even between unrelated individuals. However, some of the most compelling evidence stems from the common observation that specific "play signals" appear to have evolved independently among many different groups of animals, especially mammals.

Play Signals and the Exploitation of Individuals

Signals that appear to have the function of indicating an intention to engage in social play have been described in a wide variety of mammals (Altmann, 1967; van Lawick-Goodall, 1968; Jolly, 1972; Bekoff, 1972, 1974, 1975, 1976a, 1977b; Sade, 1973; Wilson, 1973, 1974; Wilson and Kleiman, 1974; Symons, 1978) and even in bird species (McBride et al., 1969; Keller, 1975). Sade (1973) points out that in many cases, these signals are clearly antithetical to movements associated with other moods, such as aggression. I now advocate dropping the term "metacommunication" as a synonym for the communication of play intention. First of all, it is one (of many) example of useless jargon that may be used to hide our ignorance concerning the communication process (see also Hailman, 1977 and W. J. Smith, 1977). In addition, as pointed out by S. Wilson (personal communication), the term "meta" implies "once removed." In fact, play signals are very much a part of play interactions (Wilson, 1973; Wilson and Kleiman, 1974; Bekoff, 1975, 1976a, 1977b). That these signals *actually* function to announce "play intention" has been accepted more on faith than on hard data (Bekoff, 1975, 1976a). However, if we assume for the moment that play signals do exist and serve to "state" that "what follows is play," then they may have been selected in one of two (in some cases, both?) ways. First, play signals may have evolved as signals of deception to lure selfishly other individuals into engaging in play. Second, they could have evolved simply to help to establish a "play mood" and to "insure" that the ensuing or on-going interaction will go smoothly and remain what it originally was intended to be, namely, play.

Ideas about the exploitation of individuals through communication processes have recently been considered by Wallace (1973), Otte (1974, 1975), Wilson (1975), Griffin (1976), and Rohwer (1977). Wallace and Otte discuss whether "big" or "small" lies would be advantageous (they do not agree) and how experience enters into the plausibility of misinformation. With respect to play signals, it seems that there are a number of facts available that could be used to support an argument against there being signals of deceit. First, the same play signals are used by conspecific (and also heterospecific) individuals of different ages. If experience is important in learning what a particular signal "means" (and what it does not mean), then one would expect the plausibility of a deceitful message to decrease with experience (and age). Individuals should learn *not* to respond to play invitations that use signals that previously had been associated with misinformation.

One must also consider whether or not there are two sets of very similar signals. If there are, then, according to Otte (1974, 1975), smaller lies or differences should be maintained in the repertoire. Available data do not support the idea that there are two sets of play signals, one for exploiting other individuals and one for signaling serious play intention. Rather, it appears that play signals are not used to transmit misinformation. Play signals are conspicuous, in some cases used solely in the context of play (Bekoff, 1972, 1974; Sade, 1973), and are used repeatedly without any apparent "loss" or change in function, although individuals often do not respond to the invitations of certain other individuals (Bekoff, 1977a). There are no data to suggest that the individuals who are unsuccessful in soliciting play are unsuccessful because they are, or have been, deceivers, and have used play to achieve their own selfish ends.

To summarize, it appears that selection has operated to make play signals an aid to fostering some sort of "agreement" between interacting individuals, to signal that "play is the name of the game." Indeed, this understanding is used for intraspecific as well as interspecific playful interactions. It may well be that in some cases there has been selection for signals that communicate lies of different magnitudes (i.e., the fitness of the recipient is diminished by different amounts; Otte, 1974). Yet, this does not seem to apply to play signals. In fact, play signals may actually function to decrease the likelihood that play will grade into an agonistic encounter (Bekoff, 1975), may be used to reinforce ongoing play, and are also used during self-handicapping.

Conclusion and Summary: Some Unanswered Questions

Detailed sociometric investigations of social play behavior from a comparative-evolutionary perspective should shed some light on a number of issues that currently are in vogue in behavioral biology.[2] For example, do individuals selfishly invite other individuals to engage in social play? Can play signals serve to communicate misinformation? Does play between related individuals differ (and how) from play between unrelated individuals? Or, is play something "special"? Is it, perhaps, a negotiated game (Rapoport, 1970) in which the players, related or unrelated, attempt to achieve cooperatively an outcome that is mutually preferred? How do individuals "agree" both to begin and terminate playing? Under what conditions does play lead to fighting? Does social play, especially among related individuals, represent an evolutionarily stable strategy (Chapter 19; Maynard Smith, 1974, 1976b) from which deviation will be penalized (Dawkins, 1976)? Are some players "better" than others because they use different strategies but, nonetheless, play by the same rules (Slobodkin and Rapoport, 1974)? Do all participants in play benefit, but some more than others (Symons, personal communication)? Just as there are cultural differences in the social play of humans (e.g., Blurton-Jones and Konner, 1973), inter- and intraspecific differences in the behavioral development of nonhuman animals are not unexpected, and, in fact, have been amply demonstrated. Indeed, variability in strategies of social play within and between species that are related to social grouping (age/sex ratio, number of kin available, etc.), state of development at birth (e.g., precocial, altricial), sex, and habitat, will provide critically important information.

Since this discussion views social play from a relatively novel perspective, convincing answers to all of the questions raised herein are still forthcoming. Fortunately, many of the questions can be answered by careful observation and experimentation. Although ontogenetic questions are most frequently assumed to be primary, many questions of behavioral development are, in fact, ultimate ones (Alexander, 1975). What is needed is a shift in orientation away from non-evolutionary emphases to the recognition of the fact that detailed analyses of social play behavior and behavioral development, in general, can have some bearing on current theories of the evolution of animal social behavior.

Acknowledgments

I thank Benjamin Beck, Joel Berger, John Byers, Debra Christein, Richard Dawkins, Victor De Ghett, Robert Fagen, Euclid O. Smith, and Donald Symons for their helpful comments on an earlier draft of this chapter and stimulating discussions. Nutan Pall kindly typed the manuscript. Of course, all shortcomings and oversights remain my responsibility.

Notes

1. In coyotes, *Canis latrans,* in which rank-related agonistic behavior typically occurs before the appearance of social play (Bekoff, 1974, 1977a), siblings do injure one another during fights. Injuries are most common during the first fights that occur (i.e., the early fights are unritualized). I have termed these early encounters "presocialization" fights (Bekoff, 1977c) because it appears that the infant coyotes have not acquired the necessary communicatory skills to "understand" the "meaning" of certain signals. That is, young coyotes frequently ignore threats directed toward them and do not "honor" submission. Fights observed in older coyotes are more ritualized and less intense. However, during social play by infant coyotes, injuries are extremely rare and rank relations frequently are abolished (see also Symons, 1978).

2. It should be stressed that we must not allow our anthropocentric feelings to enter into our interpretations of what our nonhuman counterparts are doing (Bekoff, 1977d). Although some nonhumans that engage in social play certainly have the ability to deceive others and to construct realities and consequences (Griffin, 1976; Mason, 1976), it does not necessarily follow that they always, if ever, use these skills in selfish or devious ways. As Mead (1971) has noted: It seems that we can pick and choose from the entire living world to find the creature whose characteristics reflect the moral wanted at the moment. In line with the present chapter, the insulting (though oftentimes unfortunately true) implications of "playing games" among humans should not be thrust on our nonhuman relatives in any haphazard way. Nor, should fashionable "socio-tautological" ideas about deceit, lying, spite, and selfishness permeate fully and be the major directive of comparative behavioral research. (At the 1977 meetings of the Animal Behavior Society at which the present symposium was held, I overheard a (frightening) conversation between two people: No. 1: "Should we go into the paper on . . . ?" No. 2: "Nah, it really doesn't have anything to do with sociobiology or kin selection.")

Although it may well be possible that some of the current thoughts about the "motivation" of behavior that stress the importance of selfishness and the like will be borne out by careful analyses and have some general applicability, the very fact that this information readily lends itself to rampant misuse by "prophets of doom" (Tinbergen, 1972) must temper our enthusiasm of accepting new, all-encompassing theories of the evolution of social behavior that may result in the premature and reckless dissemination and application of this information before solid facts are in. Dawkins' (1976) last chapter is especially important in discussing how humans (and nonhumans as well?) ". . . can rebel against the tyranny of the selfish replicator" (p. 215). In the end, the observed facts, which currently are sorely needed, must tell the story.

References

Ainsworth, M. D. S., S. M. Bell, and D. J. Stayton. 1974. Infant–mother attachment and social development: "socialization" as a product of reciprocal responsiveness to signals. Pages 99–135 *In The integration of a child into a social world,* ed., M. P. M. Richards, New York and Cambridge: Cambridge Univ. Press.

Alexander, R. D. 1974. The evolution of social behavior. *Annu. Rev. Ecol. Syst.* 5:325–383.

Alexander, R. D. 1975. The search for a general theory of behavior. *Behav. Sci.* 20:77–100.

Altmann, M. 1963. Naturalistic studies of maternal care in moose and elk. *In Maternal behavior in mammals,* ed., H. Rheingold, pp. 233–253. New York: Wiley.

Altmann, S. A. 1962. A field study of the sociobiology of rhesus monkeys, *Macaca mulatta. Ann. N.Y. Acad. Sci.* 102:338–435.

Altmann, S. A. 1967. The structure of primate social communication. *In Communication among primates,* ed., S. A. Altmann, pp. 326–362. Chicago: Univ. of Chicago Press.

Barash, D. P. 1977. *Sociobiology and behavior.* New York: American Elsevier.

Barnett, S. A. 1969. Grouping and dispersive behaviour among wild rats. *In Aggressive behaviour,* eds., S. Garattini and E. B. Sigg, pp. 3–14. Amsterdam: Excerpta Medica.

Barry, H., III, and J. M. Roberts. 1972. Infant socialization and games of chance. *Ethnology* 11:296–308.

Bekoff, M. 1972. The development of soical interaction, play and metacommunication in mammals: an ethological perspective. *Q. Rev. Biol.* 47:412–434.

Bekoff, M. 1974. Social play and play-soliciting by infant canids. *Am. Zool.* 14:323–341.

Bekoff, M. 1975. The communication of play intention: are play signals functional? *Semiotica* 15:231–239.

Bekoff, M. 1976a. Animal play: problems and perspectives. *In Perspectives in ethology,* Vol. 2, eds., P. P. G. Bateson and P. H. Klopfer, pp. 165–188. New York: Plenum.

Bekoff, M. 1976b. The social deprivation paradigm: who's being deprived of what? *Dev. Psychobiol.* 9:499–500.

Bekoff, M. 1977a. Mammalian dispersal and the ontogeny of individual behavioral phenotypes. *Am. Naturalist* 111:715–732.

Bekoff, M. 1977b. Social communication in canids: evidence for the evolution of a stereotyped mammalian display. *Science* 197:1097–1099.

Bekoff, M. 1977c. Socialization in mammals with an emphasis on nonprimates. *In Biosocial development among primates,* eds., S. Chevalier-Skolnikoff and F. E. Poirier, pp. 603–636. New York: Garland Publishing.

Bekoff, M. 1977d. "Man" and "animal": a sociobiological dichotomy? *The Biologist* 59:1–10.

Bernstein, I. S. 1964. The integration of rhesus monkeys introduced to a group. *Folia Primat.* 2:50–63.

Bertram, B. C. R. 1975. Social factors influencing reproduction in wild lions. *J. Zool.* 177:463–482.

Bertram, B. C. R. 1976. Kin selection in lions and in evolution. *In Growing points in ethology,* eds., P. P. G. Bateson and R. A. Hinde, pp. 281–301. New York and Oxford: Cambridge Univ. Press.

Bingham, H. C. 1927. Parental play of chimpanzees. *J. Mammal.* 8:77–89.

Blurton-Jones, N. G., and M. J. Konner. 1973. Sex differences in behaviour of London and Bushman children. *In Comparative ecology and behaviour of primates,* eds., R. P. Michael and J. H. Crook, pp. 689–750. New York: Academic Press.

Breuggeman, J. A. 1976. Adult play behavior and its occurrence among free-ranging rhesus monkeys (*Macaca mulatta*). Ph.D. dissertation, Northwestern Univ., Evanston, Ill.

Brownlee, A. 1954. Play in domestic cattle: an analysis of its nature. *Br. Vet. J.* 110:48–68.

Bulger, A. J. 1975. The evolution of altruistic behavior in social carnivores. *Biologist* 57:41–51.

Burton, F. 1972. The integration of biology and behavior in the socialization of *Macaca sylvana* of Gibraltar. *In Primate socialization,* ed., F. E. Poirier, pp. 29–62. New York: Random House.

Byers, J. A. 1977. Terrain preferences in the play of Siberian ibex kids (*Capra ibex siberica*). *Z. Tierpsychol.* 45:199–209.

Crook, J. A. 1971. Sources of cooperation in animals and man. *In Man and beast: comparative social behavior,* eds., J. F. Eisenberg, W. S. Dillon, and S. D. Ripley, pp. 235–260. Washington: Smithsonian Inst. Press.

Dawkins, R. 1976. *The selfish gene.* Oxford and New York: Oxford Univ. Press.

Estes, R. D. 1976. The significance of breeding synchrony in the wildebeest. *E. Afr. Wildl. J.* 14:135–152.

Evans, H. E. 1977. Extrinsic versus intrinsic factors in the evolution of insect sociality. *Bio Science* 27:613–617.

Fady, J. C. 1969. Les jeux sociant; le compagnon de leux chez les jeunes. Observation chez *Macaca irus. Folia Primat.* 11:134–143.

Fagen, R. 1976. Exercise, play and physical training in animals. *In Perspectives in ethology,* Vol. 2, eds., P. P. G. Bateson and P. H. Klopfer, pp. 189–219. New York: Plenum.

Fagen, R. 1977. Selection for optimal age-dependent schedules of play behavior. *Am. Naturalist* 111:395–414.

Fagen, R., and T. K. George. 1977. Play behavior and exercise in young ponies (*Equus caballus* L.). *Behav. Ecol. Sociobiol.* 2:267–269.

Fedigan, L. 1972. Social and solitary play in a colony of vervet monkeys (*Cercopithecus aethiops*). *Primates* 13:347–364.

Ficken, M. S. 1977. Avian play. *Auk* 94:573–582.

Frank, F. 1952. Adoptionversuche bei Feldmäusen (*Microtus arvalis* Pall). *Z. Tierpsychol.* 9:415–423.

Gould, S. J. 1977. *Ontogeny and phylogeny.* Cambridge, Mass.: Harvard Univ. Press.

Griffin, D. R. 1976. *The question of animal awareness.* New York: Rockefeller Univ. Press.

Hailman, J. P. 1977. *Optical signals.* Bloomington: Indiana Univ. Press.

Hamilton, W. D. 1964. The genetical theory of social behavior. I and II. *J. Theoret. Biol.* 7:1–52.

Hamilton, W. D. 1975. Innate social aptitudes of man: an approach from evolutionary genetics. *In Biosocial anthropology,* ed., R. Fox, pp. 133–155. New York: Halsted Press.

Harlow, H. 1969. Age-mate or peer affectional system. *Adv. Study Behav.* 2:333–383.

Hartung, J. 1977. An implication about human mating systems. *J. Theoret. Biol.* 66:737–745.

Hinde, R. A., T. E. Rowell, and Y. Spencer-Booth. 1964. Behaviour of socially living rhesus monkeys in their first six months. *Proc. Zool. Soc. London* 143:609–649.

Hornaday, W. T. 1922. *The minds and manners of wild animals.* New York: Charles Scribner's Sons.

Jolly, A. 1972. *The evolution of primate behavior.* New York: Macmillan.

Keller, R. 1975. Das Spielverhalten der Keas (*Nestor notabilis* Gould) des Zürcher Zoos. *Z. Tierpsychol.* 38:393–408.

King, J. A. 1954. Closed social groups among domestic dogs. *Proc. Am. Philos. Soc.* 198:327–336.

Konner, M. 1975. Relations among infants and juveniles in comparative perspective. *In Friendship and peer relations,* eds., M. Lewis and L. A. Rosenblum, pp. 99–112. New York: Wiley.

Kropotkin, P. 1902. *Mutual aid: a factor in evolution.* London: Heinemann.

Lamb, M. E., ed. 1976. *The role of the father in child development.* New York: Wiley.

van Lawick-Goodall, J. 1968. The behaviour of free-living chimpanzees in the Gombe Stream Reserve. *Anim. Behav. Monogr.* 1(3):161–311.

Lent, P. C. 1974. Mother–infant relationships in ungulates. *In The behaviour of ungulates and its relation to management,* Vol. 1, eds., V. Geist and F. Walther, pp. 14–55. Morges, Switzerland: Internatl. Union for the Conserv. of Nature.

LeResche, L. A. 1976. Dyadic play in hamadryas baboons. *Behaviour* 57:190–205.

Loizos, C. 1969. An ethological study of chimpanzee play. *Proc. 2nd Internatl. Congr. Primat.* 1:87–93.

Loy, J., and K. Loy. 1974. Behavior of an all juvenile group of rhesus monkeys. *Am. J. Phys. Anthropol.* 40:83–97.

Marler, P. 1976. On animal aggression: the roles of strangeness and familiarity. *Am. Psychol.* 31:239–246.

Mason, W. A. 1976. Windows on other minds. *Science* 194:930–931. (Review of Griffin, 1976.)

Massey, A. 1977. Agonistic aids and kinship in a group of pigtail macaques. *Behav. Ecol. Sociobiol.* 2:31–40.

Maynard Smith, J. 1974. The theory of games and the evolution of animal conflict. *J. Theoret. Biol.* 47:209–221.

Maynard Smith, J. 1976a. A short-term advantage for sex and recombination through sib-competition. *J. Theoret. Biol.* 63:245–258.

Maynard Smith, J. 1976b. Evolution and the theory of games. *Am. Scientist* 64:41–45.

McBride, G. 1964. A general theory of social organization and behaviour. *Queensl. Univ. Fac. Veter. Sci. Papers* 1:75–110.

McBride, G., I. P. Parer, and F. Foenander. 1969. The social organization and behaviour of the feral domestic fowl. *Anim. Behav. Monogr.* 2(3):127–181.

McDonald, D. L. 1977. Play and exercise in the California ground squirrel (*Spermophilus beecheyi*). *Anim. Behav.* 25:782–784.

McDougall, P. 1975. The feral goats of Kielderhead Moor. *J. Zool.* 176:215–246.

Mead, M. 1971. Innate behavior and building new cultures: a commentary. *In Man and beast: comparative social behavior,* eds., J. F. Eisenberg, W. S. Dillon, and S. D. Ripley, pp. 369–381. Washington: Smithsonian Inst. Press.

Mech, L. D. 1970. *The wolf.* New York: Natural History Press.

Merrick, N. J. 1977. Social grooming and play behavior of a captive group of chimpanzees. *Primates* 18:215–224.

Otte, D. 1974. Effects and functions in the evolution of signaling systems. *Ann. Rev. Ecol. Syst.* 5:285–417.

Otte, D. 1975. On the role of intraspecific deception. *Am. Naturalist* 109:239–242.

Owens, N. W. 1975. Social play behaviour in free-living baboons, *Papio anubis. Anim. Behav.* 23:387–408.

Peters, R. P., and L. D. Mech. 1975. Scent-marking in wolves. *Amer. Sci.* 63:628–637.

Pitcairn, T. K. 1976. Attention and social structure in *Macaca fasicularis. In The social structure of attention,* eds., M. R. A. Chance and R. R. Larsen, pp. 51–81. New York: Wiley.

Plooij, F. X. How wild chimpanzee babies trigger the onset of mother-infant play and what the mother makes of it. *In Before speech: the beginnings of human communication.* M. Bullowa, ed. Cambridge: Cambridge Univ. Press. (In press.)

Pond, C. M. 1977. The significance of lactation in the evolution of mammals. *Evolution* 31:177–199.

Poole, T. B. 1966. Aggressive play in polecats. *Symp. Zool. Soc. London* 18:23–44.

Poole, T. B., and Fish, J. 1976. An investigation of individual, age and sexual differences in the play of *Rattus norvegicus* (Mammalia: Rodentia). *J. Zool.* 179:249–260.

Rapoport, A. 1970. *Two-person game theory: the essential ideas.* Ann Arbor: Univ. of Michigan Press.

Redican, W. K. 1976. Adult male–infant interactions in nonhuman primates. *In The role of the father in child development,* ed., M. E. Lamb, pp. 345–385. New York: Wiley.

Redican, W. K., and G. Mitchell. 1974. Play between adult male and infant rhesus monkeys. *Am. Zool.* 14:295–302.

Rohwer, S. 1977. Status signaling in harris sparrows: some experiments in deception. *Behaviour* 61:107–129.

Rudge, M. R. 1970. Mother and kid behaviour in feral goats. *Z. Tierpsychol.* 27:687–692.

Sade, D. S. 1973. An ethogram for rhesus monkeys. I. Antithetical contrasts in posture and movement. *Am. J. Phys. Anthropol.* 38:537–542.

Savage, E. S., J. W. Temerlin, and W. B. Lemmon. 1973. Group formation among captive mother–infant chimpanzees (*Pan troglodytes*). *Folia Primat.* 20:453–473.

Sbrzesny, H. 1976. *Die Spiele der !Ko-Buschleute.* Munich: Piper.

Schaller, G. B. 1972. *The Serengeti lion.* Chicago: Univ. of Chicago Press.

Slobodkin, A. B., and A. Rapoport. 1974. An optimal strategy of evolution. *Q. Rev. Biol.* 49:181–200.

Smith, E. O. 1977. Social play in rhesus macaques, *Macaca mulatta.* Ph.D. dissertation, Columbus: Ohio State Univ.

Smith, W. J. 1977. *The behavior of communicating.* Cambridge, Mass.: Harvard Univ. Press.

Steiner, A. L. 1971. Play activity of Columbian ground squirrels. *Z. Tierpsychol.* 28:247–261.

Stringham, S. F. 1974. Mother–infant relations in moose. *Nat. Can. Ottawa* 101:325–369.

Symons, D. 1974. Aggressive play and communication in rhesus monkeys (*Macaca mulatta*). *Am. Zool.* 14:317–322.

Symons, D. 1978. *Play and aggression.* New York: Columbia Univ. Press.

Symons, D. The question of function: dominance and play. *In Primate play,* ed., E. O. Smith, (In press.) New York: Academic Press.

Tinbergen, N. 1972. Functional ethology and the human sciences. *Proc. Roy. Soc. London B* 182:385–410.

Trivers, R. 1971. The evolution of reciprocal altruism. *Q. Rev. Biol.* 46:35–57.

Trivers, R. 1974. Parent-offspring conflict. *Am. Zool.* 14:249–264.

Wallace, B. 1973. Misinformation, fitness, and selection. *Am. Naturalist* 107:1–7.

Weisler, A., and R. B. McCall. 1976. Exploration and play: resumé and redirection. *Am. Psychol.* 31:492–508.

Wemmer, C., and M. J. Fleming. 1974. Ontogeny of playful contact in a social mongoose, the meerkat, *Suricata suricatta. Am. Zool.* 14:415–426.

Wickler, W. 1976. The ethological analysis of attachment: sociometric, motivational, and sociophysiological aspects. *Z. Tierpsychol.* 42:12–28.

Williams, G. C. 1966. *Adaptation and natural selection.* Princeton, N.J.: Princeton Univ. Press.

Wilson, E. O. 1975. *Sociobiology: the new synthesis.* Cambridge, Mass.: Harvard Univ. Press.

Wilson, S. 1973. The development of social behaviour in the vole (*Microtus agrestis*). *Zool. J. Linn. Soc.* 52:45–62.

Wilson, S. 1974. Juvenile play of the common seal *Phoca vitulina vitulina* with comparative notes on the grey seal *Halichoerus grypus. Behaviour* 48:37–60.

Wilson, S., and D. G. Kleiman. 1974. Eliciting play: a comparative study. *Am. Zool.* 14:341–370.

Zajonc, R. B. 1971. Attraction, affiliation, and attachment. *In Man and beast: comparative social behavior,* eds., J. F. Eisenberg, W. S. Dillon, and S. D. Ripley, pp. 141–179. Washington: Smithsonian Inst. Press.

Zimen, E. 1976. On the regulation of pack size in wolves. *Z. Tierpsychol.* 40:300–341.

Chapter Nineteen

Evolutionary Biological Models of Animal Play Behavior

ROBERT M. FAGEN

University of Illinois at Urbana-Champaign
Department of Ecology, Ethology and Evolution
Urbana, Illinois 61801 USA

I. Introduction: Two Ethological Cultures?

Vladimir Nabokov (1900–77), writer and biologist, whose novel *The Gift* (New York: Capricorn Books, 1970) includes a limpid distillation of the joys of the field naturalist, was once asked to comment on C. P. Snow's view that a gulf existed between the literary and scientific communities. Nabokov is reported to have answered, "I would have compared myself to a Colossus of Rhodes bestriding the gulf between the thermodynamics of Snow and Laurentomania of Leavis, had that gulf not been a mere dimple of a ditch that a small frog could straddle" (Appel, 1967, p. 33). While territorial disputes, including those between scientists themselves, are not new to students of cultural history, the intellectual validity of any existing or imminent gulf between theory and data in ethology seems particularly questionable, as will be argued in the sequel. Of course, intellectual dimples may form when political gulfs yawn, and empiricists and theoreticians alike have been observed advancing to the boundaries of their respective areas of competence, inflating themselves, and vocalizing. These stereotyped vocalizations may include "Stamp collector," "Armchair speculation," "What does that mean," "Where are your data," "The theoretical relevance of this study is not clear," or "You would be well advised to put in more data and cut short the theory." The exact identification of Snow's two cultures with the two ethological cultures might likewise be contested, the disputants each claiming the exclusive right to be called scientists. Can small frogs cooperatively outcompete large toads?

The text (parts II and III) of this chapter discusses and illustrates ways in which the two cultures of developmental ethology can practice mutually beneficial cultural exchange. In part II, "Interactions Between Data and Theory," I argue that the theory-data dichotomy is as intellectually meaningless as the nature-nurture dichotomy, and in part III, "Models and Metaphors in Developmental Ethology," I present two theoretical analyses of the evolution and function of animal play behavior. As these theoretical constructions become more explicitly mathematical, they become easier to test and to interpret, suggesting that the value of mathematical formalism in ethology is precisely its ability to make vague theory operational.

II. Interactions Between Data and Theory

A. ON THE NECESSITY AND INSUFFICIENCY OF DATA

Believing as Seeing

Data on behavioral ontogeny, especially vertebrate ontogeny, have long been available, and yet when data of a particular sort are actually needed in order to test hypotheses on ontogeny it is seldom the case that appropriate information can be found in the literature. Extant theory, however implicit, and the practical constraints of any given study both tend to limit the kinds of data that are actually gathered to a degree that is not fully appreciated until one has suffered through the frustration of a literature search.

Why are useful accounts of particular phenomena hard to find in the voluminous descriptive literature of ethology? Where is Description now that we need her?

If we assume that theoretical constructs, whether explicit or unspoken, always determine the nature of the data gathered, it becomes clear that existing data are insufficient because the theory that informs current empirical concerns differs from that motivating past observational studies. Either ontogeny was not studied at all, or it was observed copiously but from a different theoretical perspective that resulted in failure to collect data relevant to current concerns. Believing is seeing.

Case History 1: Behavioral Variability

A major concern of this symposium is behavioral variability. This concept is now important in ethology, apparently because a number of theoretical streams have coalesced: first, a taxonomic concern for variation (Barlow, 1977; Bekoff, 1977a; Schleidt, 1974) parallelling an earlier and other revolution against typology (see Mayr, 1963); then the idea that behavioral variability in play could lead to new adaptive behavior (Jolly, 1966; Fagen, 1974) led to attempts to quantify variability (Bekoff, 1975) and subsequently to the demonstration, in three species of carnivores, that play was in some sense more variable than comparable nonplay behavior (domestic cats, Fagen and Goldman, 1977; coyotes, Bekoff, 1977a and Table 1; ferrets, Table 1).

Variability in ethology is theoretically important at many levels of organization and across many different kinds of sampling units. Variability of acts (of their form, duration or orientation) is of interest in ethological analyses of stereotypy (Schleidt, 1974) or modality (Barlow, 1977) and in structuralist analyses of play (Henry and Herrero, 1974; Bekoff, 1977b; Hill and Bekoff, 1977). Such comparisons may be performed ontogenetically (Chapter 18; Bekoff *et al.,* ms.), across extant species (Bekoff, 1977b), or between extant and extinct species (Chapter 1). Sequential variability is important both to structuralists and to functionalists who analyze play, for different reasons (Fagen, 1974). Behavior may vary between individuals, between dyads, between litters or families, and between larger social groups (in which case questions about origin

TABLE 1. Relative variability of play and agonistic fighting in two species of carnivores

	$H_{B/A} \div \log_2$ CATALOG SIZE		NORMALIZED VARIABILITY $(H_{B/A}/H_B)$	
	PLAY	FIGHTING	PLAY	FIGHTING
Ferrets (*Mustela furo*)	0.62 ± 0.007	0.48 ± 0.004	0.70	0.61
Coyotes (*Canis latrans*)	0.63	0.57	0.69	0.67

Source: Analysis by N. Mankovich of his ferret data, and of coyote data kindly furnished by M. Bekoff. Bounds given are ± 1 S.E.; coyote sample sizes for play and for fighting were too small to permit approximate variance computation, and expressions for the sampling variance of normalized variability are not known at this time. For background on information-theoretic measures of behavioral diversity or variability, see Bekoff (1975) and Losey (in press). In each case, higher values indicate greater sequential variability.

and maintenance of cultural variation or about ecological scaling of social behavior may be of interest). Some analyses of variability lump individuals together; others consider each individual separately; a component-based approach considering variability both within and between individuals would appear promising but cannot easily rest on use of coefficients of variation, which cannot conveniently be partitioned in this way (Dow, 1976; Lewontin, 1966; Moriarty, 1977). These authors suggest use of variance of logarithms as an alternative approach to the problem of simultaneous analysis of different levels of variability.

Studies of variability illustrate changes in the kind of data gathered by ethologists as a consequence of new theoretical developments. Before, variability, in the rare instance that it was measured at all, was one datum among many; after, it was the crux of the matter.

Quantifying and Testing Behavioral Variability

Ethologists have used various indicators to quantify variability: estimates of ethogram size (Fagen and Goldman, 1977), coefficients of variation or equivalent measures (Chapters 2 and 18; Schleidt, 1974; Barlow, 1977; Bekoff, 1977a, b), or indices of sequential dependence (Bekoff, 1974; Bekoff, 1977a). These attempts call for application of theory of a different sort, namely statistical theory, in the service of hypotheses on relative variability of behavior in different contexts or at different ages. While no stigma need be attached to the use of indicators as such (Mosteller and Tukey, 1968), it is important to recognize the difference between an indicator and a test statistic. An indicator of variability is merely a mathematical formula that embodies an arbitrary theoretical definition, based on intuition, of what it means for behavior to be variable. The meaning of differences between two measured values of the same index of variability can not be assessed a priori because of possible sampling error. An appropriate statistical procedure must be used in order to test the hypothesis that behavior in one context or at one age is actually more variable than in another. For example, the c statistic (Dawkins and Dawkins, 1973) serves to test the statistical significance of observed differences in coefficients of variation if sample sizes are sufficiently large (Dow, 1976) and provided that we meet certain assumptions about the form of the probability distributions of the variable whose coefficient of variation is at issue (note in Bekoff, 1977a). How many published ethological "tests" on coefficients of variation can be shown to meet both of these standards? The need to report results of specific prior tests of the distributional assumptions should not be overlooked in view of the variety of forms such distributions are known to exhibit in ethological data (Fagen and Young, in press). As Wiepkema (1968), Slater (1974), and others have clearly shown, the form of distributions of behavioral durations, latencies and intervals is not robust and may vary qualitatively across individuals of a given age, sex and species, through ontogeny, and between drug-treated and control animals.

B. SEEING AS BELIEVING

Cohen (1972, p. 417) discusses the classical morphological problem of selecting meaningful transitions between biological forms. As Cohen (1972, p. 420) and Bateson and Hinde (1976, p. 423) point out, this problem is no less urgent in the study of behavior. What does it mean to say that one type of act "develops from" or "emerges from" another? Before we test the hypothesis that A is the precursor of B, we must formulate a biologically meaningful definition of the precursor relation. The problem is not merely an empirical one, for what is required is fundamental theory on the kinds of transformational mechanisms that actually change the characteristics of behavior ontogenetically. One suggestion that appears promising (even if its ultimate role in ethology is to serve as a null hypothesis against which to weigh more interesting alternatives) is the allometry hypothesis (Schleidt, 1974): as the organism grows, the shape (form, duration, or other characteristic) of its behaviors changes allometrically with body size. For instance, if Y is the duration of a feeding bout, the length of a stride, or the depth of a play-bow, and X is the animal's snout-to-vent length (one measure of body size), then the hypothesis $Y = bX^a$, where b and a are fitted parameters, means that observed changes in behavior result solely from size changes and cannot be interpreted as reflecting anything more "interesting" (e.g., refinement of skill). For additional ethologically relevant information on size and shape in biology, see Gould (1968, 1971, 1977) and McMahon (1973, 1975).

The study of ontogeny may prove to be an area of biology where ordinary (human, adult) "concepts and everyday logic are no longer competent" (Cohen, 1972, p. 417; Cohen, 1976). If so, without theory relating observations on behavioral transformations in ontogeny to some underlying invariants, seductive juxtapositions of subjectively similar behaviors observed in animals of different ages will continue: seeing is believing.

Case History 2: Transitional Play

The study of play behavior furnishes an excellent example of the need for meaningful theory on behavioral transformations in ontogeny. Mammals and some birds playfight, especially as juveniles. That is, they perform behavioral sequences containing many acts identical to or resembling acts used in escalated fighting. When animals fight agonistically, they seek to control a conspecific's position or velocity, or even to injure or intimidate that conspecific, in order to gain access to a valuable, defendable resource (Deag, 1977). Temporal sequences of these same acts are different in play (Poole and Fish, 1975), playfighting sequences are more variable than agonistic ("earnest") fighting (Bekoff, 1977; Table 1), play has its own signals and displays (Bekoff, 1974, 1976, 1977b; Chevalier-Skolnikoff, 1974; Fagen, 1974) not seen in agonistic fighting (and vice versa [Symons, 1974]), and play is decorated by characteristic capers, gambols, and frisks, i.e., Locomotor-Rotational Movements (S. Wilson and Kleiman, 1974). Moreover, playfighting includes frequent role reversal, is easily interrupted, induces loose body tone, and is not felt to result in injury to an opponent, enduring dispersal, lasting possession of a disputed object, or any of the other benefits of aggression cited by Deag (1977, p. 471) (Aldis, 1975; Fagen, ms 1; West, 1974).

Unfortunately, but not surprisingly, the above categorization is not always of use. Biological categories are based on intersecting sets of phenomena, and intermediate cases always exist (cf. discussion in Hinde, 1974, pp. 7–8). Skilled field observers sometimes find it impossible to discriminate between playfighting and agonistic fighting (e.g., Barash, 1973). Playfights may "escalate into" serious fights, both in young, inexperienced animals who have not yet learned to recognize play-signals (humans, Blurton-Jones, 1967) and in older animals at an age when agonistic fighting becomes increasingly frequent (squirrel monkeys, Baldwin and Baldwin, 1974, p. 310).

A single play bout may "change into" agonistic fighting; ontogenetically, play itself may "become" increasingly rough or aggressive as animals age (Aldis, 1975). The terms "escalate into," "change into," and "become" above are placed in quotation marks because it is not clear a priori just what sort of behavioral transformation is implied. In the absence of theory, one procedure has been arbitrary selection of measurable characteristics of behavior and subsequent comparison of their frequencies, durations, and intensities in interactions of different types. If we knew what sorts of transformations made sense theoretically, we would be in a better position to select and to quantify relevant observable characteristics of the transition between playfighting and agonistic fighting, or indeed of any ontogenetic transition. Three theoretical views, social comparisons (Fagen, 1974) nonzerosum game theory (Fagen, mss 1, 2), and general social theory (Chapter 18) all suggest definition of indices of mutuality, reciprocity, or even cooperativeness of a play interaction. One such index might be the percentage of play-wrestling time each individual spent upright and in a supine position, as in rhesus macaques (Altmann, 1962; Bernstein and Draper, 1964; Sade, 1972; Symons, 1974) and in domestic cats (West, 1974), or the percentage of time spent in the roles of pursuer and pursued in chases (Owens, 1975a). According to Aldis (1975, pp. 35–36), such reciprocity has been verified empirically and the hypothesis of reciprocity in play cannot be rejected. Whatever the specific measure used, age comparisons of form or intensity could benefit from allometry correction (see above). Perhaps the most general approach to the problem of intermediate behavior categories is that of Boorman (1970), who presents measures of distances between complex entities. To my knowledge, none of Boorman's measures has yet been applied to behavior.

B. ON THE NECESSITY AND INSUFFICIENCY OF THEORY

The preceding section offers a standard, if low-key version of the traditional theoretician's sermon: without theory, seeing is believing, believing seeing. In the face of such criticism, empiricists, like Darwin at Captain Fitzroy's table (Gould, 1976), could but maintain silence. A frequent debating tactic under such circumstances is to shift the battlefield to one's home ground and to criticize theoreticians for being poor empiricists. As the unfortunate Fitzroy's fate (Gould, 1976) suggests, this tactic can be most effective. However, ethologists will find that two excellent papers speaking for the power of models in ethology (Cohen, 1972; Colgan, in press) both stress the following points:

(1) Mathematical models are not always necessary. (The authors then specify conditions under which models are worth the considerable effort required to formulate and to test them.)

(2) Modelers need to understand "what is important and what is feasible" in field work (Cohen, 1972, p. 436). Cohen further points out that modelers can gain such appreciation by participating actively in field work, "or by consulting frequently and for a long time with field workers, or by trusting his armchair insight and luck. The first option is best. The armchair is an ideal vantage point from which to overlook the obvious and make impossible demands: when estimating [population] density by choosing random points and measuring the distance to the nearest neighbor, the field worker must have alternative procedures if the random point is in a patch of six-foot-tall grass occupied by an elephant" (Cohen, 1972, p. 436).

Embattled empiricists in search of new weaponry might consult the two articles cited. Because both authors are qualified applied mathematicians, these criticisms of "modeling for its own sake" are far more telling than the competitor's product.

Is the separation between formal theory and empirical study in behavioral development a dimple or a gulf? Indeed, does such formal theory yet exist to be attacked? Such diverse areas of animal behavior as motivation, aggression, mating systems, social insect caste, foraging strategies and techniques, parental care, and play have all been profitably subjected to formal modeling (McFarland, 1976; Wilson, 1975). Moreover, the work of a number of investigators (Chapter 13; S. Altmann, 1977; Barash, 1975; P. P. G. Bateson, 1976a, b; G. Bateson, 1963; Cohen, 1976; Gould, 1977; Hansen, 1974; Katz, 1974; Krebs, 1976; Levins, 1968; Marler, 1975; Maynard Smith, 1976; Slater, 1974; Teitelbaum, 1977; Trivers, 1974) furnishes bases, at various levels of rigor, for formal models of behavioral development.

III. Models and Metaphors in Developmental Ethology

A. TWO METAPHORS

Analogies and metaphors based on mathematics are sometimes confused with mathematical models (Cohen, 1971). For instance, Thom (1975, p. 303) "explains" animal prey capture by using a mathematical metaphor:

> [A predator perceives its prey.] Then, as soon as an external form is recognized as a genetic form, a perception catastrophe takes place, and the "ego" is recreated in an action, in the motor chreod (of capture or flight) that the genetic form projects onto the external form.

The metaphorical nature of the above reasoning should be self-evident.

A moderately suggestive, if less than rigorous metaphor is the dynamic programming hypothesis (E. O. Wilson 1976). P. P. G. Bateson 1976a, p. 402) urges that theory on behavioral development be based on "more than vague appeals to fashionable technologies" and cites as an example of such vagueness the reference to "internal programming" "without any indication being given of what the programme might be like and what precisely is being controlled." The dynamic programming hypothesis is one possible answer to Bateson's challenge. To give the concept of dynamic programming substance, imagine an organism that requires environmental feedback in order to refine fighting skill, and consider some relatively efficient and some relatively inefficient mechanisms for obtaining necessary feedback that might be programmed into such an organism. E. O. Wilson (1971, 1975) discusses mechanisms by which feedback from the environment molds complex behavior sequences in simple organisms.

In this framework, behavior develops, as is generally agreed (cf. Hinde, 1974, p. 190) through experience brought about by the organism's own behavioral tendencies. This developmental mechanism is feasible because relatively little information need be specified in advance. A detailed blueprint is not necessary. What is essential, however, is an error correcting mechanism (Bateson, 1976a).

Escalated fighting in mammals is a complex sequence of acts that contains numerous decision points where responses might conceivably be learned and refined in a way that permits continued flexibility. Because mistakes during fighting can result in serious injury, it would be suicidal for a young, inexperienced animal to attempt to develop tactics and maneuvers by engaging adults in escalated fights, but it would be equally suicidal to rely wholly on genetic information for specific guidance at each instant while fighting. How could we design an organism that would act so as to have the experiences necessary for refining these fighting techniques, if we allow it only a few simple behavioral predispositions?

I propose that the following three rules might be sufficient to produce the required experience. Aldis (1975, p. 32) notes the importance of the first two rules for structuring play interactions but does not discuss the developmental dimension.

(1) Behave in a manner that maximizes your probability of achieving a certain harmless physical relationship ("positional goal") with another organism (it could be "sniff face of conspecific," or "mouth neck of conspecific").

(2) Behave in a manner that minimizes the other organism's probability of achieving this relationship with you (e.g., try to keep the conspecific from sniffing your face).

(3) Behave in a manner that minimizes your probability of being injured by your partner.

Initially, organisms following these rules will either reach the goal (1) or will take simple defensive actions that cause the partner's attempts to fail. The organism whose direct tactic has now been stymied will attempt to overcome its partner's defensive move in some way. A network of tactics is eventually built up in this reverse, stepwise fashion. Rule 1 predicts that the probability of behaviors that cause physical pain will be low, since they tend to decrease the chance of achieving the goal (the animal who is hurt will interrupt the bout and may even leave the scene of the interaction).

The developmental mechanisms that actually link these sequences of behavior together are not specified by the dynamic programming hypothesis. Simple chaining based on a conventional learning paradigm may or may not suffice. Particularly in view of P. P. G. Bateson's (1976b) interesting discussion of levels of specificity in behavioral development, it would be premature to restrict the determinants of this process to an arbitrary point in what is most likely a complex continuum.

The following example illustrates the principle involved. A novice who is familiar with the moves of each piece in chess and who understands that the object of the game is to achieve a checkmating position is not yet a skilled chess player. Practice in simple endgames (queen versus two rooks, or knight and bishop versus rook) teaches skills that a beginner can master before tackling the complexities of midgame situations. It would be easier to master possible strategies involving a queen and two pawns after prior experience with a queen and one pawn.

If responses are stimulated by a maneuver aimed directly at the goal and/or reinforced when this maneuver is successfully parried, then moves and countermoves ought to be brought into the sequence in retrograde fashion, and ontogeny of a sequence of skilled behavioral acts should progress antiparallel to the direction of the behavior sequence under development. This prediction is not concerned with the age of first appearance of the acts themselves, but rather to the order of integration into sequences of behavior. Acts may appear some time before their integration into sequences (e.g., ontogeny of prey-catching behavior in the domestic cat, Leyhausen, 1973, pp. 62–63). (In the literature on ontogeny, it is not always made clear whether ages cited for first appearance of acts referred to individual acts or to their first performance in combination with other acts, a distinction which is nevertheless important.) This argument suggests a dynamic programming optimization algorithm in which optimal sequences of action are constructed by mapping routes backward from the goal (Bryson and Ho, 1969). All such paths are considered, the relative benefits and costs of each calculated, and optimal sequences identified for future use. One consequence of dynamic programming is that complex behavioral modalities in which the number of choice points or alternative moves is sufficiently high to warrant sophisticated developmental programming should progress ontogenetically from the consummatory to the appetitive act, i.e., in retrograde order. Of particular interest are non-stereotyped interactions with other animals: escalated fighting or predation are prime candidates, for in both of these cases the interest of the two participants is diametrically opposed and there is no a priori reason for future behavior to be either predictable or reciprocal; the need for skill, and accordingly for previous experience, ought to be relatively great in such cases.

Two questions are now apparent. At what level of organization should unit behaviors be defined for this purpose (i.e., coarse-grained—chasing and wrestling—vs. fine-grained—alternative moves, tactics and positions or even different orientations of the same act)? And what sequential order is to be used for comparisons?

Level of Organization

Because the same act can appear in several positions in a sequence of behavior, and because the same act can be used in several different types of sequences (in intraspecific fighting and in predation, for example), serious conceptual difficulties hamper fine-grained analysis. In addition, since development is presumed to be at least in part hierarchical (Dawkins, 1976a), it makes sense to think of blocks of behavior including several constituent acts being brought into ontogeny as a unit. For instance, in domestic kittens two acts used in playfighting first appear on days 21–26, and six other acts appear on days 32–48; no new acts appear during days 27–31 (West, 1974). The cluster of "late" acts (days 32–48) appears further subdivided into acts that appear on days 32–35 and those that appear on days 40–48. One act appears between these intervals. An interesting ancillary question (compare P. P. G. Bateson, 1976b, p. 10) concerns the order in which these blocks of behavior ("stages"?) are integrated into the existing repertory. Do they appear as integral blocks in existing sequences, do they form their own sequences separate from those that had already existed, are the individual acts interspersed among existing sequences, or can more than one of these possibilities occur?

Sequential Order

Poole and Fish (1975) in rats, and Bekoff (1977a) in coyotes found that sequential orders of playfighting and agonistic fighting differed. It would seem logical to compare the order of appearance of movements in play with the order of acts in the agonistic sequence, since by one hypothesis it is precisely this sequence whose performance is being improved. On the other hand, if play is performed for its own sake and for reasons having nothing in common with the function or motivation of serious fighting (play that helps form or maintain a social bond, for example), then the order of appearance of play acts should be compared with the play sequences themselves. If it were already known from independent evidence that dynamic programming rules governed play, we could then use these alternative predictions to test the skill development and socialization hypotheses of social play separately and independently. Unfortunately, as should be clear from the foregoing, such a test would not yet be justified.

Müller-Schwarze (1971) presents data that seem to support the dynamic programming hypothesis. "One striking peculiarity of ludic [= "play"] behavior," writes Müller-Schwarze (p. 243), "is that the movements which occur at an early age belong to the end of a chain of responses in the adult animal. For instance, the young animal mounts but does not show any courtship patterns; it strikes but does not threaten. During maturation the various phases of appetitive behavior start to precede the original ludic components".

A review of existing literature on play for additional data that could be used to test the dynamic programming hypothesis illustrated once again that (as mentioned earlier) there is no guarantee that even the most "detailed" and "intensive" ethological observations on animal play will include data relevant to novel theory. Poole and Fish (1976) was the only source found to present quantitative data on ontogeny of play *and* on sequences of agonistic fighting, and these data reflected samples taken at only two different ages. At the earlier age, moreover, all acts but one had already appeared. West (1974) reports complete ontogenetic data and incomplete data on sequential ordering. In both studies cited above, individuals were lumped. Schaller (1972) describes the ontogenetic order of appearance of behavioral acts in a home-reared lion cub but does not quantify sequences. In general, authors who report ontogenetic data on play seem not to perform sequential analysis (and vice versa).

Domestic cat social play (West, 1974) lacks simple relationships between ontogeny and behavioral sequencing at the level of molecular acts. On the average, domestic cat social play sequences begin with pounces or sidesteps, proceed with belly-ups and stand-ups, and end with a horizontal leap or with a chase; pounce and sidestep develop at age 32–35 days, belly-up and stand-up at age 21–26 days, and horizontal leap and chase at age 38–46 days (West, 1974). The final acts both develop latest, but the earlier portion of the sequence develops in retrograde order.

Because the preceding analysis may have been excessively fine-grained, I again asked the question "Is there a consistent relationship between ontogenetic order of appearance and sequential order of performance" at a more molar level. Two recognized categories of social play are contact and noncontact play (wrestling and chasing, or rough-and-tumble and approach-withdrawal). In this case a regular pattern in order of appearance emerged across species: in ontogeny, noncontact play emerges later than contact play in all species for which data were available (e.g., domestic cats, West, 1974; laboratory rats, Poole and Fish, 1976; ponies, personal observation; olive baboons, Owens, 1975b; rhesus, Hinde and Spencer-Booth, 1967). Precocial ungulates are particularly interesting in this context because they run in a solitary context weeks before they begin social play (Müller-Schwarze, 1971; Fagen and George, 1977). However, contact play precedes chasing play in these animals; the observed ontogenetic order is not simply due to maturation of necessary motor patterns.

Unfortunately it is not known whether chases or wrestling matches are more likely to occur in the first position in serious fights in all the species cited. Domestic cat fights typically begin the body contact and often end in a chase (Leyhausen, 1973, pp. 114–140). The same rule holds for laboratory rats (Grant, 1963). Sade (1967) describes a similar outcome in rhesus monkeys.

The dynamic programming hypothesis has therefore been given preliminary evaluation in three separate contexts. It appears to hold for the relationship between social communication patterns and the motor acts associated with them. It does not agree with the only available data on molecular acts, nor does it explain the apparent parallel ontogenetic and sequential order of wrestling and chasing. West's (1974) data agree with the above generalizations on molar acts: chasing and certain signals both develop after wrestling, but in a given agonistic bout signals precede body contact and chasing follows it. Because (Symons, 1974) there is little if any evidence for the social communication practice hypothesis of play behavior, the sequential and ontogenetic regularities described by Müller-Schwarze (1971) and those involving chasing and wrestling, both empirically well founded, call for a different kind of theoretical explanation. Perhaps a future dynamic programming model will succeed where metaphor failed.

B. A MODEL

I conclude by briefly reviewing a formal mathematical model of the evolution of animal play behavior (Fagen, ms 1).

In analyzing possible ontogenetic strategies it is important to specify how such strategies could evolve. Ontogenetic strategies that involve reliance on another animal for protection, for food, for information or for stimulation involve cooperation, competition, and conflicts of interest. They are therefore subject to all constraints on evolution of any social strategy. For example, they may be vulnerable to cheating by the partner, in which case evolutionary stability of such strategies is an open question awaiting coordinated theoretical and empirical scrutiny. When an individual's caregivers or playmates are also its relatives, evolution of ontogenetic patterns by kin selection becomes important.

The question of evolutionary stability of cooperative behavior like play can be colorfully restated (Chapter 18) by asking why animals do not cheat in a playfight. Rephrased, the question implies two levels of analysis: here-and-now mechanistic questions (what behavioral tactics might enable animals to enhance their own future reproductive success through play at the playmate's expense? Are such tactics actually observed in nature?) and evolutionary questions (under what circumstances will genes predisposing such cheating behavior be selected or counterselected?). In order to answer the latter question I formulated a formal nonzerosum game model of animal playfighting (Fagen, ms 1) in which cheating was possible and in which encounters could be nonrandom with respect to strategy type (Fagen, ms 2). Encounter nonrandomness is extremely important. With any amount of encounter nonrandomness, however small, wholly symbolic agonistic contests can evolve (Fagen, ms 2), and it is therefore incorrect to claim, even in theory, that symbolic aggression is effective only because the possibility of escalation exists in species capable of inflicting injury. A "conspiracy of doves" (Dawkins, 1976b) may indeed be evolutionarily stable.

I modeled playfighting as a reciprocal social interaction from which both participants obtain equal, small benefit B. Playful animals gained no benefit in non-reciprocal encounters, but the nonplayful "cheater" gained benefit $V > B$. Possibly nonrandom encounters between nonplayful animals followed the standard Hawk-Dove formalism (Maynard Smith, 1976): Hawks exhibit escalated, potentially dangerous fighting behavior. Doves fight conventionally by displaying and do not use anatomical weapons to inflict damage even if such weapons are potentially available for use.

Under the above assumptions I investigated existence, uniqueness, and evolutionary stability of pure and mixed strategies based on playful and nonplayful fighting tactics. Playfighting was always evolutionarily unstable when animals were assumed to encounter each other randomly with respect to strategy type. But for significantly positively assortative encounters (like encounters like with high probability), playfighting was evolutionarily stable under a wider range of conditions. These results suggest that playfighting is most likely to have evolved and should be observed most frequently in family-living animals, where positively assortative, nonrandom encounters between playful young are inevitable. Moreover, although I did not explicitly model sequences of contests, social play (unlike dominance fights or mating season contest displays) is actually repeated day after day by the same individuals. It is not the single brief encounter that the model assumes. Because the model's predictions tend toward pessimism, factors other than those already considered can help explain evolution of social play behavior. Social play in repeated encounters between individuals can result in skill development, physical training, or even in a "friendship," a social bond involving individual recognition, prediction, and cooperation. This history may be marred by a serious fight over a valuable, defendable resource. When such fights occur, the winner may reap immediate gains, but in so doing it could lose some or all of the investment made in friendship formation (but not in skill development or physical training) up to that time. Demographic factors, including availability of conspecifics of suitable age and sex, the length of the juvenile stage and age at sexual maturity, and the probability that both friends will live to reap the benefits of friendship, should all influence the tendency to play socially.

In summary, playfighting should be stable against cheating (will evolve) in family-living animals in environments that select for life histories with a prolonged period of immaturity permitting a very large number of play-interactions to occur. This model, like an earlier one resting on quite different premises (Fagen, 1977), helps explain the fact that while solitary or companion-oriented play are found in almost all mammals, including voles (S. Wilson, 1973) and laboratory mice (Poole and Fish, 1975), truly reciprocal social play is only found in relatively long-lived mammals and birds having slow development and a long period of parental care. This model therefore offers specific predictions about the degree of sociality of play as a function of the overall life-history strategy of an animal, and these predictions may be tested in groups like rodents, insectivores, corvids and parrots, on previously unstudied species, or across populations within species in different environments. It suggests that play at its most cooperative will involve repeated encounters each of which has a relatively minor effect on future reproductive success. If a single encounter can have major effects, the temptation to cheat will be greatly increased. Further theory can be developed along these lines, paralleling the theory of mate desertion in parental care (Dawkins and Carlisle, 1976; Maynard Smith, 1977).

IV. Conclusions

This paper argues that data and theory, like the American colonies, must hang together or they will hang separately. The point is illustrated in several ways: by suggesting that theory colors all ethological observation whether or not it has received a formal invitation to participate; by citing two mathematical ethologists, enthusiastic practitioners of the modeling approach, who are visibly concerned about possible abuse and misuse of models in ethology; by presenting a metaphorical relationship between play and other behavior that helps organize data on ontogeny of play in mammals of several orders; and by discussing a model that helps explain how social play might have evolved and predicts what kinds of mammals will exhibit social play.

Acknowledgments

The editors of this volume supplied valuable critiques of an earlier draft. I thank E. O. Wilson for his suggestion that dynamic programming could develop complex behavior in relatively simple organisms. J. E. Cohen's papers on modeling behavior were influential. Data for Table 1 were kindly furnished by M. Bekoff and by N. Mankovich, and analyzed by N. Mankovich.

References

Aldis, O. 1975. *Play fighting.* New York: Academic Press.

Altmann, S. A. 1962. Social behavior of anthropoid primates: analysis of recent concepts. *In Roots of behavior,* ed., E. L. Bliss, pp. 277–285. New York: Harper & Bros.

Altmann, S. A. 1977. Demographic constraints on behavior and social organization. *Paper presented at a symposium on Ecological Influences on Social Organization: Evolution and Adaptation. Burg Wartenstein, Austria, August 12–21, 1977.*

Appel, A. 1967. An interview with Vladimir Nabokov. *In Nabokov: the man and his work,* ed., L. S. Dembo, pp. 19–44. Madison: Univ. of Wisconsin Press.

Baldwin, J. D., and J. I. Baldwin. 1974. Exploration and play in squirrel monkeys (*Saimiri*). *Am. Zool.* 14:303–315.

Barash, D. P. 1973. The social biology of the Olympic marmot. *Anim. Behav. Monogr.* 6:171–245.

Barash, D. P. 1975. Behavior as evolutionary strategy. *Science* 190:1084–1085.

Barlow, G. W. 1977. Modal action patterns. *In How animals communicate,* ed., T. A. Sebeok, pp. 94–125. Bloomington: Indiana Univ. Press.

Bateson, G. 1963. The role of somatic change in evolution. *Evolution* 17:529–539.

Bateson, P. P. G. 1976a. Rules and reciprocity in behavioral development. *In Growing points in ethology,* eds., P. P. G. Bateson and R. A. Hinde, pp. 401–421. Cambridge: Cambridge Univ. Press.

Bateson, P. P. G. 1976b. Specificity and the origins of behavior. *In Advances in the study of behavior,* Vol. 6, eds., J. Rosenblatt, R. Hinde, E. Shaw, and C. Beer, pp. 1–20. New York: Academic Press.

Bateson, P. P. G., and R. A. Hinde. 1976. Editorial: 6. *In Growing points in ethology,* eds., P. P. G. Bateson and R. A. Hinde, pp. 423–424. Cambridge: University Press.

Bekoff, M. 1974. Social play and play-soliciting by infant canids. *Am. Zool.* 14:323–340.

Bekoff, M. 1975. Animal play and behavioral diversity. *Am. Naturalist* 109:601–603.

Bekoff, M. 1976. Animal play: problems and perspectives. *In Perspectives in ethology,* Vol. 2, eds., P. P. G. Bateson and P. H. Klopfer, pp. 165–188. New York and London: Plenum.

Bekoff, M. 1977a. Quantitative studies of three areas of classical ethology: social dominance, behavioral taxonomy, and behavioral variability. *In Quantitative methods in the study of animal behavior,* ed., B. A. Hazlett, pp. 1–46. New York: Academic Press.

Bekoff, M. 1977b. Social communication in canids: evidence for the evolution of a stereotyped mammalian display. *Science* 197:1097–1099.

Bekoff, M., D. G. Ainley, and A. Bekoff. ms. The ontogeny and organization of comfort behavior in Adélie penguins, *Pygoscelis adéliae.*

Bernstein, I. S., and W. A. Draper. 1964. The behavior of juvenile rhesus monkeys in groups. *Anim. Behav.* 12:84–91.

Blurton-Jones, N. G. 1967. An ethological study of some aspects of social behavior of children in nursery school. *In Primate ethology,* ed., D. Morris, pp. 347–368. Chicago: Aldine.

Boorman, S. A. 1970. Metric spaces of complex objects. Senior honors thesis. Cambridge, Mass.: Division of Engineering and Applied Physics, Harvard College.

Bryson, A. E., and Y. C. Ho. 1969. *Applied optimal control.* Waltham, Mass.: Ginn and Co.

Chevalier-Skolnikoff, S. 1974. The primate play face: a possible key to the determinants and evolution of play. *Rice Univ. Stud.* 60:9–29.

Cohen, J. E. 1971. Mathematics as metaphor. *Science* 172:674–675.

Cohen, J. E. 1972. Aping monkeys with mathematics. *In The functional and evolutionary biology of primates,* ed., R. Tuttle, pp. 415–536. Chicago: Aldine-Atherton.

Cohen, J. E. 1976. Irreproducible results and the breeding of pigs. *Bioscience* 26:391–394.

Colgan, P. Modelling. *In Quantitative ethology,* ed., P. Colgan, New York: Wiley-Interscience. (In press.)

Dawkins, R. 1976. Hierarchical organization: a candidate principle for ethology. *In Growing points in ethology,* eds., P. P. G. Bateson and R. A. Hinde, pp. 7–54. Cambridge: University Press.

Dawkins, R. 1976b. *The selfish gene.* New York: Oxford Univ. Press.

Dawkins, R., and T. R. Carlisle. 1976. Parental investment and mate desertion: a fallacy. *Nature* 262:131–132.

Dawkins, R., and M. Dawkins. 1973. Decisions and the uncertainty of behaviour. *Behaviour* 45:83–102.

Deag, J. M. 1977. Aggression and submission in monkey societies. *Anim. Behav.* 25:465–474.

Dow, D. D. 1976. The use and misuse of the coefficient of variation in analysing geographical variation in birds. *Emu* 76:25–29.

Fagen, R. M. 1974. Selective and evolutionary aspects of animal play. *Am. Naturalist* 108:850–858.

Fagen, R. M. 1977. Selection for optimal age-dependent schedules of play behavior. *Am. Naturalist* 111:395–414.

Fagen, R. M. ms 1. Hawks, doves and harlequins: the logic of animal playfighting behavior.

Fagen, R. M. ms 2. When doves conspire: evolution of conventional fighting tactics in a nonrandom encounter animal conflict model.

Fagen, R. M., and T. K. George. 1977. Play behavior and exercise in young ponies (*Equus caballus* L.) *Behav. Ecol. Sociobiol.* 2:267–269.

Fagen, R. M., and R. N. Goldman. 1977. Behavioural catalogue analysis methods. *Anim. Behav.* 25:261–274.

Fagen, R. M., and D. Y. Young. Temporal patterns of behavior: durations, intervals, latencies, and sequences. *In Quantitative ethology,* ed., P. Colgan, New York: Wiley-Interscience. (In press.)

Gould, S. J. 1966. Allometry and size in ontogeny and phylogeny. *Biol. Rev.* 41:587–640.

Gould, S. J. 1971. Geometric similarity in allometric growth: a contribution to the problem of scaling in the evolution of size. *Am. Naturalist* 105:113–136.

Gould, S. J. 1976. Darwin and the captain. *Nat. Hist.* 851:32–34.

Gould, S. J. 1977. Ontogeny and phylogeny. Cambridge, Mass.: Harvard University Press.

Grant, E. C. 1963. An analysis of social behaviour of the male laboratory rat. *Behaviour* 21:260–328.

Hansen, E. W. 1974. Some aspects of behavioral development in evolutionary perspective. *In Ethology and psychiatry,* ed., N. F. White, pp. 182–186. Toronto: Univ. of Toronto Press.

Henry, J. D., and S. M. Herrero. 1974. Social play in the American black bear. *Am. Zool.* 14:371–389.

Hill, H. L., and M. Bekoff. 1977. The variability of some motor components of social play and agonistic behaviour in infant Eastern coyotes, *Canis latrans* var. *Anim. Behav.* 25:907–909.

Hinde, R. A. 1974. *Biological bases of human social behaviour.* New York: McGraw-Hill.

Hinde, R. A., and Y. Spencer-Booth. 1967. The effect of social companions on mother-infant relations in rhesus monkeys. *In Primate ethology,* ed., D. Morris, pp. 267–286. Chicago: Aldine.

Jolly, A. 1966. Lemur social behavior and primate intelligence. *Science* 153:501–506.

Katz, P. L. 1974. A long-term approach to foraging optimization. *Am. Naturalist* 108:758–782.

Krebs, J. R. 1976. Review of *Sociobiology. Anim. Behav.* 24:709–710.

Levins, R. 1968. *Evolution in changing environments.* Princeton, N.J.: Princeton Univ. Press.

Lewontin, R. C. 1966. On the measurement of relative variability. *Syst. Zool.* 15:141–142.

Leyhausen, P. 1973. *Verhaltensstudien an Katzen. Z. Tierpsych. Supp. 2,* 3rd ed. Berlin and Hamburg: Paul Parey.

Losey, G. Information theory. *In Quantitative ethology,* ed., P. Colgan. New York: Wiley-Interscience. (In press.)

McFarland, D. J. 1976. Form and function in the temporal organization of behaviour. *In Growing points in ethology,* eds., P. P. G. Bateson and R. A. Hinde, pp. 55–93. Cambridge: University Press.

McMahon, T. A. 1973. Size and shape in biology. *Science* 179:1201–1204.

McMahon, T. A. 1975. Allometry and biomechanics: limb bones in adult ungulates. *Am. Naturalist* 109:547–563.

Maynard Smith, J. 1976. Evolution and the theory of games. *Am. Sci.* 64:41–45.

Maynard Smith, J. 1977. Parental investment: a prospective analysis. *Anim. Behav.* 25:1–9.

Marler, P. 1975. On strategies of behavioural development. *In Function and evolution in behaviour,* eds., G. Baerends, C. Beer, and A. Manning, pp. 254–275. Oxford: Clarendon Press.

Mayr, E. 1963. *Animal species and evolution.* Cambridge, Mass.: Belknap Press of Harvard Univ. Press.

Moriarty, D. J. 1977. On the use of variance of logarithms. *Syst. Zool.* 26:92–93.

Mosteller, C. F., and J. W. Tukey. 1968. Data analysis, including statistics. *In Handbook of Social Psychology,* Vol. 2, 2nd ed., pp. 133–160. Reading, Mass.: Addison-Wesley.

Müller-Schwarze, D. 1971. Ludic behavior in young mammals. *In Brain development and behavior,* eds., M. B. Sterman, D. J. McGinty, and A. M. Adinolfi, pp. 229–249. New York: Academic Press.

Owens, N. W. 1975a. A comparison of aggressive play and aggression in free-living baboons, *Papio anubis. Anim. Behav.* 23:757–765.

Owens, N. W. 1975b. Social play behaviour in free-living baboons, *Papio anubis. Anim. Behav.* 23:387–408.

Poole, T. B., and J. Fish. 1975. An investigation of playful behavior in *Rattus norvegicus* and *Mus musculus* (Mammalis). *J. Zool.* 175:61–71.

Poole, T. B., and J. Fish. 1976. An investigation of individual, age and sexual differences in the play of *Rattus norvegicus* (Mammalia: Rodentia). *J. Zool.* 179:249–260.

Sade, D. S. 1967. Determinants of dominance in a group of free-ranging rhesus monkeys. *In Social communication among primates,* ed., S. A. Altmann, pp. 99–115. Chicago: Univ. of Chicago Press.

Sade, D. S. 1973. An ethogram for rhesus monkeys. *Am. J. Phys. Anthropol.* 38:537–542.

Schaller, G. B. 1972. *The Serengeti lion.* Chicago: Univ. of Chicago Press.

Schleidt, W. 1974. How "fixed" is the fixed action pattern? *Z. Tierpsychol.* 36:184–211.

Slater, P. J. B. 1974. The temporal pattern of feeding in the zebra finch. *Anim. Behav.* 22:506–515.

Symons, D. 1974. Aggressive play and communication in rhesus monkeys (*Macaca mulatta*). *Am. Zool.* 14:317–322.

Teitelbaum, P. 1977. Levels of integration of the operant. *In Handbook of operant behavior,* eds., W. K. Honig and J. E. R. Staddon, pp. 7–27. Englewood Cliffs, N.J.: Prentice-Hall.

Thom, R. 1975. *Structural stability and morphogenesis.* Tr. D. H. Fowler. Reading, Mass.: W. A. Benjamin.

Trivers, R. L. 1974. Parent-offspring conflict. *Am. Zool.* 14:249–264.

West, M. 1974. Social play in the domestic cat. *Am. Zool.* 14:427–436.

Wiepkema, P. R. 1968. Behaviour changes in CBA mice as a result of one goldthioglucose injection. *Behaviour* 32:179–210.

Wilson, E. O. 1971. *The insect societies.* Cambridge, Mass.: Belknap Press of Harvard University Press.

Wilson, E. O. 1975. *Sociobiology.* Cambridge, Mass.: Belknap Press of Harvard University Press.

Wilson, E. O. 1976. The social instinct. *Bull. Am. Acad. Arts Sci.* 30(1):11–25.

Wilson, S. C. 1973. The development of social behaviour in the vole (*Microtus agrestis*). *Zool. J. Linn. Soc.* 52:45–62.

Wilson, S., and D. G. Kleiman. 1974. Eliciting play: a comparative study. *Am. Zool.* 14:341–370.

Chapter Twenty

Ontogeny of Tool Use by Nonhuman Animals

BENJAMIN B. BECK

Chicago Zoological Park
Brookfield, Illinois 60513
        and
Department of Anthropology
University of Chicago
Chicago, Illinois 60637

While tool use has been reported for a wide variety of animals, data (as contrasted with speculation) pertaining to ontogeny are scarce. The literature suggests consideration of two distinct sets of questions. The first, the "questions of origin," concerns the manner in which a novel mode of tool use is acquired for the first time by a member of a population or group. The second, the "questions of dissemination," concerns the manner in which the novel pattern is acquired by other members of that population or group.

Questions of Origin

To my knowledge, the origin of a novel form of tool use in a wild population, or in a captive population under nonexperimental conditions, has never been observed. Events surrounding origin allow reasonable reconstruction in a few cases, and there are some records of origin in experimental settings. Regrettably, most of these data pertain to primates.

Some patterns of tool use undoubtedly have specific, inherited determinants with only general, environmental determinants in the sense of Bateson (1976). As Bateson has noted, in some terminologies such patterns would be called "innate." The only example supported by data is the occupancy of discarded gastropod shells by hermit crabs. While some might not regard this behavior as tool use, the crab does select and literally carry the shell, and in so doing gains increased protection from predation and from environmental extremes. Reese (1962) hatched eggs of *Pagurus longicarpus* in a laboratory and deprived the glaucothoe of all experience that could reasonably be considered as relevant to shell occupancy. When exposed to empty shells for the first time, the glaucothoe behavior in choosing and entering shells was essentially identical to that of experienced adults. I would conclude that the behavior is innate in the sense noted above, and that the behavior probably originated through mutation and selection.

Alcock (1972) enumerated patterns of tool use during feeding by insects and fish. He concluded that they are "fixed action patterns," and emphasized preadaptation in their origin. For example, larval ant-lions (*Myrmeleon* spp.) and worm-lions (*Lampromyia* spp. and *Vermileo* spp.) excavate pitfalls by removing sand with a stereotyped head-tossing movement. Alcock hypothesized that individuals that threw sand especially frequently or intensely would be more likely to strike prey and thus would be more successful at prey capture. There would be consequent selection for increased frequency, intensity, and/or accuracy of tossing in the presence of prey. The preexisting (non-tool) tossing pattern would be fixed as a tool phenotype. Other patterns of tool use show considerable interindividual and interpopulational stereotypy. Such attributes suggest but do not demonstrate that the patterns are innate. In the absence of conclusive descriptive and experimental data, reconstruction of origin is premature.

Many patterns of tool use, especially by birds and primates, are characterized by considerable interindividual and interpopulational variability. These attributes suggest that the behaviors are learned. Reconstruction and experimental investigation support this conclusion. Indeed there are genetically based morphological requisites for these patterns, and the patterns thus have, in Bateson's terms, general inherited determinants. However, while such factors are necessary for phenotypic expression, specific environmental effects are additionally required.

I studied the acquisition of a novel form of tool use by captive hamadryas baboons (*Papio hamadryas*), Guinea baboons (*P. papio*), and pigtailed macaques (*Macaca nemestrina*) (Beck, 1972, 1973a, 1976). In each experiment, a pan of food was placed out of reach and a rod was supplied with which the food could be secured. No training or shaping was used. In each case, initial solution resulted fortuitously from exploratory manipulation of the tool. The reinforcement resulting from the first correct response increased the probability that the tool user would repeat the response. With further repetitions, the probability of the response continued to increase and trial duration declined to an asymptotic value. Additionally, inefficient motor elements became less frequent since they did not result in reinforcement, and efficient elements were linked to form a skillful response sequence.

The process by which the monkeys learned to use the tool is fundamentally the same as that by which a cat learns to escape from a problem box or by which a pigeon learns to peck a key or a rat learns to press a bar in traditional operant settings. While the first correct response by the pigeon or rat is usually shaped by the experimenter, the monkeys' were not shaped. However, this difference is trivial since it affects only the time that elapses before the occurrence of the first correct response. A skillful operant practitioner can produce key pecking or bar pressing in less than an hour; the first successful use of the tool required accumulated times of between 8 and 14 h in my experiments. Because I did not shape the response, I prefer to describe the origin of tool use in my experiments as learning by trial and error, but it is important to recognize that this is essentially the same as operant conditioning. Once the first correct response occurs, whether quickly through shaping or slowly through accident, there is an identical facilitative effect of reinforcement on the probability and efficiency of subsequent responses.

Shurcliff, Brown, and Stollnitz (1971) found that rhesus macaques (*M. mulatta*) could not use a rod to get food unless they were shaped. However, they allowed the naive monkeys only five minutes to get the food before training was initiated. They concluded that shaping was necessary but I suspect that rhesus macaques could solve this problem by trial-and-error if given sufficient time.

Wolfgang Köhler's pioneering work (1927) on tool use by captive chimpanzees (*Pan troglodytes*) is often cited as demonstrating the role of insight, as opposed to trial and error, in the origin of tool use. However, Köhler was not able to specify the object-manipulation and reinforcement histories of his subjects. He also did not prevent all opportunities for a subject to observe other chimpanzees that already were skilled tool-users. As a result, when Köhler's animals attempted to solve a tool problem for the first time, we do not know how naive they actually were. Further, Köhler did not provide operational criteria for distinguishing insightful from trial-and-error solutions. For these reasons, Chance (1960) suggested caution in interpreting Köhler's results as being demonstrative of the role of insight. Insightful solution is operationally definable (Beck, 1967, 1973b, 1975) but retrospective application of the criteria is difficult.

In my opinion, some tool use by Köhler's chimpanzees did originate insightfully, but most resulted by trial-and-error learning, generalization of previously acquired responses, or observation of other colony members. Yerkes (1943) also investigated tool use by captive chimpanzees. At least one solution appears to have originated insightfully but, again, most resulted from trial-and-error learning or operant shaping.

Birch (1945) studied the ability of captive chimpanzees to procure food with a stick. Upon first exposure to the problem, only two of six subjects were able to get the food within 30 min. One of these animals had already learned to use sticks to reach distant incentives and solution resulted simply from transfer of previously acquired responses. The other got the food accidentally during manipulation of the tool. All six subjects were then supplied with sticks for three days in a free-play setting during which time they manipulated the sticks in a variety of ways. The chimpanzees were then returned to the testing situation and each used a stick to sweep in food within 20 sec. Birch felt that the chimpanzees acquired information about sticks per se while playing with them in a nonproblem setting. He identified as most relevant the information that a stick can serve as a functional extension of the arm. The animals were able subsequently to utilize this information to use a stick quickly and efficiently to get an incentive. Birch concluded that the solutions were insightful and could be distinguished from trial-and-error solutions. However, he added that insight is dependent on previous experience with elements of a problem and, in fact, may simply be a reorganization of that experience. Bingham (1929), who studied box-stacking by chimpanzees, also felt that insight represented only novel recombination of already established responses.

Jackson (1943) and Schiller (1957) found an age-related competency among captive chimpanzees in learning to use sticks as reaching tools. Such a correlation could result from older animals having more experience with sticks in nonproblem settings. Older animals would therefore have more general information about sticks and thus could more readily solve problems demanding use of a stick as a tool. This explanation would be totally consistent with Birch's findings. However, the age-related tool competency could also reflect maturation of sensorimotor abilities. Schiller noted that complexity and variety of patterns involved in playful object manipulation also increases with age. Older animals have a larger and more elaborate repertoire of manipulative responses to apply to a tool problem and thus would be more likely to succeed. A third explanation, of course, is that both experience and sensorimotor maturation contribute interactively to age-related tool ability. Schiller found, in fact, that the ability of two-year-old chimpanzees to use sticks to procure food did not increase after two weeks of play experience with sticks, while

Birch's subjects, which were four to five years old, showed considerable improvement after only three days.

Menzel, Davenport, and Rogers (1970) compared restriction-reared and wild-born chimpanzees with regard to their ability to learn to use reaching tools. The restriction-reared subjects were captive-born and were raised in impoverished physical environments until two or three years of age. The wild-born animals had lived in enriched environments since their importation as infants. Both groups were tested at six to eight years of age by which time the restriction-reared subjects had had extensive experience with a variety of novel objects and showed "normal prehensory and simple play patterns with objects." Despite this experience, the restriction-reared animals were significantly deficient on a variety of measures of ability to use reaching tools. The authors agree that sensorimotor maturation and object manipulation experience are important in the ontogeny of tool use. However, they emphasize that deficits in these regards in the restriction-reared subjects had largely been reversed by the time of testing and could not account for their deficient task performance. They conclude that early deprivation had produced a lasting deficiency in the ability to apply manipulative abilities and experience to novel situations, i.e., the re-striction-reared chimpanzees lacked "adaptability." Mason (Chapter 12) demonstrates that rhesus macaques reared in environments devoid of response-contingent stimulation are deficient in their responsiveness to novel situations. Macaques raised in environments rich in response-contingent stimulation develop "coping strategies" that allow more effective adaptation to novel situations. Perhaps Mason's coping strategies are equivalent to adaptability.

Adaptability as postulated by Menzel *et al.* may also be related to Birch's and Bingham's views of insight as the reorganization of previously acquired information, and to the interaction of sensorimotor ability and experience that was revealed in Schiller's research. Since most of the wild-born subjects of Menzel *et al.* used the tool successfully within several minutes, their solutions would also conform to Köhler's and Yerkes' concept of insight. If these equivalencies are valid, the varying and sometimes contradictory formulations of insight can be reconciled and synthesized. Among chimpanzees, it seems that individual histories will determine whether tool use will originate by trial and error or by insight. The capacity for insight may be genetically limited in other taxa but a "chimpomorphic" bias has precluded systematic investigation of this subject. I found that a female hamadryas baboon learned insightfully to bring a tool to a male that could use it to procure food for both (Beck, 1973b).

Tool use by captives sometimes originates spontaneously, i.e., without anticipation or intervention by the investigator. Understandably, the first performances are rarely documented and origin therefore can not be analyzed conclusively. A captive Northern blue jay (*Cyanocitta cristata*) learned spontaneously to use pieces of newspaper to reach food pellets (Jones and Kamil, 1973). While the authors did not witness origin, subsequent experimental probes indicated that it was learned by trial and error during exploratory object manipulation. More extensive data emerged from the spontaneous invention of ladders among a captive chimpanzee group that occupied a large corral at the Delta Regional Primate Research Center (Menzel, 1972, 1973; McGrew *et al.*, 1975). These chimpanzees were wild-born and imported at about one year of age. Ladder use first occurred when they were six to seven years old. In the intervening period the animals were kept in socially and physically enriched environments but tool use was not purposely reinforced. Early in their captive years, the chimpanzees were seen to balance long objects on end and climb to the top. The behavior remained common for five years and poles as long as three meters were balanced and climbed efficiently. Whatever reward the animals secured from this was intrinsic, i.e., the behavior seemed simply to be playful.

The chimpanzees were highly attracted to a glass-fronted observation booth that was built into the top of one corral wall. After almost five years of playful balancing and climbing of poles, one animal ("Rock") utilized a slight variant of the behavior to get close to the booth. He balanced and climbed the pole and then steadied himself by propping his foot against the wall. From this position he could look through and slap the glass. About one week later, the investigators arrived at the corral one morning to find the booth occupied by the chimpanzees. Since the break-in had occurred at night, the original mode of access was not actually observed. However, shortly after the apes were chased from the booth, Rock was seen to prop a pole against the wall beneath the booth, climb up it, and reenter the booth. It was likely that he had propped and climbed the pole during the night and thus had originated the use of ladders among this group.

Menzel (1972) notes that the chimpanzees had seen humans use ladders on many occasions but several years of observation of human ladder use elapsed before the animals propped and climbed poles. Further, McGrew *et al.* (1975) note that when previously given occasional opportunities to use human ladders, the chimpanzees did not climb them. Thus origin by imitation of humans is counterindicated. Trial-and-error learning was doubtlessly involved. It is likely that Rock, sitting atop a pole he had balanced and climbed in front of the booth, lost his balance and fell toward the booth. The pole was thus accidentally propped against the wall resulting in a ladder configuration that allowed access to the booth. Menzel emphasizes that such propping had occurred accidentally on numerous occasions in preceding years, indicating that the chimpanzees had possessed the requisite sensorimotor abilities for some time. Yet such events did not result in the chimpanzees' learning to purposively position, prop, and climb poles. Such a transition took place literally overnight, and Menzel invites explanation of why one or two such accidents should suddenly have such a dramatic effect on behavior when previous ones did not.

One might respond that previous accidental propping and climbing had never resulted in access to such powerful reinforcement as it did on the night of the break-in. However, cognitive readiness may also have been involved. The studies of the use of reaching tools (see above) suggest that older chimpanzees (especially wild-born individuals) profit more from experience than do younger ones, and the effect is not due simply to motor maturation. Menzel (1972) concludes that the critical ingredient of ladder using is "attention to and perceptual differentiation of specific consequences of the (pole's) contacts with other objects." This may be related to "adaptability" as revealed in the experiment of Menzel *et al.* (1970) to be critical in facile mastery of reaching tools. If so, adaptability may be a function of age as well as infant experience, and may have matured in Rock by the night he invented ladders.

To summarize, the origin of tool use (at least among captive chimpanzees) is dependent in part on the maturation of necessary sensorimotor abilities, object-manipulation experience, infant experiences of an unknown sort that predispose an animal to adapt sensorimotor abilities and previous experiences to novel situations, and an age-related factor that can be specified at this time only as cognitive readiness. This terminology may interface meaningfully with attempts to study primate cognitive development within a Piagetian framework (Hughes and Redshaw 1974; Mathieu *et al.,* 1976; Parker 1976; Redshaw 1975).

Whether tool use originates by trial-and-error or by insight, the behavior must be drawn from the inventor's existing repertoire of non-tool patterns. The baboons and macaques had to manipulate the rod before the initial accidental acquisition of food could occur. The chimpanzees had to have extensive object manipulation experience before they could adaptively produce the recombinations that comprised insightful tool responses. One source of object manipulation in vertebrates is play. While playful manipulation can not be said to be nonfunctional (Fagen 1976), it is not motivated by conventional types of reinforcement. Eaton (1972) describes the playful construction of snowballs by captive Japanese macaques (*M. fuscata*) and notes that this behavior does not involve the basic reward systems that are most commonly associated with tool behavior. However, he adds that it does prepare the animals to adapt to situations in which tool use would be adaptive. Whatever the motivation for playful exploration and manipulation of objects, such behaviors are important precursors to the origin of much learned tool use. As a result, any factor, e.g., age, availability of objects, amount of "leisure time," that influences the frequency of object play will affect the origin of tool use.

Alcock (1972) emphasizes the role of conflict behavior as the source of varietal responses from which tool use originates. He hypothesizes, for example, that a sea otter that could not bite open a hard-shelled mollusk might hit it against a rock or another mollusk out of frustration. If the mollusk were broken, the sea otter would be reinforced and thus might learn to repeat the behavior. Many nonhuman primates use tools in agonistic contexts (Beck 1975). Hall (1975) suggests that such behavior may have originated from aggressive displays that did not involve tools. For example, disturbed monkeys leaping animatedly through trees may accidentally dislodge dead branches and other debris that deter pursuers on the forest floor. This would be reinforcing and might predispose a monkey to break off and throw or drop branches toward subsequent intruders. It is difficult to estimate the importance of play, conflict, aggressive display, and generalization of previously-acquired responses in the origin of tool use until more cases of origin in nonexperimental settings are documented.

For origin to occur, the environment must be such that tool use is both possible and advantageous. Hermit crabs would not have evolved as hermits if empty shells were not available or if the environment were stable, nonextreme, and free of predators. By introducing a rod and a pan of attractive food beyond reach, I was able to increase the probability of origin of reaching tools by monkeys in my experiments. If the Delta corral had routinely been cleared of dead limbs, or if the chimpanzees had been allowed free access to the observation booth, it is improbable that ladders would have been invented. Thus the origin of tool use is a function not only of animal psychobiology but also of environmental characteristics. The same is true of the frequency of performance of established tool patterns. California sea otters (*Enhydra lutris*) open mussels by pounding them on stones that they balance on their chests (Fisher, 1939; Hall and Schaller, 1964; Murie, 1940). Pounding occurs less frequently among Aleutian sea otters (Jones 1951; Kenyon 1969; Murie 1940). Hall and Schaller suggest that the size of available prey is the main factor determining the difference in pounding frequency: the mussels eaten by the Aleutian sea otters are relatively small, and healthy adults can open them directly with their teeth.

These considerations reveal why learned tool use originates so rarely, especially in the wild. A potential inventor must be confronted with a particular configuration of environmental circumstances that would make tool use possible and advantageous. Further, the animal must have in its existing repertoire behaviors that could accidentally be reinforced by, or insightfully recombined to suit, the contingencies of that particular environmental configuration. Third, the animal must be prepared by sensorimotor maturation, previous experience, and cognitive readiness to perform the response. Fourth, the animal must be free of the necessity to perform competing behaviors that would preclude performance of tool use. The probability of each condition is undoubtedly low, especially in the wild, and the combinatorial probability that all conditions will be spatiotemporally coincident is even lower. It is surprising that tool use originates as frequently as it does!

Questions of Dissemination

Population genetics accounts for the dissemination of those tool patterns that have specific genetic determinants. The frequency of the genes influencing the new tool phenotype will increase, due to differential reproductive success. If the selective advantage is great, equilibrium will be attained when nearly all members of the population exhibit the phenotype with marked stereotypy. In the absence of reproductive isolation, the tool phenotype would be recognized as species-specific. This explanation, of course, is entirely theoretical and its simplicity does not reflect the complexity of genetic determination of behavior.

Many authors have commented on the dissemination of learned tool patterns, and I know of few issues in animal behavior about which so many conclusions have been drawn on the basis of so few data. Observation learning is usually identified as the means of dissemination, especially among vertebrates.

An initial problem is that there is still confusion about the nature of observation learning. The term is applied to at least three phenomena. The first, social facilitation, is the increased readiness to perform a response upon observation of a conspecific performing the same response. Since the response is established in the repertoire of demonstrator and observer(s), learning is not actually taking place. McGrew *at al.* (1975) note that promising attempts by one of the Delta chimpanzees to use escape tools often contagiously triggered other group members to renew their own efforts. Another type of observation learning is stimulus enhancement, in which reinforcement attained by a demonstrator in a specific situation influences the orientation of an observer's attention or behavior in the same or similar situation. Guinea baboons, after repeatedly observing a fellow group member use the rod to get food, showed significantly increased rates of handling the rod and hitting the food pan with the rod. However, they did not replicate his behavior and were unsuccessful in getting the food. Only the orientation of their behavior and attention was influenced by having observed the skilled tool user (Beck, 1973a). A third type of observation learning is imitative copying in which an observer replicates the topography of a demonstrator's behavior. A young female pigtailed macaque repeatedly observed an adult male use a rod to get food, and then used it in exactly the same manner. Despite ample opportunity, she had not shown the behavior before observing the male. After observing him, however, she reproduced the elements of his tool behavior in sequence and with coordination (Beck, 1976). Social facilitation, stimulus enhancement, and imitative copying should be differentiated when invoking observation learning to explain the dissemination of tool behavior.

Given this differentiation, it is still difficult to verify empirically which if any types are operative in dissemination. The basic problem is estimating the frequency with which the observers would perform the behavior in the absence of a demonstrator, and comparing it with the frequency of performance after demonstration. In the laboratory, two matched subject samples can be utilized. Experimental subjects are given the opportunity for observation and their subsequent performance is compared with that of an observationally naive control group. If the experimental subjects perform the behavior more frequently, then occurrence of some type of observation learning can be inferred.

A less rigorous paradigm is to compare the behavior of the same subject sample before and after demonstration. Baseline frequencies of tool-related behavior are established before demonstration and compared to values of the same measures after demonstration. The paradigm assumes that the observers are not reinforced for manipulating the tool during the observation period. Stimulus enhancement is indicated if there are significant increases in such measures as manipulation of the tool, orientation to the incentive, contact of the tool with the incentive, and/or success in attaining the incentive with the tool. If, additionally, detailed comparison of an observer's behavior with that of the demonstrator reveals clearly increased similarity, imitative copying is indicated.

These paradigms can rarely be used in noncaptive settings because matched naive groups are not available or baseline values cannot be established. However, differences in the tool repertoires of different chimpanzee populations provide an approximation of the matched-groups paradigm. Such differences have been enumerated by Beck (1974), van Lawick-Goodall (1973), Struhsaker and Hunkeler (1971), and Teleki (1974). For example, several west African populations open hard-shelled fruits and nuts by pounding them with sticks and stones (Beatty, 1951;

Rahm, 1971; Savage and Wyman, 1843–44; Struhsaker and Hunkeler, 1971). In contrast, chimpanzees of Gombe National Park open hard-shelled fruits by pounding them against trees (van Lawick-Goodall, 1973). Since the motor pattern of pounding is present at Gombe, it is reasonable to assume that the western and Gombe populations are homogeneous with regard to the genetic requisites of this tool pattern. Further, since the fruits are similar and sticks and stones are equally available, it is reasonable to assume that the Gombe environment would make the use of pounding tools no less possible or advantageous. The only critical difference appears to be that infants in the western populations have the opportunity to observe skilled demonstrators while those at Gombe do not. Since the use of pounding tools is more frequent among the western populations (the "experimental group") than it is in the Gombe population (the "control group"), observation learning is implicated. The case would be strengthened if a Gombe chimpanzee, whose baseline frequency of use of pounding tools is zero, were to show increased use of such tools after translocation and introduction to a western group.

Observation learning is often inferred if most members of a group or population show a similar tool pattern. The inference is especially attractive if a newly invented pattern quickly becomes a group characteristic. However, neither ubiquity nor nearly synchronous acquisition is conclusive evidence for observation learning. It is possible that an environment can change in such a way that a tool pattern suddenly becomes possible and/or reinforcing, and many animals learn the pattern independently at about the same time. Jones and Kamil (1973) invoke observation learning in dissemination since six of eight blue jays in their colony used reaching tools when tested shortly after the first case was documented. However, the authors provide no data that preclude independent acquisition. If nearly synchronous acquisition is independent, however, kinship, social affinity, and visual proximity of other group members to the discoverer should operate randomly with regard to those that acquire the behavior and the order in which they acquire it. Ladder use was acquired quickly by seven of the eight Delta chimpanzees and became a characteristic component of the group's behavioral repertoire. While this suggests observation learning, the case is strengthened because the order of acquisition was positively correlated with social affinities to Rock, the discoverer (Menzel, 1972).

Observation learning is commonly inferred if unskilled group members watch intently as accomplished group members use tools. However, while attention to a demonstrator is necessary for observation learning, it may not be sufficient. Van Lawick-Goodall (1968, 1970, 1973) and McGrew (1977) note that immature Gombe chimpanzees watch attentively as skilled adults make and use tools in feeding, drinking, and body care. Additionally, the youngsters often repeat elements of the adults' behavior while they watch. While this indicates that stimulus enhancement and imitative copying may be operative, both authors emphasize that such evidence is circumstantial. They add that while elements of tool behavior may be acquired by observation learning, trial-and-error learning appears to be necessary for integration and coordination of the elements into efficient sequences. Further, some typical Gombe tool patterns may be acquired independently by trial and error resulting from reinforcement that is accidently procured in the course of playful object manipulation or aggressive display.

I am persuaded that observation learning is operative in the dissemination of tool behavior in chimpanzees and that it accounts for the relatively high frequency and diversity of chimpanzee tool behavior (Beck, 1974). However, it is important to recognize both the paucity of conclusive data and the importance of individual trial-and-error learning. There are even fewer data pertaining to dissemination in other taxa and generalization should be tempered by apparent differences in the capacity of even closely related forms, e.g. baboons and macaques (Beck, 1973a, 1976), to learn tool use by observation.

Summary

Innate tool patterns originate through preadaptation or by mutation, and are disseminated by selective reproduction. Learned tool patterns originate through generalization of previously acquired responses, trial-and-error learning, or by insight. The origin of learned tool patterns is controlled primarily by sensorimotor maturation, object manipulation experience, and cognitive readiness (psychobiological factors) and by environmental contingencies that make tool use possible and advantageous. Observation learning is operative in the dissemination of some primate tool behavior but not to the degree that is commonly reflected in textbooks and secondary sources. Individual learning appears to be instrumental in the acquisition of even those tool patterns that are characteristic of primate groups or populations, and remains the most parsimonious explanation for nonprimates.

Study of the ontogeny of tool behavior demands no unique terminology, concepts, or strategies but hard data are scarce. This scarcity is surprising in view of the diversity of animal tool behavior and the interest it engenders, and is unfortunate in view of the importance of such information for understanding the human evolutionary complex.

References

Alcock, J. 1972. The evolution of the use of tools by feeding animals. *Evolution* 26:464–473.

Bateson, P. 1976. Specificity and the origins of behavior. *In Advances in the study of behavior,* Vol. 6, eds., J. Rosenblatt, R. Hinde, E. Shaw, and C. Beer, pp. 1–20. New York: Academic Press.

Beatty, H. 1951. A note on the behavior of the chimpanzee. *J. Mammal.* 32:118.

Beck, B. 1967. A study of problem solving by gibbons. *Behaviour* 28:95–109.

Beck, B. 1972. Tool use in captive hamadryas baboons. *Primates* 13:277–295.

Beck, B. 1973a. Observation learning of tool use by captive Guinea baboons (*Papio papio*). *Am. J. Phys. Anthropol.* 38:579–582.

Beck, B. 1973b. Cooperative tool use by captive hamadryas baboons. *Science* 182:594–597.

Beck, B. 1974. Baboons, chimpanzees, and tools. *J. Human Evol.* 3:509–516.

Beck, B. 1975. Primate tool behavior. *In Primate socioecology and psychology,* ed., R. Tuttle, pp. 413–447. The Hague: Mouton Press.

Beck, B. 1976. Tool use by captive pigtailed macaques. *Primates* 17:301–310.

Bingham, H. 1929. Chimpanzee translocation by means of boxes. *Comp. Psychol. Monogr.* 5(3).

Birch, H. 1945. The relation of previous experience to insightful problem solving. *J. Comp. Psychol.* 38:367–383.

Chance, M. 1960. Köhler's chimpanzees—how did they perform? *Man* 59:130–135.

Eaton, G. 1972. Snowball construction by a feral troop of Japanese macaques (*Macaca fuscata*) living under seminatural conditions. *Primates* 13:411–414.

Fagen, R. 1976. Exercise, play, and physical training in animals. *In Perspectives in ethology,* Vol. 2, eds., P. Bateson and P. Klopfer, pp. 189–219. New York: Plenum.

Fisher, E. 1939. Habits of the southern sea otter. *J. Mammal.* 20:21–36.

Hall, K. 1963. Tool-using performances as indicators of behavioral adaptability. *Current Anthropol.* 4:479–494.

Hall, K., and G. Schaller. 1964. Tool-using behavior of the California sea otter. *J. Mammal.* 45:287–298.

Hughes, J., and M. Redshaw. 1974. Cognitive, manipulative and social skills in gorillas: Part 1, the first year. *Annu. Rep. Jersey Wildl. Preservation Trust* 11:53–60.

Jackson, T. 1942. Use of the stick as a tool by young chimpanzees. *J. Comp. Psychol.* 33:223–235.

Jones, R. 1951. Present status of the sea otter in Alaska. *In Transactions of the sixteenth North American wildlife conference,* ed., E. M. Quee, pp. 376–383. Washington: Wildl. Manage. Inst.

Jones, T., and A. Kamil. 1973. Tool-making and tool-using in the northern blue jay. *Science* 180:1076–1078.

Kenyon, K. 1969. *The sea otter in the eastern Pacific Ocean.* Washington: U.S. Bureau of Sport Fisheries and Wildlife.

Köhler, W. 1927. *The mentality of apes.* London: Routledge and Kegan Paul.

Lawick-Goodall, J. van. 1968. The behavior of free-living chimpanzees in the Gombe Stream Reserve. *Anim. Behav. Monogr.* 1(3):161–311.

Lawick-Goodall, J. van. 1970. Tool-using in primates and other vertebrates. *In Advances in the study of behavior,* Vol. 3, eds., D. Lehrman, R. Hinde, and E. Shaw, pp. 195–249. New York: Academic Press.

Lawick-Goodall, J. van. 1973. Cultural elements in a chimpanzee community. *In Precultural primate behavior,* ed., E. Menzel, pp. 144–184. New York: Karger Press.

Mathieu, M., M.-A. Bouchard, L. Granger, and J. Herscovitch. 1976. Piagetian object-permanence in *Cebus capucinus, Lagothrica flavicauda* and *Pan troglodytes. Anim. Behav.* 24:585–588.

McGrew, W. 1977. Socialisation and object manipulation in wild chimpanzees. *In Primate bio-social development: biological, social and ecological determinants,* eds., S. Chevalier-Skolnikoff and F. Poirier, pp. 261–288. New York: Garland Publishing.

McGrew, W., C. Tutin, and P. Midgett. 1975. Tool use in a group of captive chimpanzees: I. Escape. *Z. Tierpsychol.* 37:145–162.

Menzel, E. 1972. Spontaneous invention of ladders in a group of young chimpanzees. *Folia Primat.* 17:87–106.

Menzel, E. 1973. Further observations on the use of ladders in a group of young chimpanzees. *Folia Primat.* 19:450–457.

Menzel, E., R. Davenport, and C. Rogers. 1970. The development of tool using in wild-born and restriction-reared chimpanzees. *Folia Primat.* 12:273–283.

Murie, O. 1940. Notes on the sea otter. *J. Mammal.* 21:119–131.

Parker, S. 1976. A comparative longitudinal study of sensorimotor development in a gorilla, a human, and a macaque infant from a Piagetian perspective. *Paper presented at the annual meeting of the Animal Behavior Society; Boulder, Colorado.*

Rahm, U. 1971. L'emploi d'outils par les chimpanzes de l'ouest de la Côte-d'Ivoire. *Terre et la Vie* 25:506–509.

Redshaw, M. 1975. Cognitive, manipulative and social skills in gorillas: Part II, the second year. *Annu. Rep. Jersey Wildl. Preservation Trust* 12:56–60.

Reese, E. 1962. Shell selection behavior of hermit crabs. *Anim. Behav.* 10:347–360.

Savage, T., and J. Wyman. 1843–44. Observations on the external characters and habits of *Troglodytes niger* Geoff. and on its organization. *Boston J. Natl. Hist.* 4:362–386.

Schiller, P. 1957. Innate motor action as a basis of learning. *In Instinctive behavior,* ed., C. Schiller, pp. 264–287. New York: International Universities Press.

Shurcliff, A., D. Brown, and F. Stollnitz. 1971. Specificity of training required for solution of a stick problem by rhesus monkeys (*Macaca mulatta*). *Learn. Motiv.* 2:255–270.

Struhsaker, T., and P. Hunkeler. 1971. Evidence of tool-using by chimpanzees in the Ivory Coast. *Folia Primat.* 15:212–219.

Teleki, G. 1974. Chimpanzee subsistence technology: materials and skills. *J. Human Evol.* 3:575–594.

Yerkes, R. 1943. *Chimpanzees: a laboratory colony.* New Haven: Yale University Press.

Index